CLIMATE CHAOS II

: How Natural Cycles Eclipse
Manmade Impacts

'If you can look into the seeds of time
and say which will grow and which will not,
speak then to me'

Macbeth

Professor David P Gregg (retired)

To my good friends and former scientific colleagues :
Bob Burn, Tony Bird, Pankaj Jadeja and Tony Baker
who have listened, with a degree of patience, to my
heretical ramblings on climate, and much else, yet
still choose to take a drink with me at the Green Man.

Climate Chaos II

Published by Green Man Books
15 Poulton Green Close, Spital, Wirral
CH63 9FS (davidgregg@talktalk.net)

Copyright © David P Gregg 2021

ISBN 9798735590224

Printed by Kindle Direct Publishing
Available from Amazon,
Green Man Books
And other retail outlets

CONTENTS   Page

General Introduction   6

Section 1 : A Brief History of Climate

1.1 :	In The Beginning	15
1.2 :	The Middle Ages & Other Climate Factors	26
1.3 :	Tertiary Cooling & The Present Ice Epoch	29
1.4 :	The Energy Budget of the Earth	37
1.5 :	Why Climate Change ?	40
1.6 :	Heresies: Biofeedback, Planets, Clouds & Cosmic Rays	45

Section 2 : Planetary Orbital Dynamics

2.1 :	The Solar Planets	63
2.2 :	Extra-Solar Planetary Systems	73
2.3 :	Discussion	83
2.4 :	Stellar Activity Cycles	86
2.5 :	Self Organised Planetary Spacing	89

Section 3 : Planetary Forcing of Solar Activity

3.1 :	Introduction	93
3.2 :	Solar Activity ; The Wolf Sunspot Numbers	94
3.3 :	The Hale 'Magnetic' Activity Cycle	100
3.4 :	Hale Cycle Wavelet Analysis	108
3.5 :	Solar X Ray Flare Cycles	110
3.6 :	Long Term Beryllium 10 & Cosmic Ray History	111
3.7 :	Review : Non-linear Oscillators & Phi	122

3.8 Self Organised criticality in Solar activity ?   127

Section 4 : 'Cosmic' Forcing of Earth-Moon Orbital Dynamics & Geophysical Cycles

4.1 : Introduction   134

4.2 : The Earth-Moon System   134

4.3 : Global Earthquake Energy & Polar Nutation   136

4.4 : Power Law Models & SOC   145

4.5 : Volcanic Activity Modulation   149

Section 5 : Weather Variables , The Sun & The Planets

5.1 :	Temperature Reconstructions : Tibet, Europe, Greenland, Antarctica	151
5.2 :	USA Tree Ring Records	161
5.3 :	The North Atlantic Oscillation	164
5.4 :	The Atlantic Meridional Oscillation	168
5.5 :	Wavelet Analysis of Central England Temperature	178
5.6 :	Total Ozone Variation	183
5.7 :	Northern Hemisphere Mean Annual Surface Air Temperature	184
5.8 :	Tropical Weather & the QBO	187
5.9 :	QBO & QTO Signatures in Brazilian Rainfall	189
5.10:	El Nino & the Southern Oscillation	191
5.11:	The Long Game : Oxygen Isotope Measurements	198
5.12:	The Pacific Decadal Oscillation	206
5.13:	IPCC Averaged Global Surface Temperature	211
5.14	Nile River Long Term Level Records	212
5.15	East Asia Winter Monsoon	214
5.16	Drought Cycles in the Western USA	216
5.17	Global temperature & a Solar Activity Index	218
5.18	Svalbard Temperature Record 1912 – 2010	220
5.19	Indian Ocean Dipole, Monsoons & Solar Activity	222
5.20	Global Surface Temperature Spectral Analysis Using SVD	225
5.21	Pacific Sea Temperature since 1600 AD via Coral D18 O records	229
5.22	The Relative Contributions of Anthropogenic Forcing & Natural Climate Change Cycles	230
5.23	The Relationship between CO 2 & Global Surface Temperature	234

5.24	Hockey Sticks & Ski Jumps	241

Section 6 : Climate Futures : Ice or Fire?   251

Postscript   282

Appendix 1 : The Behaviour & Analysis of Dynamic Systems   288

Appendix 2 : Long Term Solar Cycle Modulation: D Gregg, Solar-Terrestrial Prediction Workshop, Paris Observatory, 1984.   310

Appendix 3 : Consensus Bullying & Cargo Cult Science   314

CLIMATE CHAOS II

Some say the world will end in fire,
Some say in ice.
From what I've tasted of desire
I hold with those who favour fire
But if I had to perish twice
I think I know enough of hate
To say that for destruction,
Ice is also great
...and would suffice

                                  Robert Frost

General Introduction

It is five years since 'Climate Chaos' was published and much has happened since in terms of new evidence for the role of natural forces in climate change. A major El Nino event broke the long 2000s temperature hiatus in 2016 – 2019 but as this book is published global temperatures have returned to the levels in 2001. It is time to take stock.

The title of this book is still appropriate in several senses. Firstly we will see that non-linear dynamic systems and, occasionally, mathematical chaos play an important role in climate and related issues. Secondly we might reasonably consider the current debate about climate change to be chaotic in a socio-political sense, while 'climate science', despite a claimed consensus against the 'climate change deniers', is in disarray as the forecasts made with the 'official' models fail badly. Global near surface temperatures, land and ocean, measured by satellite have returned to 2001 levels, totally contrary to the Intergovernmental Panel on Climate Change (IPCC) approved model forecasts. Over the years we will see that the published estimates of Equilibrium Climate Sensitivity to CO_2 doubling has steadily fallen to much ,lower values. Meanwhile solar activity and magnetic field strength has also fallen with some astrophysicists projecting a grand solar minimum...perhaps with Little Ice Age consequences but almost certainly with a fall in global temperatures for some decades ahead.

Finally, the author admits that the climate issue is so complex, with so many twists and turns, with so much, often contradictory, research in disparate fields not properly collated, that he too sometimes finds his thoughts on the edge of chaos and is unable to coalesce them into a 'simple solution' or a simple story.

He merely promises the reader that by the end he or she will be in a better position to knowledgably question the 'simple solutions' currently being offered for sale by politicians and other snake oil vendors, who should know better. We will at least show, forensically, that certain neglected climate 'influences' are critically important and focus down to a few central questions which, if answered, would resolve the future direction of climate change and what we should do about it.

This book is about climate change past, present and future and about emerging evidence for alternative, heretical, interpretations of how it happens.

It is not about geopolitics, green ideologies, child saints and other quasi-religious cults, establishment science, geriatric TV 'nature' presenters, any view expressed by BBC 'science' reporters, dirty tricks, data fiddling, consensus bullying, post industrial angst or fairy stories (but see also App.3 for a few observations). It is about what we know, what we suspect and what we don't know. Above all it is about perspective: the need for a wider vision of the Earth as a living planet in a far from empty solar system: the third rock from a G2 main sequence star in middle age; one star in a collection of thousands in our galactic neighbourhood now known to have both planets and solar activity cycles.

To make important decisions about this world we need to build a rational, conscious, racial perception of a geologically active planet which has been turning for over 4 billion years, suffering great cycles of heat and cold long before the invention of industry, central heating and the sports utility vehicle.

The Sun provides the energy which keeps the Earth at an equitable temperature, making life possible and sustainable over billions of years. That much is universally accepted by 'science'. Beyond that 'fact' the complexity of the Earth as a bio-geophysical system, probably influenced by outside cosmic forces, challenges our understanding…to say the least. Is our ignorance of importance? Surely the Sun will keep burning and the Earth will keep turning long after our turbulent species has gone to dust? We can be certain they will and only ignorant hubris interprets our puny fumbling as an existential threat to life on this planet. However, from a parochial point of view, the continuation of our species, and our current civilisation, such as it is, might be seen as a desirable aim. If there is a slight chance that our ignorance could damage the Earth, at least temporarily, at least to our own disadvantage, understanding the planet and its component systems better should clearly be seen as a species wide priority by everybody.

It is too important to be left to politicians and jobs-worth scientists who are all too human. And when politics and big science make common, and mistaken, cause God help us all!

Some would say that our ignorance also means that we should be cautious about 'changing' the Earth in the mean time. However some would say that we must also be cautious about over reaction. Our species is sustained by our own complex, and now global, socio-economic system; a system even more complex and potentially more chaotic than the climate system we fear. If that human system fails due to instability brought about by a too rapid retreat from the global use of fossil fuels, we will regret it. Using the tool of historical 'perspective' we should recall that civilisations have fallen for lack of key inputs such an as adequate water supply. And yes, droughts occurred long before human interference with the climate.

It is a 'fact' that the global 21st century economy, dependent on high intensity energy sources, cannot be sustained by windmills. The late lamented Professor Sir David MacKay made the point most elegantly (1). In 2009 Britain was generating ~4.5% of electricity from renewable sources: mainly wind, hydro power and a little rubbish dump methane. What about the other renewable options? McKay calculated that to meet a **quarter** of our electricity demand, **75%** of our land would have to be covered in biomass plantations. 300 miles of coast line would need to be filled with wave generators…maybe not impossible. Covering 10% of the country with current solar panels would give us about half the average daily electricity usage of a typical European… again not impossible (see below). The same result, perhaps, could be got from an offshore wind farm with the area of Wales… forget it!

McKay supported renewables but realistically pointed to the necessity of also using nuclear and 'clean' fossil sources. By the way McKay kindly reminded us that electricity generation accounts for only 20% of our carbon emissions: 80% comes from heating, transport and food production. It turns out that reducing our electricity related emissions by 80% would only cut overall British emissions by 16%. On a personal level McKay urged us to 'turn the thermostat down' and 'drive less' and 'fly less'…but the media promoted green lifestyle choices, he describes, over politely, as 'greenwash'. Please note, McKay was no mad outsider: from 2009-14 he was chief scientific advisor to the Dept. of Energy and Climate Change …what a pity he wasted his great talents and time in the hopeless task of trying to educate politicians.

It would be ironic if our economies failed and civilisation fell, because of self-imposed energy droughts.

But of course China and India, building both coal and nuclear stations at an astonishing rate, and the USA with its 'fracking revolution', will ensure some vestige of 'western' civilisation survives even as 'green' Europe and the UK in particular, move back to tallow candles. In 2012-13 the margin between peak electricity demand and normal supply was ~15%. In 2016 - 17 it will be zero thanks to closure of evil coal powered stations. In 2016, for the third year running, the UK National Grid warned that 'emergency' measures would be put into effect again.

Energy conservation is sensible but beyond building fission stations, another species wide priority should be to develop and deploy fusion power and space based solar power systems along with high efficiency, cheap solar cells for the masses…remember covering 5 or 10% of the area of the UK is not impossible (and current conversion efficiencies of 12-15% can be easily doubled with emerging solar cell materials, such as perovskites) … but that is another story for another time.

Despite the IPCC political juggernaut and its insistence that it represents the 'scientific consensus' we will see that not all climatologists are in its thrall and some have spoken out. Here is one of the strongest statements of frustration I have yet noted, although we will see others later. Professor Richard Lindzen of MIT (2) said

'Future generations will wonder in bemused amazement that the early 21st century's developed world went into hysterical panic over a globally averaged temperature increase of a few tenths of a degree and, on the basis of gross exaggerations of highly exaggerated computer predictions, combined into implausible chains of evidence, proceeded to contemplate a rollback of the industrial age.'

Is that a fair reading of the situation? Lindzen was quite ready to share his views in giving evidence to a House of Lords enquiry in 2005 (3)

'If the major greenhouse substances –water vapour- and the clouds, remain fixed, a doubling of CO_2 should lead, on the basis of straightforward physics, to a globally averaged warming of about one degree Celsius.'

Is this a fair conclusion? The reader can decide after considering what follows in this book. The key problem is whether this direct effect of CO_2 is enhanced or suppressed by natural feedbacks which are not reliably quantified. This is why the IPCC models forecasts differ so widely. Nor is Lindzen, a distinguished climatologist, alone although many are slightly quieter in their criticisms.

Take Professor Lennart Bengtsson (former IPCC collaborator and head of the Max Plank Institute in Hamburg in the 1990s; head of the ISSI in Berne) who commented in 2013 (4)

'We are creating great anxiety without it being justified...there are no indications that the warming is so severe that we need to panic. *The warming we have had over the last hundred years is so small that if we didn't have meteorologists and climatologists to measure it we wouldn't have noticed it at all...* The Earth appears to have cooling properties that exceed those previously thought and computer models are inadequate to try to foretell chaotic systems like climate, where using actual observations is the only way to go. Climate change has become highly politicised. The issue is so complex that one cannot ask the people to be convinced that the whole economic system must be changed because you have done some computer simulations.'

By 2010 Robert Watson, then chief scientist at DEFRA, former chairman of the IPCC (1997-2002), was calling for the panel to tackle its 'blunders' or face loosing all credibility. In particular very strong claims about future impacts found their way into official Synthesis Reports based on 'scientific' material which was not peer reviewed or checked. Professor Chris Field the lead author on the IPCC Impacts team was equally concerned (5). We lack the space here to look at the manipulation of climate data to reinforce the case for global man made warming. Interested readers need only type in 'Climategate' to their favourite search engine to be flooded with material. Hopefully things are improving now but do not bet on it (see, for examples, App.3).

So what do we do now? In terms of critical 'ignorance' it comes down to these questions:

Are current climate change and 'global warming' primarily driven by man made changes to our planetary atmosphere, most notably by the rise in CO_2 and aerosols?

Or, are the climate changes we see due primarily to natural cyclical changes which would occur even if environment 'polluting' humanity did not exist?

Is our planetary bio - geophysical - climate system self-correcting in response to the kind and scale of environmental shocks humanity is capable of imposing on it?

If our actions may lead to some 'global warming' and given that we may be nearing the end of the current warm, interglacial interlude, before entering the next ice age, is that a bad thing or a good thing?

We will at least try to explore, if not definitively answer, these questions and others, by applying a long term historical perspective and through a detailed look at the temporal 'fingerprints' of geophysical and climate variables and the equivalent 'fingerprints' of some possible other perpetrators of climate change. We will at least name and cross examine *all* the suspects, not just CO_2.

We are repeatedly told by patronising 'climate' scientists that it is a 'simple scientific fact' that CO_2 is a 'greenhouse' gas and that increased CO_2 **must** therefore lead to global warming and catastrophe. QED!
The first statement is true: CO_2 is a powerful greenhouse gas but so are water vapour (even stronger), methane (even stronger) and the chloro-flourocarbons (amazingly strong). In isolation we might well expect warming to follow increases in greenhouse gases. However the atmosphere is not isolated. We have oceans and complex ecosystems also in play.

It is a 'simple fact' that much of the CO_2 pumped out by humanity over the last century has not stayed in the atmosphere but 'gone' elsewhere …perhaps into increased biomass in the oceans, wetlands, peat bogs, forests, grasslands and soils.

It is also a 'simple fact' that considering the troposphere and stratosphere together, there was **no** increase in global near surface temperature for over 15 years from satellite measurements. It is a fact that the Hadcrut3 surface temperature series shows a small global fall from 2001 to 2014. It is a 'simple fact', now grudgingly acknowledged, that all the IPCC promoted climate models forecast strong global warming which has **not** happened. How can this be? Well now we are told that the models **did not include** known natural cycles whose amplitudes are greater than the recently claimed, CO_2 generated, multi-decadal, trend in temperature. But if the post 1999 warming halt was down to 'natural' cycles perhaps much of the late 20th century warming was also down to 'natural' cycles?

We will see by the way, that the 'natural' cycles now adduced to explain the halt, include El Nino, the North Atlantic Oscillation, the Atlantic Meridional (multi- decadal) Oscillation and the Pacific Decadal Oscillation and that these cycles display very clear spectral 'fingerprints' of the Sun's influence! Also by the way, some hold that the important role of the clouds in climate change is not captured in a realistic way by the current climate models.

We should realise that the models are still spatially coarse in the area covered by surface calculation cells and in the depth of the atmosphere modelled, where we will see all kinds of critical processes occur. It is certainly true that the models do not in any way include the responses of the Earth's ecosystems to climate variations although some suspect that powerful feedback loops exist between bio-botanical productivity, weather and climate, to the extent that life stabilises conditions on the planet and has done so for aeons: the Gaia Hypothesis. If so it is critical that we understand the power and limits of that stabilisation process.

So we will look at evidence for the 'natural' variation of the Sun's output, measured in a number of ways, on a range of time scales from months to millions of years and compare its 'fingerprints', in the form of high resolution spectra, with 'fingerprints' of long term weather and climate variables such as temperature and rainfall and their accepted, proxy time series. The reader can judge the results but the case against the Sun as a major perpetrator, based on these 'forensic' analyses and circumstantial physical and mathematical evidence, is overwhelming.

It has always been said that the small physical variations in the Sun cannot explain large shifts in weather and climate but you have to look at several aspects of the Sun's influence, not just total irradiance, and we will argue that the non-linear, even semi-chaotic, behaviour of many of our terrestrial systems naturally enhances and magnifies small input signals under some circumstances. There is growing evidence for these contentions. Moffa-Sanchez et al (6) have recently shown that North Atlantic proxy ocean temperatures and salinity from 800 to 1780 AD are strongly correlated with total solar irradiance. In simulations low irradiance promotes persistent atmospheric blocking events. Halkos et al (7) have recently analysed reconstructed global temperatures from 704 – 2004 AD and found strong evidence for non-linear (but non-chaotic) behaviour in the climate system in addition to stochastic noise. We will see much more evidence of non-linear behaviour and the amplification of small effects through processes such as Self Organised Criticality as we proceed including confirmation of the 'astronomical' causes of the ice ages!

Even more shocking, our mathematical 'fingerprinting' techniques will also show that solar activity is modulated by the gravitational influence and dynamics of the planets in their orbits on a range of time scales and that we can see the same 'fingerprints' in terrestrial geophysical variables such as earthquake energy release. This is perhaps the ultimate climate heresy but the evidence is clear. Assuming these analyses are correct what is the outlook for climate change in this century and beyond?

And how would the continued growth in man made atmospheric CO_2 at various levels affect the situation? Will we overheat as the 'warmists' fear? We are 11,000 years into the current warm, interglacial period. For the last several ice ages such 'interglacials' lasted on average no longer than ~10,000 years. So should we fear the fire or the coming of the ice? We can at least ask these questions against the perspective of the last million years and much longer, instead of merely the last hundred. That can surely do no harm and may actually help us to come to more rational decisions about the future of our planet.

Introduction References

1. MacKay D; 'Sustainable Energy – Without Hot Air', UIT Cambridge, 05/02/2009. Also 12 Mb pdf download : www. withouthotair.com/download.html

2. Lindzen Quote : Calder, Daily Telegraph 25th October, 2009.

3. Lindzen Quote : Minutes of Evidence, House of Lords Economic Affairs Committee; 25th January 2005.

4. Climate Depot article; February 3rd, 2013. www. climatedepot.com/2013/02/03

5. 'Top British Scientist says UN panel is loosing credibility', Times Online, The Sunday Times, February 7th 2010.

6. Moffa-Sanchez P et al; 'Solar forcing of North Atlantic surface temperature and salinity over the past millennium', Nature Geoscience, doi:10.1038/ngeo2094, 2014.

7. Halkos G et al; 'Nonlinear time series analysis of annual temperatures concerning the global Earth climate', www. mpra.ub.uni-muenchen.de /59140/1/MPRA_paper_59140.pdf

Section 1: A Brief History of Climate

> A thousand ages in thy sight
> Are like an evening gone;
> Short as the watch that ends the night
> Before the rising Sun
>
> Isaac Watts

1.1 In The Beginning

Climate refers to the physical state of the Earth's surface, solid, liquid and gaseous and primarily with respect to variables such as temperature, velocity of fluid flows, that is winds and ocean currents, and precipitation, that is rain and snowfall. Climate also refers to conditions considered on a longer time scale but what that means is increasingly unclear. Even some scientists seem ready to tell the media that some extreme weather event in one year, in one location, like unusual flooding, is 'probably' an indicator of global climate change. I suggest we think of climate on decadal time scales at least. Most of the cycles and processes we will examine operate on such scales and longer...much longer. We can say that 'climate' so defined has been changing since the beginning of the Earth, 4.5 billion years ago, and will continue to change until the end. That will come in four billion years as the Sun exhausts its nuclear fuel and expands into a red giant star, probably in the process, consuming the Earth: by then, a red hot cinder.

What happened in the beginning? About 4.6 billion years ago a large interstellar cloud of gas and dust was compressed by one or more shock waves from nearby supernovae explosions: the death throes of giant neighbours. These giant, short lived stars, were probably born in the same stellar nursery, perhaps like that in today's familiar Orion Nebula. Our cloud was sufficiently disturbed to begin a collapse under its own gravity, which produced a rotating, flattened disk of material consisting mainly of hydrogen and helium along with the ashes of the supernovae. Those ashes contained all the other elements necessary to create planets and, potentially, life. At the centre of the disk, gas density eventually reached a critical level where nuclear fusion could begin and our new Sun began its long burn. Across the disk small random variations in density led to clumping in places. Some of the clumps, the largest, would form major planetesimals, the cores of the planets, which would attract more and more material.

We have some confidence now about this sequence of events for three reasons. Firstly in recent decades astronomers have directly observed proto-planetary disks around young stars, complete with forming planets. Secondly super-computer simulations have allowed astrophysicists to model these disks and planet formation in detail. Thirdly we can now look directly at the end result of this process in general because we have identified hundreds of planetary systems around sun-like stars in our local galactic neighbourhood. We have dozens of multi-planet star systems to compare with our own. In section 2 we will take a closer look at some of these systems since planetary orbital dynamics will turn out to play a significant role in climate modulation.

By ~4.5 billion years BP (before the present) the Earth, other planets and various rings of debris were in place. Powerful solar winds from the new Sun rapidly blew away the unused gas and dust, halting the formation of other bodies. Young, sun like stars we now know, may have winds 100 times the strength of the Sun's current wind. These winds we will also see, played a major role in modifying the early climate of the Earth, so setting a pattern many believe still exists today.

In these early days the Earth's surface was an ocean of magma and for hundreds of millions of years that surface was bombarded by leftover planetesimals of all sizes. The largest collision involved a body somewhere in size between Mars and the Earth now given the name 'Theia'. It is now believed that both bodies were liquefied in a near head on collision and the materials of both worlds intermixed. Masses of magma thrown into space eventually condensed into the Moon in an orbit much closer than today.

At this point the planetary system was not yet stable and resonance between the orbital periods of Jupiter and Saturn led to the migration sunwards of both giant planets. The ice giants Uranus and Neptune migrated outwards in response. Later we will explore the consequences of these interactions for the spacing of the planets overall. The interactions also produced major consequences for the small rocky worlds near the Sun. The outer ring of debris, the forerunner of today's Kuiper Belt, was disturbed and for tens of millions of years untold numbers of rocks and icy comets fell into the inner system. The results of this quaintly termed 'Late Heavy Bombardment' can still be seen on the battered faces of Mars and the Moon. The Earth was similarly battered and any early atmosphere would have been stripped away…along with any early attempts at the 'creation' of life. Nothing remains of that early surface thanks to plate tectonics and plate subduction over four billion years.

This fall of the icy comets from the outer system may have delivered most of Earth's water although this is disputed on isotope composition grounds. Very probably the comets did deliver a good supply of pre-biotic organics to the surface. Water vapour and other gases like CO_2 and SO_2 may have also come from volcanic outgasing over a long period. This hellish time of surface instability from ~4.5 to ~3.8 billion years BP is appropriately known as the Hadean era. One surprising thing is clear: soon after the close of the Hadean the Earth's surface had stabilised and our planet had acquired both an atmosphere and oceans. We know this because in some places ancient sedimentary rocks, rocks laid down in water, have been found. Distinctive pillow lavas which only form beneath water have also been found for that period. Even more remarkable, the ancient sedimentary rocks at the well named, Godthab in Greenland are peppered with minute globules of carbon. The levels of Carbon 13, C_{13}, relative to C_{12} and other chemical arguments about lead isotopes, convince many geologists that the carbon is from the remains of bacteria caught in seabed sediments nearly 3.8 billion years ago.

Darwin famously imaged life beginning in a 'warm little pond'. However the current existence, near the root of the reconstructed bio-chemical tree of life, of extremophile microbes with remarkable tolerance to high temperatures, salinity and acidity suggest that life may have begun in the deep, wet crust where there was plenty of heat energy and mineral nutrients. It may be that life got an early start even before the surface stabilised. Given the long Hadean era perhaps life began several times, only to be destroyed, until finally, one strain got lucky. Some also claim that the seas contained a little free oxygen by 3.8 billion years ago which if true, suggests that some bacteria had already discovered the process of photosynthesis. By 3.6 billion years BP the presence of life is undisputed with many examples of bacterial cells clearly preserved in the ancient rocks.

The early existence of life and the liquid seas pose a severe puzzle: the first of many in exploring ancient climates. Generally speaking current microbial life thrives in a narrow temperature band between 0 and 50 degrees C. There are still extremeophile exceptions which can live in near boiling hot springs or in extremely saline or acidic waters. The deep ocean bed volcanic vents, the 'black smokers', are teeming with life under immense pressures, with food webs depending upon the extremeophiles extracting energy from the hot mineral fountains. Even so the temperature range for most life is narrow and bounded below by the freezing point of water. Here is the problem: stellar physics insists that 'main sequence' stars like the Sun follow a stable and predictable evolutionary pathway (until the red giant stage begins).

The radiation output of the early Sun must have been ~30% less than today. By 3.8 billion years BP it would still have been at only 75% of today's brightness. If today's Sun dimmed by 25 – 30% temperatures would fall by at least 30 degrees C. At a mean global temperature of -15 degrees C we could expect extreme glaciation and frozen oceans. But various geophysical estimates suggest a temperature of 20 – 30 degrees C and possibly more. We know that there were liquid oceans then and no sign of extensive glaciation in the surviving rocks. How can this paradox be resolved? Two solutions have been proposed which are still in the frame as we discuss current climate change over three billion years later. The first solution involves greenhouse gases: water vapour, CO_2, methane and ammonia. However the composition of the early atmosphere is unknown. The only certainties are a dim Sun and liquid oceans at up to 30 degrees C, which implies a great deal of water vapour. The rest is assumption including the very high levels of CO_2 aided by the presence of molecular nitrogen, N_2. The lowest estimate of CO_2, with appreciable N_2, is some 200 times today's level. However, such a level would make the early oceans extremely acidic…perhaps too acidic even for extremeophile bacteria. That is a problem.

The second solution to the dim Sun paradox is, well, the Sun itself. Although dim in radiation terms, the fast rotating young Sun had a powerful magnetic field and a massive solar wind. Recently a young twin of the Sun, Kappa Ceti has been closely studied. Kappa is only ~500 million years old with a powerful magnetic field and a solar wind 50 x stronger than the Sun's today. The fast rotating star has massive surface spots and also puts out super flares. Recall that the Sun's wind had quickly cleared dust and gas from the early solar system. The magnetic field and the wind created a protective bubble around the solar system preventing most galactic cosmic rays from impacting our upper atmosphere. This may have led to low amounts of cloud cover and a low albedo (see below). Fortunately the Earth's magnetosphere must have protected us to some extent from the turbulent early Sun. What role did this early irregularity of the Sun play in the early days of life? Perhaps life was initially confined beneath the surface in the hot, wet upper crust as some have suggested ….or later, it sheltered deep in the early seas.

We will see later that certain energies of cosmic rays can seed low cloud formation and low clouds reflect solar radiation back into space. The clouds reflect 90% of solar energy; the dark oceans, 3%. Before the rise of the continents in the Archean the albedo contrast was stark. Zero cosmic rays implies a world of dark oceans ready to efficiently soak up the dimmer sunlight with only 3% reflected away compared with the Earth's current average reflectivity of 36%.

The result of moderate greenhouse warming and low albedo was equitable temperatures even in the Archean. Recent work seems to confirm this. De Wit et al (26) based on Delta oxygen 18 levels in Australian greenstone beds 3.5 billion years old, suggest cool ocean temperatures. They found evidence of hydrothermal vents which biased older samples towards an explanation of high general temperatures of 30 degrees C or more…falsely. CO_2 did not need to be as high as once thought.

An even more interesting climate issue from the Archean onwards is the role of life in influencing climate via various feedback processes. Lovelock and others (1) have proposed models which link the rise of anaerobic bacterial populations and the rise of methane with a fall in CO_2 as carbon was incorporated into biomass, some of which was sequestered in ocean sediments (Figure1).

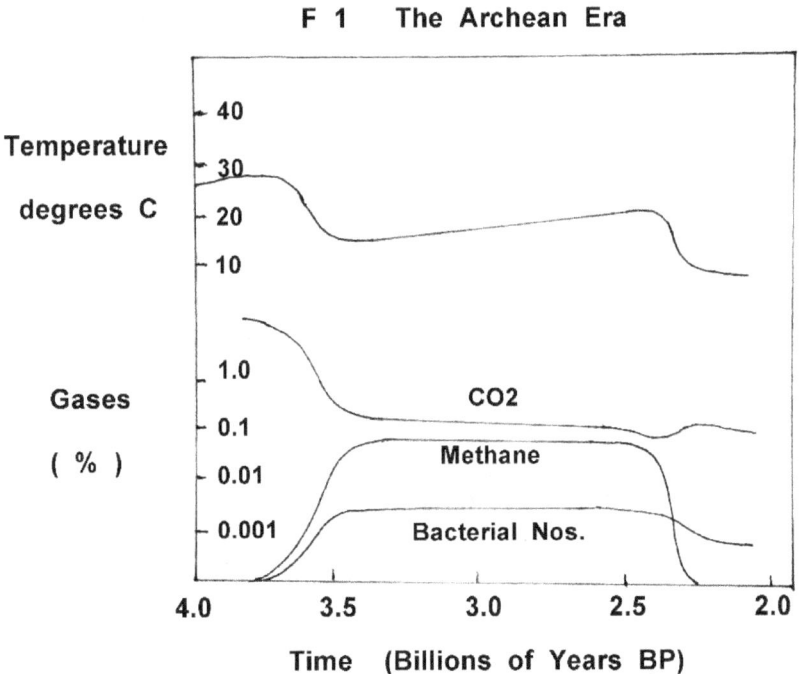

F 1   The Archean Era

Time scales are uncertain but through some feedback mechanism, temperatures remained stable and suitable for life as the Sun steadily increased in brightness over a billion years through the Proterozoic era. No doubt there were disturbances and from time to time, left over planetesimals, asteroids and comets, struck the Earth. But the Earth recovered and life continued to evolve with massive consequences for climate.

At some uncertain point photosynthetic bacteria appeared at first, it has been suggested, using sunlight to dissociate hydrogen sulphide. Then came the cyanobacteria, the blue-green algae, which used sunlight to metabolise CO_2 and water to make sugars and as a by-product, oxygen. Initially, probably for hundreds of millions of years, free oxygen in any quantity immediately reacted with iron, nitrogen and sulphur compounds in the oceans. Initially the anaerobic methanogen bacteria widely coexisted with the photosynthetic bacteria, acting as scavengers on their corpses and returning methane and CO_2 to the atmosphere. In the long term the composition of the oceans changed and oxygen was no longer buffered. Oxygen oxidised the methane (a very powerful greenhouse gas) to CO_2 (a less powerful greenhouse gas).

Methane and CO_2 levels fell and O_2 rose to some significant but unknown level. O_2 is not a greenhouse gas. This is known as the 'Great Oxygen Catastrophe'. The greenhouse effect was supposedly greatly reduced and around 2.3 billion years BP the Huronian ice epoch, Earth's first, began and lasted for two or three hundred million years. Oxygen levels were still only a few percent and kept under control by land surface absorption. Other causes for this long epoch have been proposed and we will explore them later but the reduction of greenhouse gases probably played a part. Some propose a major collapse of CO_2 during the rise of oxygen and photosynthesis as more biomaterial was buried in ocean sediments.

The Huronian ice epoch may have hosted the first Snowball Earth episodes where land and oceans were glaciated almost to the equator. The loss of the very high levels of CO_2 proposed by some, further suggests huge reserves of carbon are hidden in the crust…far beyond those so far discovered. This can be viewed as good or bad news according to taste. Over many millions of years some carbon trapped in the crust is recycled by volcanic action in the subduction zones at tectonic plate boundaries. The state of plate tectonics before, during and after the Huronian is uncertain but life may have played a role in enhancing the system. Some geologists claim that the plates were moving by 2,500 million years BP, just before the Huronian began; some say that tectonics has been with us since 3,500 million BP. Huge quantities of calcium carbonate may have been formed in the first burst of photosynthetic life in the seas and that limestone, if subducted, can change the mineral composition of the Earth and lubricate plate movement.

Perhaps life helped to accelerate the pace of 'continental drift' and we will see that those movements played a major role in determining climate in later ice epochs.

There are still many uncertainties and unknowns about these early periods but early life undoubtedly played a role in climate and geophysics as it did later.

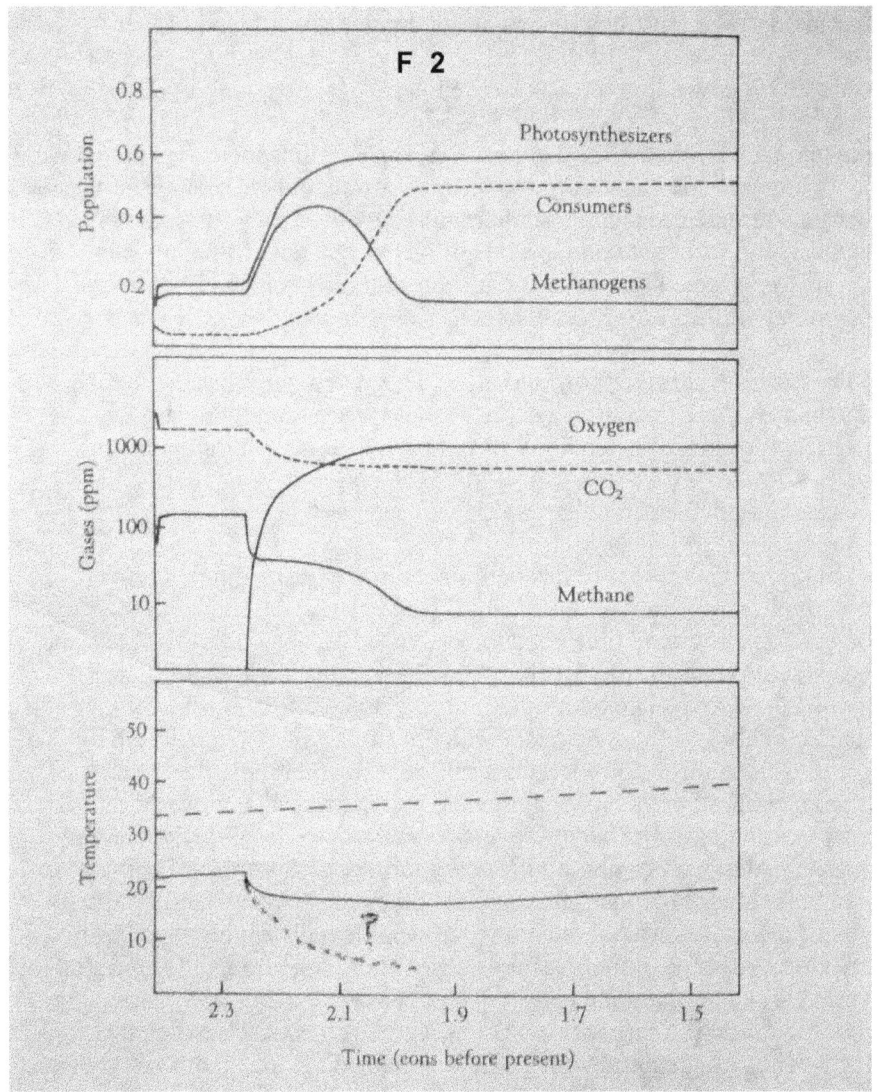

That should be remembered when considering the complete absence of biology and botany in current climate models and their predictions of climate change (not to mention the over simple treatment of clouds and their effects).

So what happened next? According to sparse evidence in Canada, South Africa, Finland and Australia the Huronian series of ice ages ended at ~1.9 billion years BP (Figure 2). 'Sparse' means for example that the positions and sizes of the proto-continents are unknown at this time.

The Earth now entered the middle Proterozoic era, a long middle age of relative stability from ~2 billion to ~1 billion BP. Over that long period oxygen may have stabilised at just 2 or 3%. The biosphere was dominated by photosynthetic bacteria in the form of stromatolites : mounds at the margins of the oceans built of alternating layers of bacteria, sediments and calcium carbonate on the scale of metres. The earliest abundant stromatolite remains go back 2.8 billion years BP, 'shortly' before the beginning of the Huronian ice epoch.

The Sun was stronger now and after the 'great oxygen crisis', the ice epoch ended and the climate stabilised and was generally unstressed …except for the occasional fall of minor and major asteroid / comet strikes, the scars of which can still be seen on the ancient continental shields (stripped clean by later ice ages) such as that of Canada.

Microbial life appears to have continued not withstanding, and the ubiquitous stromatolites continued to feed calcium carbonate to the plate boundaries providing limestone to be scraped off the colliding plates and heaped into limestone mountains …such were the volumes of material involved. Were plate tectonics as active as today? If so it is interesting that no major continent drifted over a pole and stayed long enough to precipitate another ice epoch for a billion years. Perhaps it was a case of good luck. During this long summer, life itself was changing. So far our anaerobic and aerobic bacteria were prokaryotic cells. Genetic material was spread in clumps around the cell interior which remained 'simple' in structure. Such cells 'eat' material and other cells by surrounding their meals, the process called phagocytosis, and then dissolving them with enzymes. At some point possibly shortly after the Huronian (~2 billion years BP?), a prokaryotic cell 'ate' another cell but for some reason digestion failed …perhaps some cells learned a new trick which turned off the eater's enzyme production. The ingested cell gained a protective home. What was the benefit to the inadvertent host cell? Perhaps in many cases nothing. But in some cases the co-opted cell was useful and a stable symbiotic relationship formed. One kind of cell definitely co-opted was the mitochondrion. We know because the mitochondria have their own genetic material which after two billion years or more, is still passed on separately from the host DNA.

In fact mitochondria still hitch a ride inside human eggs and those of every other 'animal' on Earth. In return for our cells feeding and protecting them they provide concentrated chemical energy without which we would not survive a day. Chloroplasts, organelles descended from free living cyanobacteria, perform a similar service in host plant cells. Within the new eukaryotic cells several species of former bacteria constituted a symbiotic colony and the genetic material of the original host became walled off into a protective nucleus....presumably in case the hungry lodgers became confused.

The eukaryotic cell was complex enough to support all kinds of new possibilities including the emergence of multi-cellular life....eventually. Once the trick of co-operation within cells was learned it could be extended between cells with each cell type evolving responsibility for particular roles. So how far back does multi-cellular life go? Nobody knows but the first certain, extensive signs of experiments in multi-cellular life occur in the Ediacaran formation of Australia maybe 700 million years ago. Now similar biota assemblages are known from elsewhere. The Ediacarans included worm like and echinoderm like life forms which survived to later times but also flat, quilted forms which have no recognised descendents. However survival of small, soft bodied animals from so long ago requires a great deal of luck. Such life may be older. In 2010 Nature (2) published a report of flattened, finger sized creatures from Gabon in West Africa, in rocks reliably dated to 2,100 million years BP. Two hundred and fifty of these 'animals' were collected with a variety of body shapes: this was a working ecosystem. However some have claimed the forms are mineralogical in origin, disliking the early date. It is interesting that the date is during the Huronian ice epoch which we speculated followed the 'catastrophic' rise in oxygen and the fall off in CO_2 levels. Does an ice epoch somehow catalyse evolution through environmental challenge? Well it seems to have happened again more than once. For a conventional but superb book on the history of life from the beginning see (3).

The long summer of the later Proterozoic era ended in a new ice epoch from ~1 billion to ~600 million years BP. Evidence for massive episodes of glaciation comes from all over the current world including vast deposits of glacial tills from the repeated advance and retreats of the ice. The length of this ice epoch was such that at current drift rates, two major rearrangements of the continents across the Earth could have occurred. At some point over a period of half a billion years all the continents spent some time at one or other of the poles.

Some geologists / climatologists claim that around 620 – 590 million years BP we endured another Snowball Earth era when most of the surface was glaciated except, perhaps, for a band of equatorial ocean.

At that time a supercontinent occupied the southern polar region providing space for a huge ice cap. Perhaps by misfortune the upper Proterozoic was an era when successive movements of the continents always left a major land mass at a pole (just as in the long, Proterozoic summer the opposite may have happened by chance). Curiously from ~800 million years BP onwards it appears that free oxygen rose rapidly from its long standing ~3% level towards roughly current levels, perhaps 15 – 20%, despite the ice ages. One story is that all the gas sinks of land and ocean had finally filled so that oxygen could accumulate in the atmosphere. Another is that a burst of high biological productivity in cool conditions boosted oxygen and further reduced CO_2 levels, enhancing the cooling.

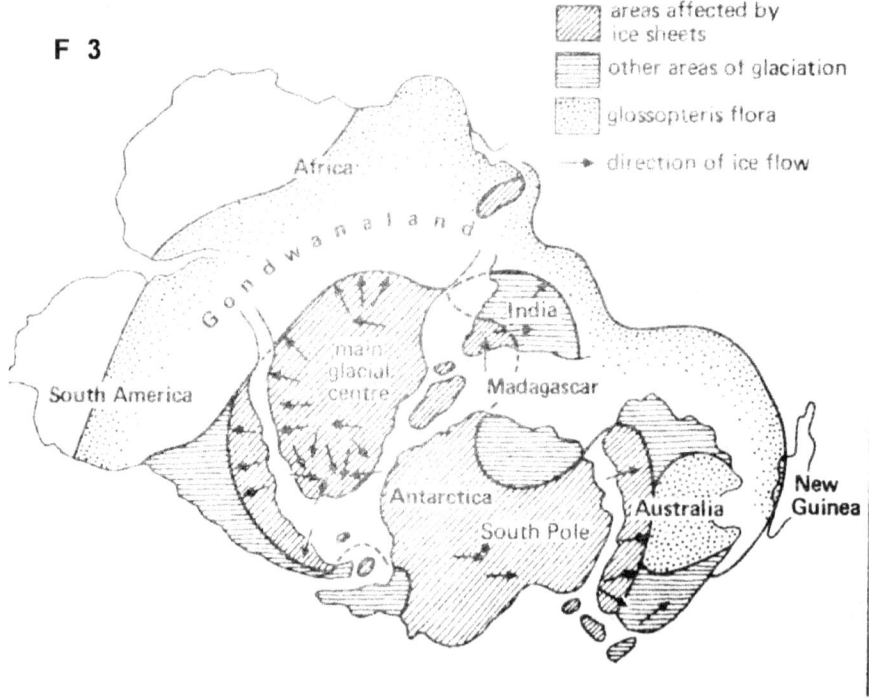

F 3

A feature of ice age conditions is aridity and dry winds producing vast movements of mineral dust. It has been suggested that iron rich and other dust blown into the mineral poor tropical oceans produced massive plankton blooms responsible for lowering CO_2 and more cooling.
This may be one of many bio-feedbacks.

Experiments have shown that 1,000 lbs of iron dispersed across 5 square miles of tropical, mineral constrained, ocean produces a massive increase in plankton productivity. This may have some relevance to the supposed current 'global warming' crisis. Ocean seeding may provide a solution if normal planetary bio-feedbacks fail, although we will see that is unlikely.

The position of continents over the poles featured prominently in all later ice epochs after the Cambrian period: the late Ordovician, 460 – 410 million years BP, the great Permo-Carboniferous epoch, 340 – 240 million years BP (Figure 3) and the Plio - Pleistocene epoch from 3 million years ago until today. Today, and for the last 40 million years, we have been in the same 'polar landmass' position thanks to the movements of Antarctica near the south pole.

The positioning of a continent at a pole may have been a necessary condition for an ice epoch (since the Cambrian) but is it a sufficient condition? Alternative or complementary theories include plausible 'cosmic' explanations for ice epochs which we will look at later. It seems true that within each long ice epoch a sequence of ice ages of alternating cold glacial and warm interglacial periods follow each other in sequence. That is what we see currently (NB : currently, being the last million years or so). During the great Permo – Carboniferous ice epoch incredible deposits of glacial till were laid down in Antarctica and South America by successive ice advances (Figure 3). In Antarctica deposits are 1,000 mts deep; in Brazil up to 1,600 mts. At least 20 sub-periods of intense glaciation have been identified but if the ice ebbed and flowed in the current ~100,000 year rhythm there may have been several hundred ice ages in that epoch with repeated partial retreats of the vast Gondwanaland super continent ice sheet at the south pole.

During this long epoch Earth's major coal measures were laid down, not in tropical swamps, but in the then peri-glacial regions in low lying forests on large exposed continental shelves; thanks to very low sea levels. Any partial melting of the huge ice caps would raise sea levels by many tens of metres flooding the low lying forests rapidly before biodegradation could occur in the cold water. Coal is often found in bands of several layers separated by thicker deposits of limestone or shale. There must have been repeated changes in sea level as the ice retreated and advanced again and again. It may be, to explain the preservation of the pre-coal deposits, that at the end of glacial periods temperatures rose very rapidly (along with sea level) just as they did at the end of the ice ages in our current sequence: in no more than a century or two.

During this long interval average temperatures varied from 12 degrees C (a little colder than today) to 20 degrees C with oxygen reaching over 30% and CO_2 at perhaps ~800 ppm. The unprecedented oxygen levels enabled the evolution of arthropod giants, with dragon flies hunting prey on one metre span wings.

After the Permo – Carboniferous ice epoch came another long summer stretching through the Triassic, Jurassic and Cretaceous periods : the so-called 'Age of the Dinosaurs' could equally be called the age of chalk and limestone given the richness of sea life and its sedimentary fall out in that period. Average global temperatures fell briefly at the Jurassic–Cretaceous boundary but no full ice age resulted; just a few glaciers in Australia. Now for 'modern' times and a look at other causes of climate change.

1.2   The Middle Ages and Other Climate Factors

We noted that there is good evidence by the Earth's middle age that ice epochs are themselves modulated into a long series of ice ages. Why this more high frequency modulation? Before we attempt to answer that it may be a good idea to consider other suggested climate forcing factors so far neglected. We did briefly discuss the possibility of continuing major asteroid strikes disturbing climate at least on the scale of decades, centuries and millennia, although the ecosystem fall out of such strikes may be much more profound and long lasting in its consequences. We left our climate briefing earlier in the long Mesozoic summer. That summer ended most famously with the 'Death of the Dinosaurs' when they and many families of sea creatures became extinct. All genera of the Dinosauria vanished but only half of land reptile genera became extinct overall. Mammals increased by 13%. Of all terrestrial organisms only 19% became extinct. Of all freshwater organisms (fish, amphibians, reptiles) only 3% of genera were lost. The seas faired worse. 70% of genera of swimming organisms died out. 42% of floating microorganisms became extinct along with 49% of bottom dwelling genera (6). Large animals on land and some sea life families were hardest hit for reasons still not clear.

The perpetrator of this mass extinction was almost certainly the giant rock some 6 miles in diameter which smashed into the Yucatan Peninsula (as it now is) some 66 million years ago. At that time the rock fell into a shallow sea. There is little doubt that massive global damage followed including the ejection of a huge volume of pulverised rock into the upper atmosphere and a regional fire storm. The blanket of dust particles quickly circled the globe as volcanic clouds do today, reducing solar radiation reaching the surface for a period probably measured in years rather than months.

It has been suggested the effect on the seas was intensified by the high level of sulphur in the rocks in the area of the strike. Was SO_2 produced in large quantities and did sulphuric acid acidify the oceans and destroy susceptible marine families: 91% of coccolith and 83% of foraminifera genera became extinct. Remarkably the radiolarians and diatoms suffered no losses (6). What is interesting is that the mass extinctions were not followed by decline into an ice age despite the popular 'nuclear winter' scenario of the media…and some alarmist scientists. This tells us that under 'normal' conditions, when the upper oceans hold say a century or more of heat reserve in warm water, major cooling shocks can be absorbed. Today dear reader our oceans (after the long temperature decline in the Tertiary era from ~50 million years BP onwards) have only a ten year reserve of heat. This is why your author fears the threat of ice as much as the fire.

Falling rocks are not the only concern. We suggested that the drifting continents have played a role since the Proterozoic in setting up the pre-conditions for ice epochs. Deep below us the forces that move the continents may manifest in more destructive forms. Occasionally the deep, turbulent mantle creates hot spots and plumes beneath a section of the crust. Vast volumes of flood basalts may break thought the surface taking the Earth (locally) back into Hadean times. Two flood basalt events are of particular interest because of their timing: those which produced the Deccan Traps of Central India and the earlier Siberian Traps.

The Deccan Traps consist of multiple layers of basalt over 6,000 ft thick, covering an area of 193,000 square miles giving a total volume of 123,000 cubic miles of material. Before erosion the area is believed to have been three times larger. The Traps are one of the largest volcanic features on Earth…but not the biggest as we will see. Their timing is the point of interest. They began 66 million years ago and lasted an estimated 30,000 years. Masses of SO_2 and other gases were also released and temperatures may have fallen by 2 degrees C from aerosol effects. Attentive readers will recall that the date of the 'Dinosaur Killer' was also 66 million years ago (the dates coincide within ~50,000 years, well within the dating uncertainties). Did the unprecedented earthquakes and shock waves generated by the asteroid fall disturb the Deccan mantle – crust hot spot and trigger the flood basalt breakthrough? The energy release of the strike was huge: the equivalent of billions of our largest thermonuclear weapons released in a few seconds. Did the massive SO_2 release from the flood basalts acidify the oceans as we suspected earlier for the asteroid strike? Perhaps.

So now we have two coincident, incredibly violent events disturbing the climate system…yet nothing drastic happened so far as we know. The climate system bounced back from any temporary fluctuation. There was no temperature runaway in either direction despite the mass extinctions in the oceans. By the way only 10% of land plant genera died out and there were plenty of species of marine photosynthetic microbes left in the seas to take up the slack there. The global O_2 and CO_2 gas cycles continued as before. That is reassuring.

The formation of the Siberian Traps may have been a different matter. That series of flood basalt eruptions still covers 2 million square kilometres of Siberia. The original pre-erosion area was ~7 million square kilometres …the size of Western Europe. Volume estimates vary from 1 to 4 million cubic kilometres. A major mantle plume is again implicated but there is a complication. The Wilkes Land crater beneath the Antarctic ice cap was found as a large gravity anomaly within an outer ring some 300 miles in diameter, discovered by deep radar scanning. The crater is at the antipodes of the centre of the Traps on the other side of the planet. The crater is claimed to be of a similar date to the Traps event, about 230 million years ago. We see that the eruptions, triggered by the Wilkes Land strike or not, spanned the Permo – Triassic boundary, and the mass extinctions at that boundary. This was the time of 'The Great Dying' when ~90% of all species on Earth became extinct. Land life may have taken 10 to 20 million years to recover in terms of diversity. Estimates from sea bed oxygen isotope sequences suggest a period of global warming, not cooling, at the peak with mean equatorial temperatures of ~35 degrees C compared with today's 25 degrees C. Remarkably the climate and life 'recovered'.

Major flood basalt events are rare although the remains of several are known. More frequent are lesser mantle plumes powering super-volcanoes which are still very much with us. The well known Yellowstone caldera is ~34 x 45 miles across. Yellowstone has exploded three times recently : at 2.1, 1.3, 0.63 million years BP. The hot spot currently beneath Yellowstone is believed to be ~18 million years old and the North American plate has moved across it travelling west – southwest leaving a string of previous super volcano sites with a rate of one or two caldera forming eruptions per million years. (The Hawaiian island chain in mid Pacific was formed in the same way as the Pacific Plate drifted eastwards over another mantle hot spot) The last Yellowstone eruption is estimated to have put out 250 cubic miles of ash and dust. The latest models suggest that a similar event would produce an ash cloud that could reach eastwards to Chicago and create a temporary 'nuclear winter' with plunging temperatures. After a few years most of the dust would be washed out of the atmosphere.

Vast amounts of volcanic gases including CO_2 and SO_2 would also be produced, no doubt with further disturbances. SO_2 would produce sulphuric acid and enhance formation of cooling clouds but CO_2 would be expected to promote warming. The event model predicts ~100,000 American deaths. Is there any direct evidence for such a scale of effect? There is, from 'ordinary' eruptions. In April 1815 the volcano Tambora produced the most powerful eruption in historic times. An estimated 10 cubic miles of ash, minute compared with the super volcanoes, went into the atmosphere. The dust cloud quickly spread around the world and fine ash stayed in the stratosphere for two years producing spectacular sunsets. 1816 was called 'eighteen hundred and frozen to death' in North America and the 'year without a summer' in Europe as crops failed. The global climate anomaly was estimated to be -0.51, -0.44, -0.29 degrees C in 1816, 17, 18; roughly equivalent to the scale of global warming in the 20th century. Polar ice cores for the period show a huge peak in sulphate concentration lasting over two years. The volcanic SO_2 appears to have formed sulphuric acid which seeded cooling cloud formation as we proposed above. Did CO_2 mitigate the cooling? We do not know. In Asia the Indian monsoons failed along with harvests for three years. Famine in Asia magnified the impact of a new cholera strain which appeared in 1816 and caused havoc in weakened populations. Overall it took at least three years for the world to recover from the Tambora eruption.

The severity of effect of super volcanoes in the Yellowstone class, with a much greater volume of ash and gases, must have been many times greater but what may be critical is the duration of the effect. The cooling caused a blip in climate, no more. The major flood basalt events must have caused cooling periodically throughout their duration but an ice epoch did not follow.

1.3 Tertiary Cooling & the Current Ice Epoch

The Tertiary era began as the world recovered from the Cretaceous extinctions. We now have a good picture of temperature and climatic conditions thanks to oxygen isotope composition measurements of sea floor sediment cores containing fossil foraminifera plankton. The first 10 million years of the Palaeocene was a period of warm and steadily rising temperatures which continued into the next, Eocene period. The reason for the long warming is unknown but this was a time of high CO_2 in the atmosphere, perhaps 2000+ ppm compared with today's 400 ppm and temperatures several degrees warmer than today. What happened next towards the Palaeocene – Eocene boundary has some relevance to the global warming debate.

Beginning around 55.5 million years until ~52 million years BP the planet experienced a series of extreme, hyperthermal, warming events of diminishing scale. The first, the PETM, was associated with an increase of 5 degrees C over several hundreds of years along with evidence of ocean acidification and a massive release of carbon into the atmosphere. Several causes have been suggested including the destabilisation of clathrates, methane hydrates, trapped in the sea beds of the continental shelves but the estimated volumes are said to be too small to explain the scale of warming and the repeated sequence of hyperthermal events over three million years.

The most interesting recent hypothesis involves the release of carbon dioxide from land based permafrost deposits (7). At that time Antarctica was moving towards the South Pole but was largely ice free with large forests and tundra. The proposal is that the steady rise in temperatures in the Palaeocene reached a critical point where the permafrost deposits destabilised. What is interesting is that the temperature spike was brief (a thousand years at most) and temperatures returned to former levels. Other, smaller spikes followed the first at intervals. **There was no long term runaway greenhouse effect on temperature despite these massive carbon injections.** In fact temperatures slowly peaked long after the events, in the mid Eocene and began a slow fall (of about 6 degrees C) from ~50 million years BP to the Eocene – Oligocene boundary at 34 million years BP. We should compare these hyperthermal events with the current 'tipping point' argument which tells us that if the temperature goes up 'two degrees' (from the pre-industrial base) biblical disasters will immediately follow! The two degrees limit was invented by an economist many years ago as a political target to aim for but it has taken on a life of its own… which scientists, who should know better, have done little to condemn. In the 2015 Paris Climate Conference it featured yet again at the top of the press briefings.

    The other point of interest of the PETM events is the suspected cause. It turns out that there is a very good match between the thermal spikes and extremes in the eccentricity of the Earth's orbit and the polar obliquity, that is, the angle of the rotation axis to the plane of the solar system. The climate forcing effects of these changes are said to be small but we will see later that these orbital changes, still small, are currently modulating the ice ages (see section 5.11 on non-linear forcing responses of the climate system). In fact we can see the series of Paleocene hyperthermal events as 'negative' ice ages.

Whatever triggered them some (feedback?) processes soon returned temperatures back to normal (for that era). We will look at these and other 'astronomical' effects in detail later.

With ice caps forming on Antarctica two effects may have slowly cooled the surrounding ocean. Firstly calving ice bergs drifting northwards would cool adjacent water bodies towards freezing point which is -2 degrees C for salt water. This water became very dense and would sink to the ocean bottom. The unsolved question is what caused the long, slow upward march of temperatures over 66 to 50 million years BP and the long, slow fall, from 50 to 34 million years BP. What is clear is that over this period and on to our own times, the temperature of the deep ocean was also falling steadily and the heat reserves near the surface declining (Figures 4 & 5).

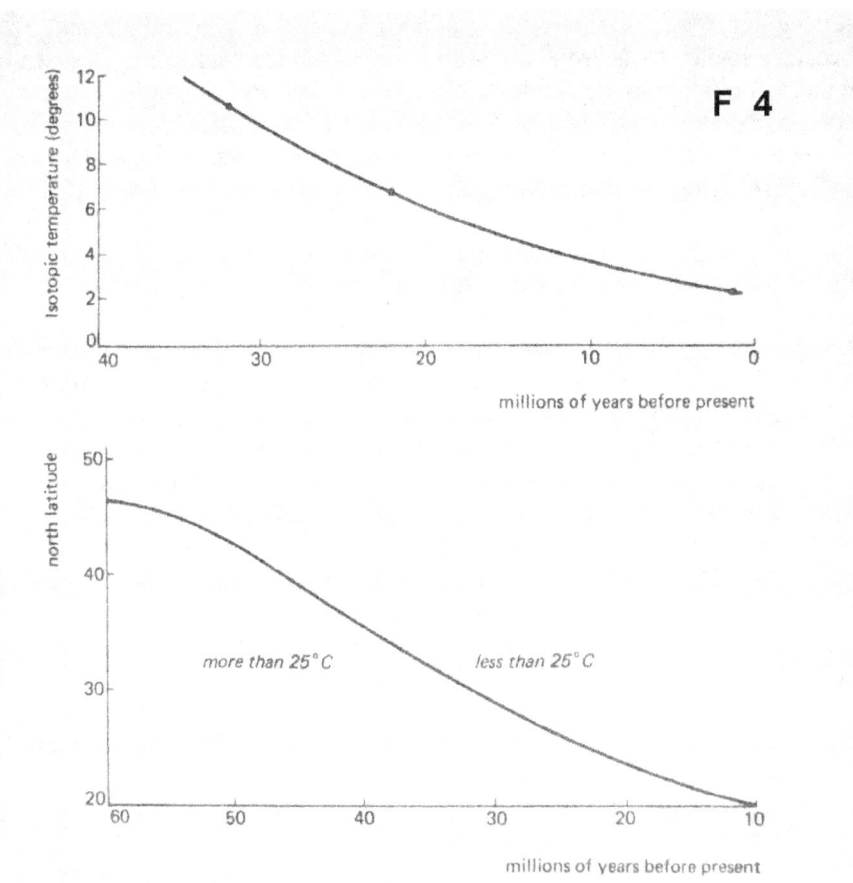

Secondly in winter seasons sea ice would form around the continent as it does today, freezing out of the salt solution of the sea and leaving dense, cold, salty water which would also sink to the bottom.

These processes over 40 million years were sufficient to cool bottom water from perhaps 12 to 3 degrees C. Conditions were being set for the next ice epoch, our own. Surface water temperatures also fell as shown by the estimated latitude at which Pacific coastal temperatures reached 25 degrees C. By the end of the Eocene CO_2 had fallen to an estimated level of 800 - 1200 ppm, still up to three times the current level.

It is probable then that these long, slow swings in global temperature followed the slow shift in the position of the continents and the ocean currents around them. Despite the high CO_2 levels at the Eocene - Oligocene boundary, at 34 million years BP, there was a sharp fall of several degrees C as Antarctica finally settled over the South Pole and became isolated by a strong circumpolar ocean current and strong westerly winds, as South America and Australia drifted away northwards. These currents and winds are still with us. The ice cap of Antarctica also increased the albedo of the Earth significantly. Temperatures were still 2 or 3 degrees higher than current interglacial periods such as today's.

At this time India was already colliding vigorously with the Eurasian plate pushing up new mountains. It has been suggested that more active mountain building during this period increased the rate of chemical weathering to a significant degree, reacting calcium silicate with CO_2 to produce silica and calcium carbonate. Perhaps this explains part of the long fall in CO_2 during the mid Eocene and Oligocene. By the end of the Oligocene CO_2 had fallen to ~500 ppm. At about 30 million years BP there was a flood basalt event in Somalia and Ethiopia covering 500,000 square kilometres which just shows up as a blip in temperature. At about 25 million years BP temperatures appear to have recovered somewhat but not back to Eocene levels. The picture is confused and geographically variable, with claims for both partial thawing and extensions of the Antarctic ice cap. Calving from the ice cap has certainly helped keep the deep ocean cold.

The Miocene, beginning at 23 million years BP was a time of change for life. The spread of the grasslands was a major development since these C4 plants were efficient users of CO_2 and good retainers of water and carbon in their deep soils. The atmosphere became more arid and the forests retreated. In the seas kelp forests proliferated and the cetaceans diversified. The land fauna became recognisably modern and the genetic line of forest and plains apes became successful. By the end of the Miocene our ancestors had split from the chimpanzees. In the mid Miocene, 18 – 16 million years BP there was a brief rise in temperature with some extinctions, quickly followed by a rapid decline which has continued into our own times.

By now the major northern landmasses surrounded the Arctic Ocean largely isolating it from the inflow of warmer southern waters. The Himalayas and other ranges like the Andes continued to rise promoting chemical weathering. The high plateaus of Tibet and Colorado acted similarly as giant 'CO_2 scrubbers'. At some point Tibet became glaciated adding 930,000 square miles of reflective ice fields at otherwise sub-tropical latitudes and increasing the regional albedo by an estimated 70%. Antarctica was fully glaciated and ice caps had formed in Greenland although forests still survived in places. CO_2 in the later Miocene was ~300 ppm, similar to today. The 60 million year Tertiary summer and autumn were over.

Now we enter the Plio – Pleistocene ice epoch. This epoch is characterised by repeated cycles of glaciation where the glacial temperature minima grew colder over the last 3 million years while the interglacial warm periods reached roughly the same temperatures as today. We have returned to conditions not seen since the Permo-Carboniferous ice epoch which lasted for ~100 million years.

F 6

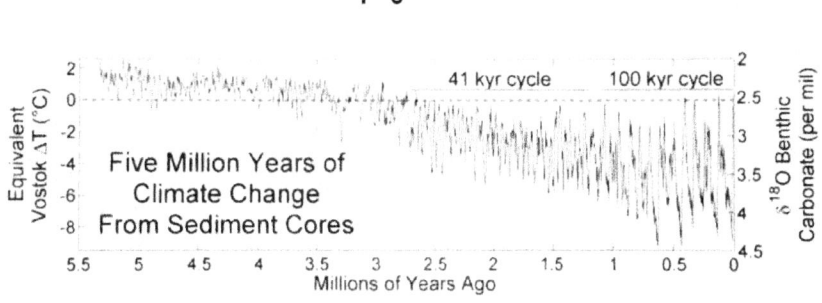

Figure 6 shows us the last 5 million years of climate change reconstructed from ocean sediment cores. Note that for 500,000 years the maximum and minimum levels of estimated temperature have remained remarkably constant. We see the same picture in several proxy climate time series. This strongly suggests that (currently at least) we have a stable 'limit cycle' in global temperature and ice volume. (We look at the maths of this non-linear behaviour in Appendix 1 and section 5.11). The climate is 'bouncing' between two quasi-stable conditions and this has implications for our search for explanations of climate change and predictions of future climate.

Figure 7 shows the most recent ice ages including proxy temperature, CO_2 and sea level. Note once again that our current, supposedly unique, 'man made', temperature peak is no higher than at the last interglacial maximum around 120,000 years ago, nor the previous ones, nor in Figure 6, the long 2 million year period before the ice epoch proper began. Note also that the last five glacial cycles have become less complex in form with a duration of about 100,000 years and brief interglacial maxima of about 10,000 years duration. For the reader's interest: hold on to the fact that our current interglacial has already lasted for ~12,000 years. We will see later the 'cosmic' origin of the 100,000 year cycle. These observations perhaps have some relevance to our later discussions of current 'global warming'.

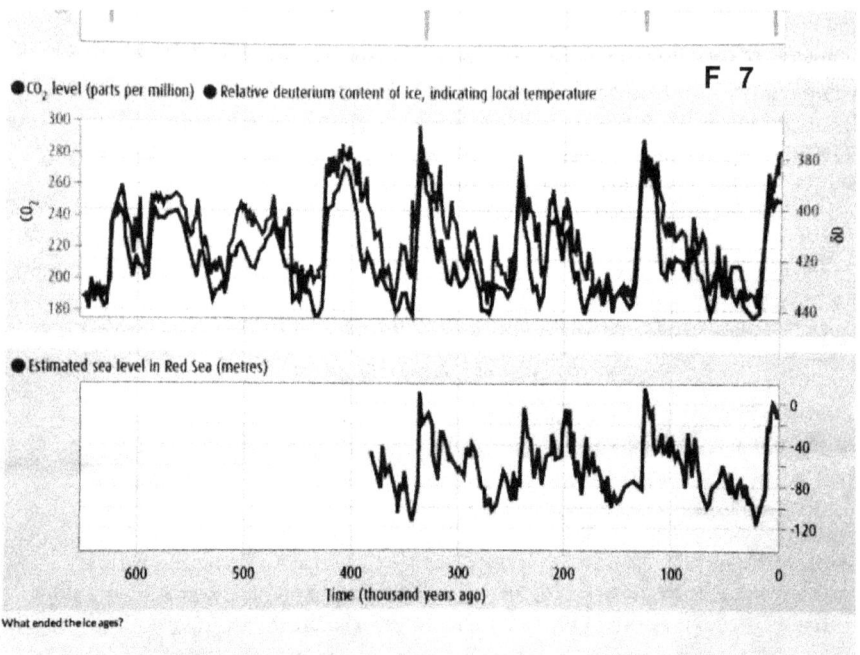

What ended the ice ages?

Hopefully this brief look at the history of climate provides several lessons. Firstly it is nothing new: the Earth has continued to change since its beginning 4.55 billion years ago. The Earth has been much warmer than today with global mean temperatures of up to 30 degrees C, oxygen levels at 30% and CO_2 levels of 2000+ ppm (during the long late Mesozoic / early Tertiary summer) and many times this level in the earliest days of the dim Sun. At times the Earth has been ice free with temperate forests at the poles and feathered dinosaurs stalking the sub-polar twilight. However several times, for reasons unclear, the Earth has fallen into ice epochs lasting tens of millions of years. Occasionally there is evidence that glaciations becomes all but complete with both land and oceans covered in ice except in a band near the equator. In the Huronian perhaps the Earth *did* become a snowball with life surviving thanks only to volcanic heat sources. Our own ice epoch is a spring chill by comparison. Secondly, through it all complex life has survived from near the beginning, perhaps 3.7 billion years ago, until today, evolving and diversifying without end.

It now seems highly likely that life itself has influenced the climate and geophysical cycles of the planet in major ways such as in the 'Great Oxygen Catastrophe', in the ongoing, vast sequestering of carbon in biomass, soils and ocean sediments and perhaps even in the carbonate 'lubricating' of the engine of the tectonic plates. There are good hints as Lovelock has claimed that some biological – climate feedbacks have co-evolved which tend to stabilise the climate (except under extreme disturbances) through processes such as cloud seeding by sulphuric acid droplets derived from dimethyl sulphide released by ocean plankton. More low cloud cools the Earth. Even so, occasionally something goes wrong: the system suffers a sudden change too large to absorb, the climate shifts in a major way, and extinctions follow.

Why does it happen? Our journey through time tells us we live on a violent planet in a violent universe. Over hundreds of millions of years the continents drift across the surface of the planet like rafts on a turbulent river. They drift from pole to equator and back again in irregular cycles. Periodically they collide to form super-continents like Rodinia, Gondwanaland and Pangea, changing the patterns of the oceans, and their currents and winds in major ways. If a major land mass should linger over a pole, glaciation and an ice epoch will follow. Mountain ranges rise and fall as the plates collide and the winds and rainfall change as when India collided with the Eurasian plate, raising the Himalayas and the high plateau of Tibet. We saw that beneath the tectonic plates the liquid mantle is restless and capable of creating long lasting mantle plumes and hot spots. When these breach the thin crust, basalt floods occur bringing vast volumes of lava, ash and most importantly, climate altering fountains of gases including SO_2 and CO_2. The climate may be disrupted periodically for tens of millions of years.

On shorter times scales but more frequently, smaller plumes feed super volcanoes like Yellowstone and ordinary plate boundary volcanoes like Tambora and Toba. We have seen that even these can produce major regional disasters and temporary global climate shifts for a period of several years. The Earth formed from the collisions and merger of primordial planetesimals and smaller rocks and comets are still falling some 4.6 billion years later.

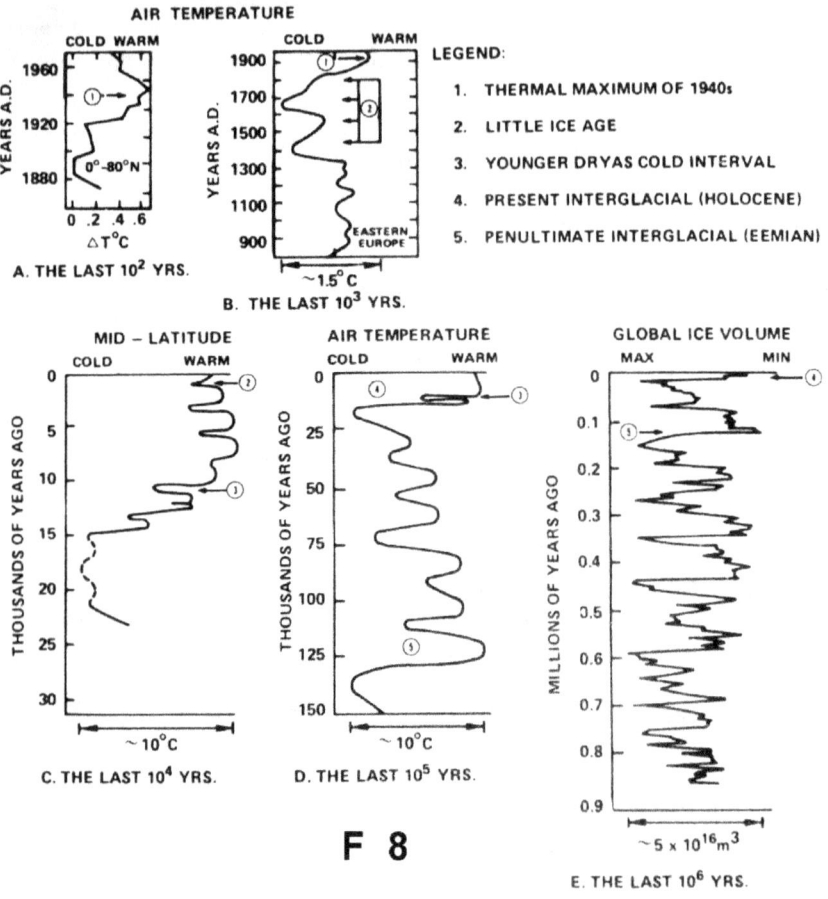

F 8

The largest falls are undoubtedly responsible for some mass extinctions, such as the Cretaceous – Tertiary boundary event, and sudden climate shifts... and the idea that major impact shock waves may set off pre-stressed areas of the crust or mantle plumes is a real possibility. All these forces are still in play and the reader will note the high level of uncertainty about when and how they interact to shape climate. And these are not the only suspects of disruption as we will see. At this point, considering the Earth's violent past and present and the global palaver about 'one degree' of climate change, the reader might well ask : **why panic now?**

Finally to complete our 'new' perspective consider Figure 8 which summarises 'recent' climate change on several time scales of one million, 100,000, 10,000, 1,000 years and the last century. Change is a constant.

To cheer ourselves up recall that since the end of the fall of the planetesimals, some 3.8 billion years ago, the Earth's average surface temperature has apparently stayed in the range 273 to 298 degrees Kelvin (0 – 25 degrees C), a variation of only 10%, with no obvious long term trend, and despite everything, including a 25% increase in the radiation output of the Sun.

Our ability to reconstruct detailed climate proxy histories from measurements of ice caps and seabed sediments, through the current ice epoch, teaches other lessons. The ice age cycles tell us that other external forces, on shorter time scales are acting to modulate climate and so far we have not considered these. To do so we will have to look beyond the Earth into the solar system and the wider galactic environment. Before we can do that we need to understand a little more about the energy balance and thermodynamics of the Earth.

1.4 The Energy Budget of the Earth

The Earth is a ball of rock 7,900 miles in diameter with a thin veneer of salty liquid water, no more than ~6 miles deep, covering 70% of the surface, all embedded in a shallow atmosphere currently consisting of 21% molecular oxygen and 78% molecular nitrogen with a very small amount of Argon, CO_2, methane and other trace gases. The planet is in a relatively stable orbit around the Sun at a mean distance of ~93 million miles. It shares the inner solar system with the other small rocky planets, Mercury and Venus, and beyond Earth, Mars. Further out is a major Asteroid Belt of left over debris which causes us occasional problems as we have seen. Beyond the Belt are the well named gas giants, Jupiter and Saturn and beyond them, far, far away are the distant, less massive, ice giants Uranus and Neptune.

Beyond the ice giants lies the Kuiper Belt which hosts a large but unknown number of dwarf planets of which Pluto is the best known. There may be many bodies bigger than Pluto further out. The Kuiper Belt lies roughly in the plane of the solar system but with some bodies in more inclined orbits. From time to time planetary dynamicists see an unexplained pattern in the known orbits and propose a distant massive planet to explain it. In 2016 New evidence emerged for 'planet 9'. This is not impossible but so far extensive infra-red telescopic searches have not located such a body.

Far beyond lies the Oort Cloud of cometary bodies, a large sphere of objects stretching up to a light year away, a quarter the distance to the nearest star, Proxima Centauri.

We remember the Kuiper Belt as the source of the Late Heavy Bombardment of asteroids which blasted the Moon and the inner planets in the Hadean era. Occasionally also, a passing star will disturb the Oort cloud bringing a similar destructive shower of comets towards the Sun. This then is our solar system, at least in terms of solid bodies. Gravitationally the system is dominated by the Sun, Jupiter and Saturn. Although Jupiter has only 1 / 1050th the mass of the Sun it has over 90% of the system's angular momentum and we will find that this has massive consequences in our climate quest. So here we are floating in an apparent vacuum at a fixed distance from our primary heat source, the Sun. (The small secondary source being internal heat generated by the long term radioactive decay of unstable elements in the body of the planet and mainly the core). What could be simpler? The energy budget of the Earth must be a simple matter to understand and model. OK, let us start with the simplest physical considerations and build gradually towards a fuller picture. The climatologists, after all, tell us they like simplicity.

1. Assume that the earth is a perfect, spinning sphere without an atmosphere.

2. The sphere absorbs all the solar energy which falls upon it.

3. The sphere is a perfect radiator of heat.

4. The sphere has good thermal conductivity and is able to rapidly transfer heat to achieve a steady average temperature. Spinning also helps in this.

If the sphere orbits the Sun at a distance of 93 million miles in a circular orbit what equilibrium temperature should it achieve? The great physicist, Professor, Sir Fred Hoyle, did the calculations thirty five years ago (6). He obtained a figure of 280 degrees Kelvin. This 'absolute' temperature is better known as 280 − 273 = 7 degrees Centigrade. This compares with a current, global average temperature of ~14 degrees C. Isn't that remarkable? A simple calculation (for Hoyle) gets us within 3% of the correct answer. Clearly understanding climate is child's play just as the climatologists tell us. It's so easy, surely anything they tell us must be correct …like that CO_2 increase must lead to massive temperature rises of 3, 4, 5 or even 6 degrees C in this century! Unfortunately nature is not as friendly when dealing with real, complex systems. We need to remember humbly, that simple laws can lead to complex phenomena. Let's review some of our assumptions. The Earth is not a perfect absorber of radiation. Fresh snow, of which we currently have quantities in Antarctica, Greenland and the Arctic Ocean, reflects about 90% of incident sunlight.

One can already see that on a glaciated Earth the picture is not simple. Low clouds reflect about 50% of light on average. Any satellite photograph will show the large areas covered by cloud. Old ice and deserts reflect 35%; depending on vegetation cover locally, land areas only reflect 10 to 20%; 70% of earth's surface is ocean which reflects only 3% of incident sunlight. Only the oceans come close to our initial perfect absorber assumption.

By the way satellite images emphasise that much cloud cover occurs over the oceans. The contrast of the bright clouds to the dark blue oceans is startling and hints at the effect that variation in cloud cover may have. Overall, taking into account the relative areas of these surface types, the Earth currently reflects about 36% of the Sun's radiation back into space. It only has 64% of available solar energy to keep it warm. Keeping our other initial assumptions this leads to a calculated surface temperature of 250 degrees K or -23 degrees C…definitely in Snowball Earth territory. By being more physically realistic our climate model is now in error by 13%. Clearly we have left out some critical features of the real Earth. Well the real planet has seas and an atmosphere and it is not a perfect radiator as we assumed: it has 'radiation traps' of several kinds. The water vapour trap blocks the escape of all heat for wavelengths longer than 20 micrometres. The famous CO_2 trap absorbs very strongly from 14 to 16.5 micrometres. The fraction of radiation caught within these traps is 42%. Note carefully that 27% comes from the water vapour trap and 15% from the 'notorious' CO_2 trap. The Earth is a wet planet which must be remembered when considering claims about CO_2 forced 'global warming'. We will see that recent spectroscopic work suggests the effect of water has been underestimated. The radiation efficiency overall is only 58% so our Earth cannot cool itself as efficiently as our model sphere.

There are also other, partial radiation traps. Water has a partial trap between 16.5 and 20 micrometres and CO_2 between 13 and 14 mmts. These additional traps increase radiation blocking from 42% to 63%. However water vapour plays another key role. Evaporation over the warm oceans can carry water vapour in rising air columns to very great heights. When this vapour eventually condenses or forms ice crystals it frees a great deal of latent heat above the radiation traps which then escapes into space. By bypassing the traps this process reduces blocking from 63% back down to ~40%. So we have several processes in play, some increasing, some decreasing the energy balance of the Earth. So where did Hoyle's calculations leave us? We first found that the surface receives 36% less of incident solar radiation because of reflection. But radiation trapping and latent heat escape add up to a 40% reduction in radiation losses compared with the perfect radiator we first assumed.

There is thus a small overall energy gain of 4% compared with our simple sphere. This gives an equilibrium temperature for the more complex Earth model of 283 degrees K or 10 degrees C compared with the actual 14 degrees C, an error of just 1.4%. We can see that alterations in the radiation traps, latent heat transport by water vapour or the reflectivity (albedo) of the Earth will alter this result…as would variations in the Sun's output of course. The key point is that CO_2 level is not the only player and perhaps not the main player in climate change. What other factors, have, are and will change?

1.5 Why Climate Change?

'On balance, the effect of increased carbon dioxide on climate is almost certainly in the direction of warming but is probably much smaller than the estimates that have commonly been accepted'

Professor H H Lamb, father of historic climate analysis (8, page 666).

Our look at long term history of climate change over 4.5 billion years identified several candidate causes. On time scales of tens and hundreds of millions of years the disposition of the drifting continents relative to the poles and periods of mountain building appear to be important. On similar time scales mantle plumes and hot spots can wreak havoc if they break through the crust. It is possible also that some flood basalt episodes are triggered by major asteroid or comet collisions which in themselves cause only blips in climate. We have seen that the Sun itself, according to modern stellar theory, has increased in brightness by 30% since its birth and will continue to brighten. Some theorists also claim that the Sun may have periods of mixing instability in the core every few hundred million years, with temporary falls in luminosity of ~5% (8; page 286). It is now known that of the thousands of sun like stars studied a very few have shown sudden falls in luminosity of a few percent but evidence is sparse. We will look at strong evidence that the boundary between the core and convective mantle of the Sun is affected by planetary gravitational effects in another context (see below)…but this should alert us to possible core disturbance effects. It is also certain that large asteroids and comets fall into the Sun from time to time: perhaps large enough to disturb the dynamo.

Later we will look in detail at 'solar' activity cycles on scales of 10 to 1,000 years duration but note now that total solar irradiance appears to vary by only ~0.1% over these cycles. It is this observation that leads to the dismissal of the Sun as an important source of climate change.

More recently it is suggested that what matters are variations in particular regions of the solar spectrum such as ultraviolet light acting on the stratosphere. This variation too was dismissed as too small but recent work proves that the variations are several times larger than previously believed based on NASA's SORCE satellite data (9). These variations appear to have a significant effect on European winter temperatures. Significantly this research was undertaken by the British Meteorological Office and may perhaps open closed minds about the role of solar influences.

However our Sun is not just a distant, isolated, radiator of energy. We live within its extended, variable atmosphere. Our spinning star has a powerful internal dynamo generating a massive magnetic field within which we orbit. The visible signs of that dynamo are the magnetic disturbances on the surface which we call sunspots. The variation of sunspot activity with a primary cycle of ~11 years has long been recognised. The peak amplitude of activity also varies over centuries and sometimes these cycles turn off completely for periods (so far observed) of many decades. Repeatedly these cycles have been linked, for a period, to weather variables on Earth only for the good correlations to disappear. Consider for example the link which caught the author's eye many years ago in Nature: a remarkable correlation between cycles in UK growing season length and sunspot numbers over some decades (Figure 9). Because these periods of correlation evaporate many deny any sunspot connection. There are two reasons for this problem. Firstly sunspots are only one measure of solar activity, the first readily observable to astronomers. We will see that there are better (and longer term) proxies for solar activity.

Secondly we will see that the variation of the Sun is complex and it interacts with the non-linear climate systems in complex, non-intuitive ways which hide the links most of the time. By 'fingerprinting' of the solar activity time series and climate time series and a bit of 'nonlinear' detective work the actually strong links will emerge.

For our climate quest the key factors turn out to be variations in the Sun's magnetic field and the strength of the powerful solar particle wind. The whole solar system sits within a gigantic magnetic bubble. The breathing in and out of this bubble determines how well protected we are from incoming interstellar material. This material includes energetic galactic cosmic rays and radiation. Cosmic ray showers reaching the upper atmosphere liberate electrons which aid the formation of nuclei around which water vapour condenses to form cloud droplets. Some energy levels of cosmic rays at low levels of the atmosphere catalyse low cloud formation, increase reflection and so alter the radiation balance of the Earth as we will see.

F 9

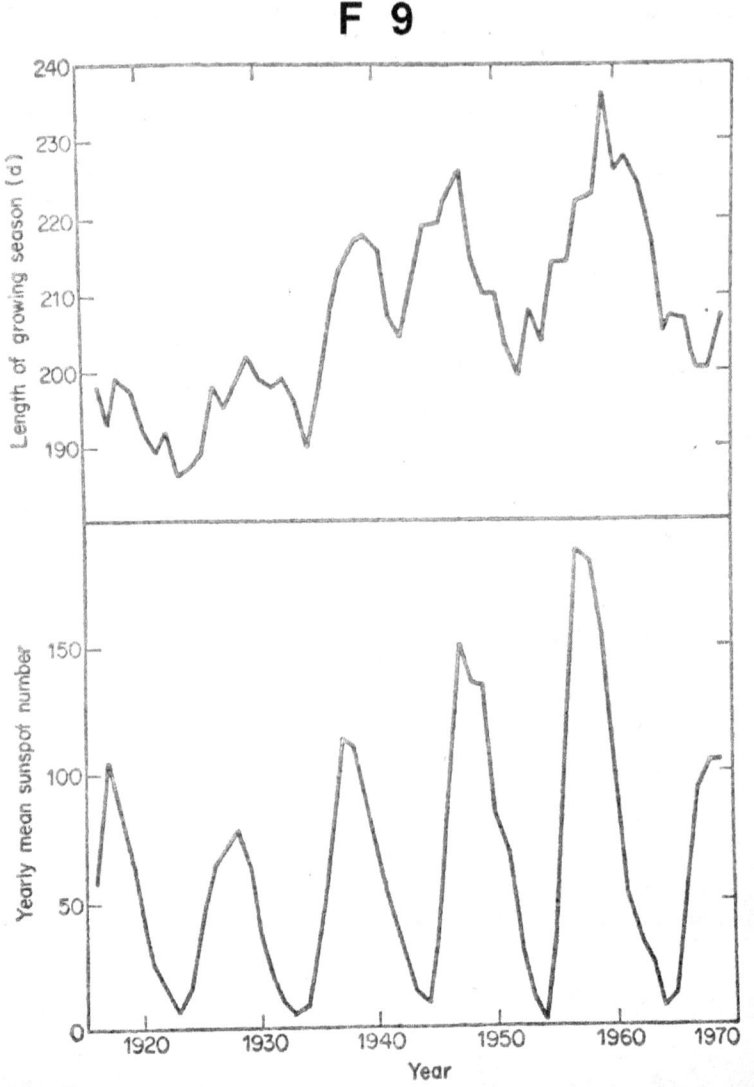

This cosmic ray flux is modulated by the strength of the solar magnetic field and the solar wind. In this way the Sun affects climate. This 'cosmic ray – cloud' theory of Svensmark (10) has proved controversial and has been ruthlessly attacked by the 'climate change' establishment. Nevertheless supporting evidence is increasing as we will see. Systematic changes in cloud cover and the Earth's albedo must be important. This is a conclusion Hoyle drew over thirty years ago in considering the radiation traps, latent heat transfer and albedo.

This analysis holds today and carries important lessons. For a major climate excursion one or more of these factors must change if the input solar radiation is constant. With a mean temperature today of 14 degrees C and 25 degrees C in the topics, the rate of evaporation should always be much higher than the minimum necessary to preserve the water vapour radiation trap and to carry latent heat above the traps. This leaves the CO_2 traps to consider. If CO_2 was somehow completely removed from the atmosphere the radiation efficiency of the earth would rise by 15%, increasing from 60% to 75%. Calculating as before Hoyle finds an equilibrium temperature of -3 degrees C or 270 degrees K. If nothing else changed this could precipitate an ice age.

However very low CO_2 levels would very quickly lead to a catastrophic reduction in land and ocean plant life and thence animal extinctions. There is nothing like this in the fossil record during the decline into recent ice ages. Ice age CO_2 levels were around 180 ppm compared with 400 ppm now. CO_2 was low but not negligible and life responded. During the long decline towards cooler conditions in the Miocene a new class of plants, the C4s, evolved including many grasses, with biochemistries still efficient at lower CO_2 levels (1; chp. 6). In the longer term as the Sun continues to brighten and less CO_2 is required to keep the Earth warm, the C4s or something new, will dominate land ecosystems.

Back to the present. Hoyle expressed clear views about the CO_2 trap based on 'simple physics' (6; chp. 7).

'The efficiency of the CO_2 radiation trap is insensitive to the amount of carbon dioxide in the atmosphere : increasing the amount fivefold would scarcely change the trap in spite of the stories that are currently being circulated by environmentalists. Only if the amount of CO_2 was enormously increased (many hundreds of times more than at present, as was hypothesised for the Archean era) would the trap widen its influence significantly'

If increasing CO_2 is a threat, this observation at least extends the window for remedial action into centuries. We should also remember that at the Palaeocene-Eocene thermal maximum when temperatures were several degrees warmer and CO_2 level was several times higher than now, the major hyperthermal spike events related to sudden large release of CO_2 from permafrost, were short lived: the system responded and some feedback process brought temperatures and CO_2 back to previous levels. There was no runaway 'greenhouse' disaster. Note also that radiation heat loss from a body goes up exponentially with temperature.

If some runaway process raised temperatures, increased heat losses would soon balance the books. Perhaps Hoyle should be listened too more carefully.

Today the CO_2 level has just reached 400 ppm and during recent glacial periods it has fallen to ~180 ppm. Mean temperatures in the other warm interglacial periods have been very similar, around 14 degrees C. At the ends of the last several ice ages temperatures and CO_2 have risen rapidly within several centuries between these CO_2 and temperature limits. Why did these rises not continue? Mathematically global temperature and CO_2 time series have the appearance of a 'limit cycle' bouncing between two (temporary) equilibrium conditions. (What powers the bounce we will discuss later). To the author this speaks of a set of negative and positive feedback processes acting together. If so perhaps Hoyle is correct, but not only is the efficiency of the CO_2 radiation trap insensitive to CO_2 levels, something else is acting to stabilise the system. Will CO_2 be 'allowed' to rise much further? In section 5.11 we will look in detail at long climate proxy records covering the recent ice ages and show that recent research strongly supports an underlying non-linear process with two highly stable, limiting temperature states.

Lovelock (1) makes the case that the Earth's bio-geophysical system may indeed self-regulate if left to itself. But we are not leaving it to itself. It may be extremely unwise to interfere with a natural system which pumps down CO_2 and has been doing so for a long, long time. The soils in areas with vegetation contain 10 to 40 times the equivalent amount of CO_2 as exists in the atmosphere. Huge amounts are also tied up in land biomass and ocean life forms and sediments. There is an inverse relationship between the abundance of CO_2 and the abundance of vegetation. It is therefore a very bad idea to chop down the tropical and temperate forests. That is probably where the real climate threat lies.

The question of why CO_2 rose rapidly at the end of the ice age (and previous ones) is instructive. Was CO_2 the cause of the temperature rise? When the 'global warming' palaver first emerged this was claimed while ignoring the millennium long lag between the rise in (proxy) temperature series and the rise in atmospheric CO_2. Miraculously the effect preceded the alleged cause! More recently this 'inconvenient' lag has been reduced by 'adjustments' to the data but a lag is still there. However contrary evidence keeps popping up 'inconveniently'. Professor Stott, et al, of USC (11) based on Western Pacific seabed cores, now say that deep sea temperatures began to warm at ~19,000 BP about 1,300 years before the tropical surface ocean and well before the rise in CO_2.

This coincided with the retreat of Antarctic sea ice attributed to increased solar input from the Milankovitch orbital / axis cycles and then changes in sea surface albedo. The warming came from the bottom up according to Stott and increased CO_2 production followed later. Interestingly enough Stott is well respected and has in the past been a scientific reviewer for the IPCC.

Now the 'official' story is that something suddenly caused a temperature increase (after 100,000 years of cooling and ice age conditions) which led to a CO_2 increase, which led to a further temperature increase, and so on in a runaway, positive feedback process. Such a feedback is plausible. CO_2 is less soluble in warmer water and so on. Higher CO_2 would also normally suggest lower vegetation productivity. Did initial rising sea levels as some ice melted, flood swamplands and forests on the continental margins, releasing CO_2 into the atmosphere? Nobody knows. Recent accepted models of the ice age 'explain' that 40% of the rapid warming came from rising greenhouse gases (23) so the initial warming at the end of the ice age and 60% of the warming overall are down to some other cause ... supposedly unknown. The rapid CO_2 'runaway' ended at ~300 ppm because of some strong negative feedback process (or a sudden change in some external forcing factor) which has also kicked in for several previous ice ages and left the Earth at close to the same global mean temperature, ~14 degrees C, during interglacial periods. Given this million year history of stabilisation why do we expect the current slow increase in CO_2 to disturb what appears to be a natural equilibrium state of the climate system? We already know that so far over half the man made CO_2 put into the atmosphere in the last century or so has disappeared...somewhere. Is the stabilising feedback process already in action? Below we will look at some of the likely candidates which, following Hoyle, relate to the reflectivity of the Earth and changes in albedo.

1.6 Heresies : Biofeedback, Planets, Clouds & Cosmic Rays

> 'So its shift boys shift
> Of that there is no doubt
> It's time to make a shift
> To the stations further out.'
>
> The Drovers Song

Having looked at the history, and more or less conventional interpretations of climate change, it's time for a little heresy, or rather a linked double heresy and some other problematical matters.

If the water vapour and CO_2 radiation traps are essentially fixed in efficiency, there is sufficient latent heat transfer beyond the traps in our current ice epoch, and the disposition of the continents and the patterns of mountain ranges are fixed, Hoyle's energy budget model forces us to look more closely at albedo changes; at the reflectivity of the Earth. We noted the very wide range of reflectivity of the various surface types earlier: from dark oceans to white snow and clouds. Trying to learn lessons from the rapid temperature rises at the end of the ice ages we require rapid, sustained changes in albedo. That points us to climate sub-systems that can change rapidly in locations where albedo changes can be extreme: the low clouds over the oceans. Cloud can reflect 90% of sunlight back into space; oceans reflect as little as 3%.

Let's look more closely at clouds. The physics is very complicated which is the excuse used by some climatologists for not including what *is known* about clouds in their models. The proper course would be to try out all possibilities and look at how the results differ. God knows, enough competing climate models have been built in recent decades. But why bother when you know the politically correct answer is CO_2? Water vapour rises in warm air columns above the oceans until it condenses into cloud droplets and in the right circumstances these will freeze. However even supersaturated water vapour will not condense in the absence of nucleation particles. Small salt particles produced from wave spray can serve as nuclei but there are other, more powerful sources.

Sulphuric acid and methane sulfonic acids are the main source of cloud nuclei. These acids are formed by the action of oxygen on sulphur compounds. This includes SO_2 released by volcanoes and industry. But the main suppliers are the ocean phytoplankton. As early as the 1980s it was known that dimethyl sulphide generated in the oceans was the main, windborne, carrier of sulphur from the oceans to the land. The quantities of dimethyl sulphide and carbon disulphide released are immense. So most cloud nuclei derive from ocean life and clouds have a large effect on the Earth's albedo and energy budget. Why do ocean plankton emit dimethyl sulphide? Too high a salt concentration in living cells is lethal. Salt can be buffered in cells by a class of compounds called betaines.

The primary betaine used by our ocean plankton is dimethylsulfonio proprionate. When plankton die or are eaten this compound decomposes into acrylic acid and dimethyl sulphide. Lovelock and others (1; chp 6) have calculated the rate of natural emission of dimethyl sulphide and its influence, via sulphuric and sulfonic acid nuclei seeding, on cloud formation above the oceans.

They showed a magnitude of effect comparable with that of CO_2 'greenhouse warming' but in the opposite direction. The ocean plankton have the capacity to cool the Earth and plankton have been around for over 3 billion years. Are they part of a bio-feedback system maintaining optimum conditions suitable for them and incidentally other life forms? There is an optimum set of conditions for growth in a given species of 'plant'. Does plankton productivity increase with CO_2, increasing dimethyl sulphide output and clouds, so limiting further temperature rises? It is not likely to be that simple. In fact there could be two or more local optima for productivity with different combinations of environmental conditions. There may be other effects. Ocean clouds increase local wind speeds and the winds mix deeper nutrient layers with the surface photosynthetic layer, to the advantage of the plankton. In early times species which promoted cloud formation, comfortable temperatures and nutrient mixing would be favoured by natural selection. It is also notable that more cloud, overall, implies more rain falling on adjacent land areas, increasing chemical weathering and increasing nutrient flows into the adjacent ocean...again to the benefit of plankton.

These ideas are difficult to test but intriguing. Lovelock (1; chp.6) reports the discovery of sulphuric acid and methane sulfonic acids in Antarctic ice cores over a 30,000 year period. There is a strong negative correlation between isotope proxies of global temperature and the concentration of these acids in the cores. Sulfonic acid levels were 2 to 5 times higher during the last ice age. Sulphuric acid may derive from volcanic SO_2 but the sulfonic acids can only come from living sources. Oceans become saltier when water is locked up in polar ice caps so did the need for salt buffering in ocean plankton lead to more dimethyl sulphide emissions, more cloud and further cooling? Is the current 'favoured' state of the climate a slowish slide back to cooler conditions? If so what is it that suddenly ends an ice age and causes a rapid temperature increase ...before CO_2 increases in response? Is there an external trigger or some failure in the global ecosystem? We do not know.

Let's look back at the long term again. It is interesting that, like Hoyle, Lovelock realised the dominance of Earth's albedo in controlling the energy budget of a wet planet. His famous, elegant, Daisyworld models (1; chp 3) demonstrated vividly how a bio-geophysical self-regulation system could evolve naturally without a hint of the dreaded 'teleology' word. In his model several species of daisies of various colours from white (like the clouds) to black (like the oceans) evolve and coexist. Over the aeons, as we earlier examined, as the Sun became brighter, the mix of daisies evolved to reflect more light into space so maintaining good conditions for daisies.

Lovelock troubled his world with several kinds of disasters and plagues and demonstrated the ability of the system to absorb punishment and bounce back…just like the Earth over 4 billion years of climate change. Eventually of course Lovelock's self-regulated Gaian world failed as his Sun reached the end of stable nuclear burning. In our real world there probably exist many positive and negative feedback climate loops and it is certain we have not identified, let alone modelled, them all. We do know that one recognised factor, clouds, are not fully understood and not taken into account realistically in current climate models. That is a deeply troubling omission as are the many missing, possible roles of Earth's life forms.

This is not the end of the story of albedo changes and the clouds. It is time to consider a second heresy. We saw that clouds only form in the presence of seeding nuclei: the water vapour condensation nuclei provided by tiny sulphuric acid and methane sulfonic acid particles and to a lesser extent salt particles in the atmosphere. That is accepted. The interesting question is what else is necessary to build nuclei of the correct size from the molecular level to something much bigger? The ultra-small proto-nuclei formed from gas molecules are only a few millionths of a millimetre across and are too small to aid drop formation. The proto- nuclei must somehow clump together to form particles at least 100 times larger to be effective. The conventional model assumes brute force, that is, lots of acid and time. However observations of atmospheric 'nucleation burst' events suggest something else is happening. The additional factor is ion seeding: electric charges encourage the building of larger condensation nuclei. Where did these charges come from? Svensmark and others (12) have argued and demonstrated that electrons generated by secondary cosmic ray collisions with air molecules are the catalytic villains of the story. These ideas were and still are controversial and strongly (some would say fanatically) opposed by the man made 'global warming' lobby and many scientists.

The animosity goes back to 1992 when the IPCC **rejected** a Danish proposal to include the Sun in a list of future climate research topics. The cosmic ray militant tendency has worked over many years to build evidence for the case of the clouds on several time scales. Most impressively a series of physical (not hypothetical computer model) experiments, beginning with SKY and culminating in CLOUD at CERN, used artificial cloud chambers, natural cosmic rays and simulated rays (from particle accelerators) to test the idea of electron catalysis of cloud nuclei formation. The cosmic rays strongly catalysed the formation of sulphuric acid and water aerosols of the right size in the cloud chambers as predicted.

By the way they also used ultra violet light to convert SO_2 into sulphuric acid as happens in nature. Recently it has been found that solar ultraviolet varies far more than originally thought over the solar activity cycle. Perhaps this is another source of cloud nucleation variation down to the Sun. In 2011 the eminent journal NATURE published the results of this research (13). Despite the positive results and the successful passing of the paper through peer review, the leaders of CERN ensured a politically correct report of the outcome so as not to offend the global warming lobby. Nevertheless the author's still managed this

'[cosmic ray action] will manifest itself as a steady production of new particles...and the effects could be quite large when averaged globally over the troposphere [lower atmosphere]'

Readers who want to follow the development of the Svensmark Heresy should read his (and Nigel Calder's) book, 'The Chilling Stars' (10). The range of evidence collected by Calder and Svensmark from many other workers is impressive. Evidence continues to accumulate but I mention only one recent item of particular interest. Many attempts have been made to mathematically link cosmic rays and global temperature variation. It is a difficult task for technical reasons explained in Appendix 1. New and powerful mathematical techniques are helping. Recently 'convergent cross mapping', CCM, has been used to identify a causal (not just statistical) relationship between the detrended HadCRUT4 global temperature record and cosmic ray intensity through time (14). The authors rejected other global temperature records because of known (ie notorious) 'data infilling, spatial over-averaging and interpolation': adjustments known to obscure non-linear effects often dominant in this arena (as we will see repeatedly in our later analyses). The authors proved definitively the causal link between inter-year variations in temperature and cosmic ray intensity. **The link is proven.**

The paper was published by the National Academy of Science of the USA and the authors are reputable. The lead author is Hao Ye from the eminent Scripps Institute of Oceanography. However of particular interest to me was the fourth author listed : Professor, the Baron May of Oxford, president of the Royal Society 2000 – 2005 and former Chief Scientific Advisor the Her Majesty's Government in the 1990s. May is a pioneer of non-linear mathematical modelling in ecosystems and a creative explorer of the seminal new ideas in Complex Adaptive Systems Theory ... popularly called Complexity. I had the pleasure to briefly cross verbal swords with him at the Santa Fe Institute and I can attest to his sharpness both scientific and political. I see it as highly significant that May has put his name to this paper.

It may tell us that the tide of evidence and opinion on the Svensmark heresy is turning. However this analysis was not all good news for the heretics. The temperature and cosmic ray data was de-trended and so the established causal relationship was confined to year on year variation …not trends. This means that this particular analysis could not explain global warming trends by cosmic ray variation. But as usual nothing is simple and the plot thickens. For example the authors used the aa geomagnetic index as a proxy for cosmic ray intensity. The aa index is computed from data at just two stations : Hartland in the UK and Canberra, Australia. The index is known to reflect the effects of : solar radiation, the solar wind (which should be anti-correlated with cosmic ray flux), interactions with our magnetosphere, interactions within the ionosphere and interactions between them, etc. This example illustrates the difficulty of obtaining pure test signals to explore geophysical relationships. The aa index reflects *all* cosmic ray flux and other things but Svensmark's theory fingers only a sub-set of rays, in a particular energy range, as the modulator of low cloud formation.

A direct measure of cosmic rays of the right energy range is really needed to test the theory. A few direct neutron count time series for cosmic rays are about but only date from the 1960s or later. I have not found any analyses of this type using direct cosmic ray data. On longer time scales Beryllium 10, Carbon 14 and Oxygen isotope series from sediment cores, as proxies for cosmic rays and temperature, go back much further but again what exactly do they measure? No wonder proving causation is difficult. However we can at least look in detail now at the spectral 'fingerprints' of climate proxies and geophysical and astronomical time series and show identical features, particularly if we accept the idea that these signals are being filtered though strongly non-linear physical systems.

Recently Svensmark et al (27) has published new, direct evidence of significant solar-cosmic ray-cloud effects. Occasionally the Sun suffers major eruptions and the enhanced solar wind blows galactic cosmic rays away from the Earth. Such events a known as Forbush decreases. Svensmark has now looked at the effects of the five largest events since 2000 on liquid cloud cover. Over a week or so atmospheric ions reduce by ~20-30 % after the event. Corresponding cloud fraction falls by ~2%. These large effects are short lived but the authors suggest that similar, slower variations in solar output over decades or centuries have a similar scale of effect. It is claimed that a 2% increase in cloudiness could offset a 100% increase in CO_2. Is this credible? Let us assume for the moment that cosmic rays and bio-feedback do have an influence on climate via the clouds and the Earth's albedo.

How big could the effect be in principal? If large changes make no difference we are on the wrong track. The cloud cover is calculated currently to reduce incoming solar energy by 8%. Keeping the other factors constant, removing the clouds completely would raise mean global temperature by several, possibly 10, degrees C. This is not trivial. An increase in low level clouds by a few percent would yield significant cooling. Svensmark (10; page 89) used data from the Earth Radiation Budget Experiment satellite to calculate cloud sensitivity at various latitudes. A 4% reduction in clouds means a 1 degree C rise in temperature at the equator and vice versa. In the Antarctic the cooling sensitivity was less at 0.5 degrees C. This is because the albedo difference between ice and cloud cover is less than between cloud and ocean, or typical land surfaces. So overall an increase in cloudiness of 2% could offset an increase of 100% in the carbon dioxide level. If correct this is highly significant.

Lockwood et al (15) were able to estimate from the aa Index and historical records that over the 20th century the Sun's magnetic field strength had increased by a factor of 2.3 X. Svensmark and Marsh used this and recent cosmic ray data to calculate that this was equivalent to a 8.6 % reduction in low level cloud cover as the Sun became more active. This is a large effect if correct. They pointed out that this implied a forcing of 1.4 watts / square metre, a rather politically incorrect figure since it is the forcing strength attributed by the IPCC to global CO_2 increase. If the Sun's magnetic field is weakening (towards a Grand Minimum) as many astronomers now believe we may soon discover who is right. In fact in his 2014 paper Lockwood noted that the open solar flux had fallen sharply since ~1990 and was now approaching 1900 levels again. If a significant proportion of 20th century warming was due solar flux variations we would soon expect significant cooling…but surely CO_2 must have some effect?

So the next questions are: how does the galactic cosmic ray flux change over various time scales and why? The answers are fascinating and may have relevance to climate change on all the time scales we considered in our history of climate….over 4 billion years. Firstly, just where do cosmic rays come from in space? Primary 'galactic' cosmic rays consist of protons and alpha particles with a small amount of anti-matter, positrons and anti-protons, thrown in. These particles are generated in supernovae explosions as stars die in our own galaxy. We also receive X rays and Gamma rays from galactic and extra-galactic sources. Very occasionally, fortunately, we receive the tail end of gamma-ray bursts via the relativistic jets of forming black holes and distant active galaxies. It is now thought that the mass extinction event at the end of the Ordovician period, 444 million years ago was due to a GRB stripping the ozone layer and exposing surface life to heavy radiation.

Recent modelling work suggests a 90% probability of a lethal (100% surface life gone) burst over the life of the solar system and a 50% probability of one in the last 500 million years (i.e. since the Cambrian period) (24). Perhaps the long lag in the development of complex surface life after its early start is down to a GRB event…or two?

Back to ordinary cosmic rays. On collision with the atmosphere these high velocity particles smash into air molecules and create an air shower of secondary cosmic rays including muons, pions, electrons and neutrons in addition to the original protons. Muons (a kind of heavy electron) have sufficient energy to reach the surface or even penetrate into shallow mines. The energy range of the cosmic rays is wide and recent evidence suggests a 2 : 1 flux variation on roughly thousand year time scales, sustained over tens of millennia. The strength of the solar wind is sufficient to strongly modulate the cosmic ray flux. Far beyond the planets the heliopause marks the edge of a magnetic bubble reducing the ray flux at lower energies by 90%. The Earth's magnetic field also deflects cosmic rays of lower energies that get thought the Sun's magnetic bubble shield. Only some energies of the cosmic rays that reach the earth are relevant to low cloud formation so the picture is complicated.

On longer time scales of tens of millions of years, cosmic ray flux approaching the solar system is believed to vary considerably as our star system circles the galactic centre with a period of ~240 million years. However individual stars move at different speeds relative to the rotation of the spiral arms of the galaxy. It is here that dust and gas is concentrated and the leading edges of the arms are the sites of star formation. Here we find concentrations of large, short lifetime stars and supernovae …the generators of cosmic rays. It follows that as the Sun transits a spiral arm it will encounter high cosmic ray flux. Some physicists like Professor Shaviv (16) suggest that periodic spiral arm passage means more clouds, more cooling and the start of long ice epochs such as we discussed in our climate history. He attempted to reconstruct cosmic ray flux history and found a strong correlation with climate variations over 500 million years. We can see that cold periods marked in greys coincided closely with high values of delta oxygen 18 used as a cosmic ray proxy. The upper panel shows various reconstructions of CO_2. Notice the wide variation of the CO_2 estimates. He concluded that 66% of the variance in the palaeo - temperature trends could be attributed to variations in cosmic ray flux with the rest possibly due to CO_2 variation, although he could see little correlation with the climate record. He believed that CO_2 forcing on these time scales must, at least, be much smaller than the IPCC claims.

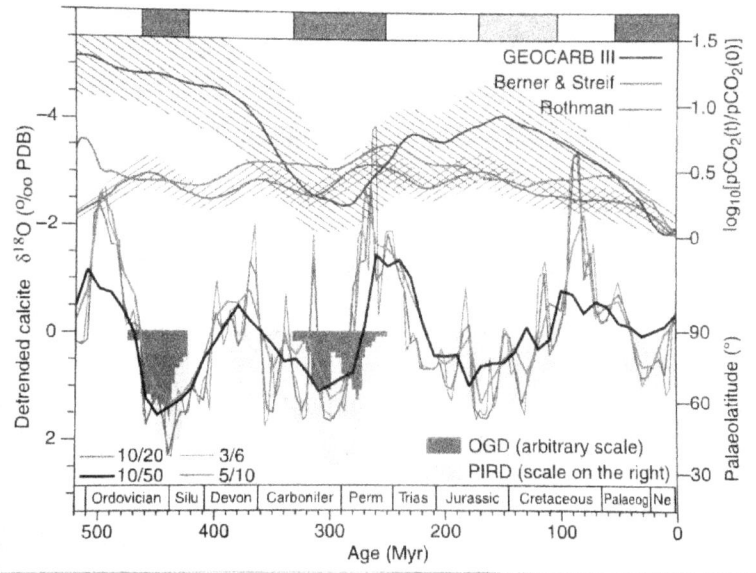

Shaviv concluded that if the Sun explains about 1 degree C variation in the short term then a passage through a galactic arm should lead to changes of the order of 10 degrees C...enough to start an ice epoch. Naturally Shaviv's extensive work on 'cosmic' climate change was widely attacked by the CO_2 lobby since in essence it adduces the cloud modulation process proposed by Svensmark... and so the raucous debate continues.

The long term history of CO_2 level is uncertain and analyses differ. However looking at the usually accepted estimates above we can see that the Cambrian levels were between 10 and 20 times current 'quaternary' levels. This was the time of the great explosion of complex life in the oceans. CO_2 fell to a minimum, roughly at current levels, during the long ice age epoch of the Carboniferous and Permian. It then rose again during the Triassic, Jurassic and Cretaceous to levels between ~500 and 2000 ppm reaching 15 time current levels at the Jurassic peak. This was the age of the dinosaur ascendency with rich, diverse life on land and in the oceans. The dinosaurs and their relatives thrived for over 100 million years.

Further work and quieter voices suggest a role for cosmic rays and CO_2 variation on a range of time scales. Taking climate sensitivity as the temperature increase caused by doubling the current CO_2 level, the IPCC has reported values up to 6 degrees C.

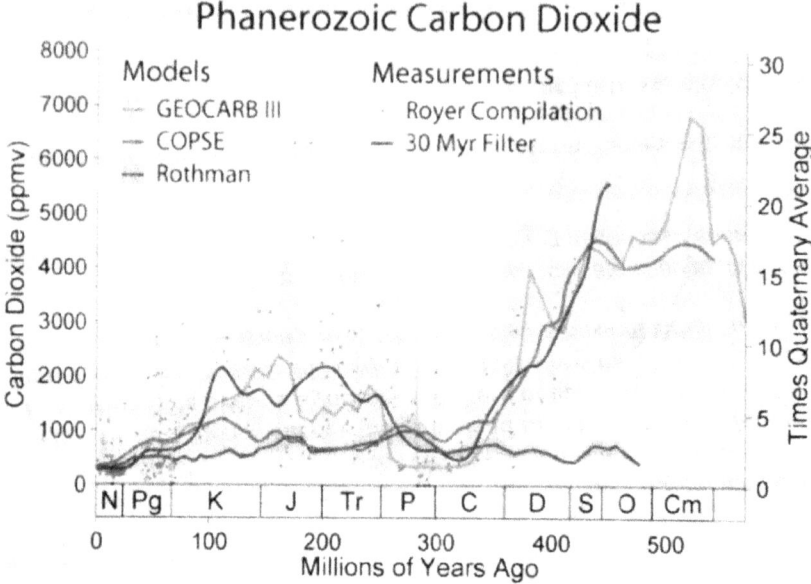

Shaviv and some other (respectable) climatologists now tentatively accept ~1 degree C, but remember we still have not explained why interglacial temperatures rose quickly and suddenly stabilised at the same level following the last four ice ages ...along with CO_2. Forcing strength may not be constant through time. We are still missing some stabilising feedback process which must still be in operation, or some change in external forcing of the system... in which case modest rises in CO_2 may be irrelevant. Under current conditions (solar output, continent positions, etc) temperature dynamics are acting as though they are described by a strange attractor with two quasi-stable fixed points: ice age conditions and interglacial conditions (see Appendix 1 for a discussion and examples of strange attractors). They have so behaved for at least a million years. Looking at the movements of ice volume and temperature proxies, the mix of frequencies dominating the ups and downs has shifted slowly but there is no sign of a disruption of the attractor. The temperature dynamic attractor over the last half a million years has been remarkably stable. Recently studies have confirmed, that for this period at least, the global temperature record can be reconstructed by defining a strange (non-chaotic) attractor driven by a 'cosmic' signal we will look at shortly. The CO_2 issue comes down to this: the 'global warming' lobby fear in effect, that the climate system will somehow move to a third, far away and hot, fixed point of attraction.

If so, they have to say what they believe has changed in the system to alter the stabilisation process (invisibly) since the last ice age. It almost certainly cannot be a small change in CO_2 around a level we have seen before and higher in several other interglacials over the last million years or so. Temperature reconstructions by NOAA from ice cores show peak interglacial temperatures for the last 4 ice ages of + 2.5 d C, 2.1 d C, 3.2 d C and 2.1 d C compared with the 1960-1990 mean baseline. Incidentally, during the early phase of this current interglacial, temperatures also briefly (for a few centuries) reached 2.1 d C above the current level. Interglacial peak CO_2 levels also remained surprisingly constant at ~ 280 ppm. Going back four hundred thousand years we see no sign of a runaway temperature event.

Of course so far we have only looked at part of the 'cosmic' story. Let's clear the decks of other long term effects before we proceed. In addition to circling the galactic centre in about 230 - 250 million years, the solar system also follows a slow oscillation, rising above and then falling below the galactic plane. A crossing occurs every ~32 million years. Star density is highest in the plane and the galactic magnetic field concentrates cosmic rays near the plane so we might expect cooling episodes on Earth every 32 million years and this has been claimed by Shaviv and others. The story never ends! Recently a remarkable new theory has been published which accepts the switchback passage of the Sun but posits a concentration of particular dark matter particles in the galactic plane sufficient to gravitationally disturb our outer Oort Cloud of comets as we cross the disk, which then fall into the inner solar system and create havoc, including climate change and mass extinctions (17). However the reconstructed cratering rate record of the Earth does not definitely indicate any strong periodicity although there is a (statistically non-significant) hint of a cycle near 30 million years in some studies.

Even more remarkable it has been claimed by Shaviv and others that near by 'star burst' episodes in, for example the Magellanic Clouds, coincided with the ice epochs (and Snowball Earths?) of ~2300 million and ~700 million years BP. Irrespective of dates it is true that the Galaxy is the gravitational centre for a dozen or so dwarf galaxies and vast streams of stars which are the remains of collisions between us and other dwarfs in the past. Such collisions compress interstellar dust and gas and greatly enhance star formation...including the short lived supernovae whose deaths fill space with cosmic rays. Star burst events also occur in other galaxies in the local group and energetic cosmic rays can travel inter-galactic distances to be captured by our Galaxy. The solar system and the Earth, despite interstellar distances, are not isolated entities.

Perhaps now we should turn back to the recent past and the present and to 'cosmic' processes which are acknowledged, reluctantly, as real against all the odds. Later our detailed forensic 'fingerprint' analyses will demonstrate these processes beyond doubt. Let us begin. The Earth swims in a complex ocean of gravitational currents defined by the Sun, Moon and planets. We should say at the beginning that on times scales of tens of millions of years that ocean is turbulent and unpredictable…some say chaotic.

On shorter time scales of decades, centuries and millennia, the currents are more regular but still complex. We are talking about time scales from those of a human to those of a civilisation.

In the 19th century the brilliant 'amateur', James Croll, without benefit of modern computers, calculated the effects on climate of the precession of the equinoxes and its effects on radiation received on different parts of the Earth from the Sun. The Earth spins like a top and the direction of the axis of rotation sweeps out a circle in the sky with a period of ~25,800 years. Additional planetary effects also change the orientation of the orbit so that radiation received by north and south hemispheres rises and falls by several percent on a cycle of ~23,000 years. The cycle alters the times of year at which the Earth is nearest the Sun in its eccentric orbit. The distribution of land and sea and ice and hence albedo is not the same in the two hemispheres and this may be significant. Later Milutin Milankovitch added further effects to the story. He pointed to the periodic change in the angle of inclination of the rotation axis to the ecliptic plane between 22 and 24.2 degrees. Calculation of solar radiation at mid summer at a latitude of 65 degrees varies by 7% across this axial tilt range. The tilt cycle has a period of ~41,000 years. The periods are non – rational so the two cycles interact to produce a more complex climate forcing pattern. The orbit of the Earth also changes in eccentricity…it stretches and relaxes in a cycle of ~100,000 years and there is another longer cycle of 413,000 years.

These 'cosmic' cycles look like prime candidates for climate forcing factors but the calculated effects at face value are too small, it is argued, to 'cause' the current ice ages. It is embarrassing for the climatologists that the three Croll – Milankovitch cycles show up clearly in global temperature proxy times series for the last million years despite being 'negligible' under conventional thinking. We will show further that most of the temperature and ice volume variation within the ice ages can be simply related, assuming a non-linear system, to these cycles. This is backed up by the recent study showing how the cycles can match the ice volume record we mentioned above (18). These cycles certainly modulate the ice age temperatures even if they are not the ultimate 'cause'.

We will see that the cycles turn up in the climate 'fingerprints' as harmonics and interaction frequencies between them: classic indicators of a forced non-linear system. As usual things are not so simple. Recent work by Sharma (19) used proxy data for solar activity (cosmic ray produced Beryllium 10 levels) and Earth's magnetic field intensity to reconstruct the Sun's long term magnetic activity history. He then compared this with (oxygen isotope) proxy temperature data and found a strong correlation. His conclusion was that the current series of 100,000 year glacial cycles was down to variation in the Sun itself, not the 100,000 year Milankovitch cycle in the Earth's orbital parameters. Ironically Sharma and Croll-Milankovitch may both be correct in that planetary orbital dynamics appear to drive both solar activity variations and changes in the Earth's orbit, at least over some time scales. We will show in section 3 that these effects are clear in the various spectral 'fingerprints' of the several suspects along the paths of connection, making the unravelling of the details of causation challenging. Nevertheless the clear links are there.

Coming down finally to human time scales and looking ahead: comparing the frequency 'fingerprints' (or spectra) of geophysical, global and regional climate variables, solar activity and planetary orbital dynamics we will demonstrate in sections 3, 4, 5 clear and significant links …provided we look though non-linear system spectacles. All this sounds uncomfortably like astrology and magic!

So that the reader is not prematurely put off I will summarise one recent, published study which provides a credible physical mechanism for planetary 'influences' on the Sun (20).The angular momentum of the four giants, Jupiter, Saturn, Uranus and Neptune is so large that the centre of mass of the solar system sometimes lies *outside* the Sun. As a result the Sun experiences a complex quasi-periodic swing around the barycentre and very real torque effects, just like a looping skater (Figure 10). The tachocline is the region where the Sun's powerful magnetic field is generated by the solar 'dynamo'. Differential rotation in the outer layer shears and intensifies the rising magnetic field in certain places, generating the famous sunspots on the surface. The varying torque caused by the swings of the Sun somehow modulates the dynamo to produce variations in magnetic field strength, the solar particle wind and the sun spots.

Whatever the details of this process, periods directly related to the orbital periods of the planets show up clearly in proxies of solar activity like the 'solar modulation potential' times series and their spectral 'fingerprints'. The torque time series and the SMP time series are identical in the main periods appearing in both, although this is disputed as we will see.

Even more telling a closer look by this author shows other significant periods in the SMP spectrum which are harmonically related to the planetary spectrum peaks. (See section 3.6). The Sun, probably through the proposed dynamo modulation mechanism, is responding to planetary gravitational forcing as a non-linear system. Most of the spectral energy in the SMP variation is straightforwardly related to the planets. We will look at these results in detail in section 3.6 but the reader can be reassured: this result is indisputable; the planets are guilty as charged.

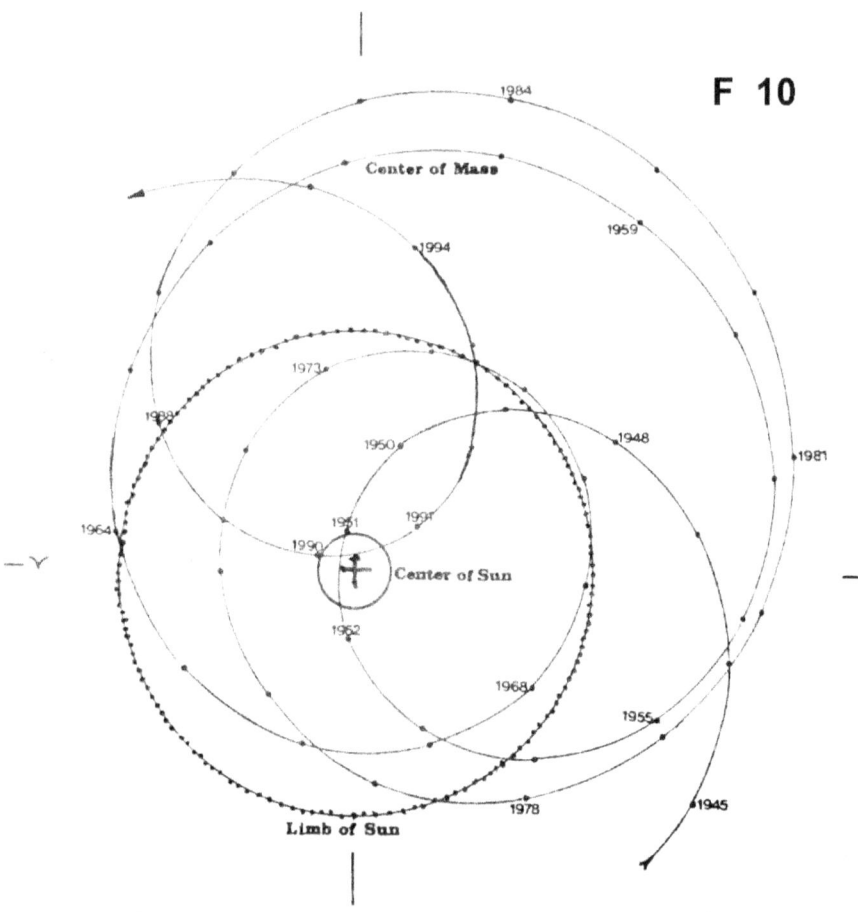

F 10

The planetary torque effect (even though as usual the forces are said to be 'too small') is probably the dominant short term influence on solar activity variation. However many decades ago my attention was first drawn to the question of solar variation by a paper that claimed that this variation was influenced by planetary tidal effects...although again the forces involved seem to be too small. However we will discover that certain non-linear systems which display Self Organised Criticality can respond strongly over some frequency ranges to very weak energy inputs. We will find the fingerprints of SOC all through our story.

Calculations show that the main tides, such as they are, are generated by Venus, Earth and Jupiter with a high frequency contribution from fast orbiting Mercury. Professor Wood (21) famously computed the VEJ tide back to 1800 and compared it with Wolf sun spot numbers (Figure 11). The height of the planetary tides over this period was fairly constant although the shape of the tides changed somewhat while the sun spot peak intensity changed much more. The interesting point here is that the tidal cycle minimum began in phase with the solar minimum and was still in phase 200 years later. In phase minima occurred in 1801, 1844, 1888, 1954 and 1965. We can say that the two systems were phase locked and we will see more evidence in later sections. It follows that the mean cycle periods of the two phenomena must be identical. The tidal cycle over this time window had a period of ~11.1 years. The long term mean of the underlying solar activity cycle is 11.16 years as we will see.

Recently Edmunds (25) revisited the effect of Mercury tides on sunspot areas during cycle 23 and found strong indications of Mercury's orbital period and harmonics in the solar spectrum, providing the north and south hemispheres were treated separately. There may be a lesson here for looking at climate variables regionally rather at than global averages when investigating possible causes of climate forcing.

These tidal period and phase correspondences seem beyond coincidence and we have to add to this the strong link between solar activity and the torque cycle. If these links are real it would be useful to understand something of planetary dynamics, the main external forcing signal acting on our Sun, and particularly since the planetary orbits are not random but harmonically related (again through an SOC process).

The spacing of planets and their periods are the result of gravitational interactions during planet formation and later migrations and have a lot to teach us about non-linear effects and resonance. Now we are in a position to study the dynamics not only of our own solar system but those of many other systems of planets and we will do so. Many sun like stars turn out to have activity cycles similar to our own along with multiple planets and soon, with several decades of data, we will be able to test out the hypothesis of planetary forcing of stellar activity on multiple examples. That is what is needed to convince the doubters of genuine 'cosmic' effects.

There is one last twist discussed in detail in section 2.4 on stellar activity cycles where we compare the Sun with emerging evidence from an increasing data base of sun like stars in the nearby galaxy. The primary solar activity cycle acts on a time scale of 11 years or so. We have seen that there may be cycles in the Sun on time scales of 100,000 years.
In between there is good evidence that on the scale of centuries and millennia the magnetic activity of the Sun declines to the point where some indicators of activity like sunspots, disappear for decades or even centuries.

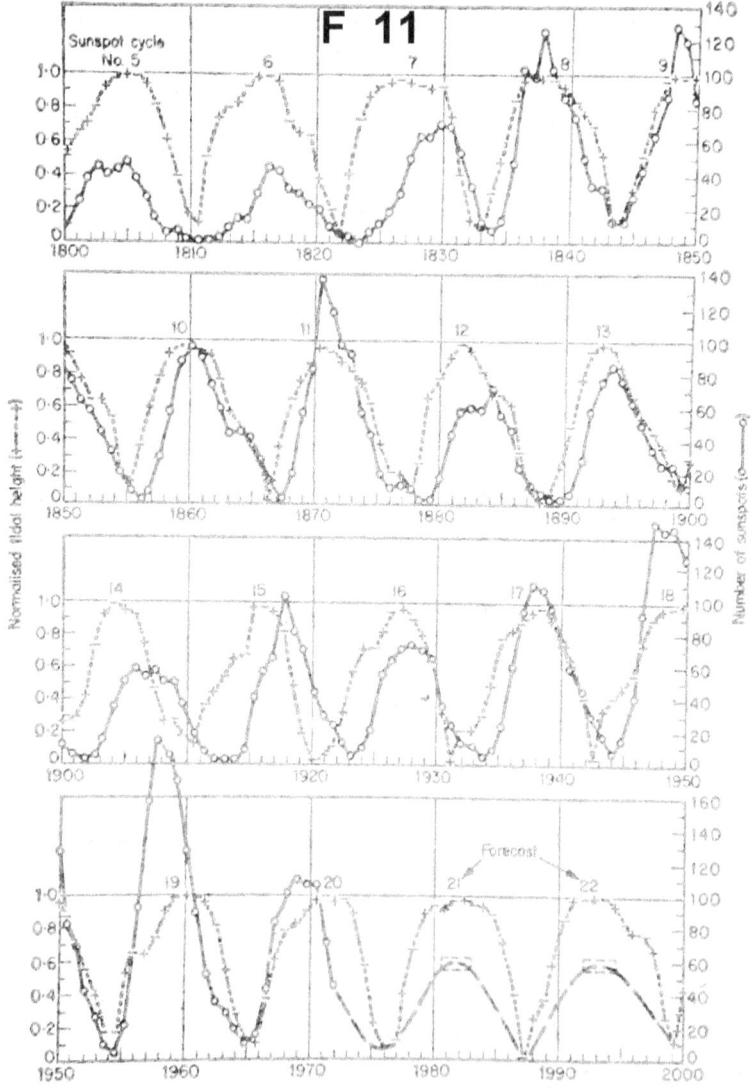

F 11

These 'Grand Minima' in activity may be associated with falling temperatures on Earth. The last 'Maunder Minimum' in the seventeenth century coincided with a period of intense cooling, several decades long, called the 'Little Ice Age'. This event is not popular with the 'global warming' lobby and is often dismissed as an artefact of poor observation. However solar proxy data suggest there have been many such minima, some of long duration, and now we see that many sun like stars, show such minima, in which state they reside, like the Sun, for ~20% of the time.

One wonders whether the longer grand minima could act as triggers to shift us out of the warm quasi-stable interglacial state and into the slow decline towards the next ice age. It would be most interesting to know if 'super - grand minima' occur on 100,000 year time scales.

This section has been about heresies. In science unusual ideas, contrary to the consensus mainstream, need unusually strong evidence to be considered seriously. What is not acceptable is when the 'establishment' and those with political influence refuse to even look at unconventional ideas or rabidly attack the proposers, even to the point of misrepresentation and data fiddling. We have noted that some scientists are willing to put their heads above the surface and cry foul…even some climatologists. In reading many papers in preparing this book it is interesting that these days authors go out of their way to say 'of course I believe in man made global warming' before presenting, quietly, good evidence to the contrary. This tiptoeing around should not be necessary (see App. 3 for a discussion). It warms the heart to see an apparent shift in attitudes as the limits of the climate models become clearer. In 2010 the prestigious Royal Society produced a 'guidance' on climate change and in particular, Climate Change Consensus, which reported the establishment 'warmist' position without discussing the dissenters and heretics, nor the growing concerns about the problems with the models (22). 43 Fellows of the Royal Society complained about this exposition and it was modified. Lord Rees, the president, accepted that the case for man-made global warming had been 'exaggerated' and much published information (from the IPCC) had been 'misleading'. Of course he also repeated that there was nothing wrong with the science 'underpinning' climate change. We must all agree with that: CO_2 is a greenhouse gas…that is certain. What has followed, on the basis of a long string of assumptions and guesses, is not. We must, nevertheless, be very grateful to the 'RS 43'. That is progress.

Section 1 References

1. Lovelock J; 'The Ages of Gaia', Oxford University Press, 1990.

2. El Albani et al; Nature 466,100-104,(2010).

3. Dawkins R; 'The Ancestors Tale', Weidenfeld & Nicolson, 2004.

4. Gould S J; 'Wonderful Life', Hutchinson Radius, 1990.

5. Morris S C; 'The Crucible of Creation', Oxford University Press, 1999.

6. Hoyle F; 'Ice', New English Library, 1982.

7. DeConto R M; 'Past extreme warming events linked to massive carbon release from thawing permafrost', Nature Letters, Vol 484, page 87, 5th April, 2012.

8. Lamb H H ; 'Climate : present, past and future: Vol 2', Barnes & Noble, 1977.

9. Scaife A et al; ' Solar forcing of winter climate variability in the Northern Hemisphere', Nature Geoscience ; 4, 755-757, 2011.

10. Svensmark H & Calder N; 'The Chilling Stars' Icon Books, 2007.

11. Stott L et al; 'Southern hemisphere deep sea warming led deglacial atmospheric CO2 rise & tropical warming', Sciencexpress / 27 September / 2007 / p1-10.

12. Svensmark H et al; 'Experimental evidence for the role of ions in particle nucleation under atmospheric conditions', Proceedings of the Royal Society A, Vol. 463, pp385-396, 2007.

13. Kirkby J et al; 'Role of sulphuric acid, ammonia & galactic cosmic rays in atmospheric aerosol nucleation', Nature 476, pp429-433, 25th August 2011.

14. Hao Ye et al; 'Robustness of causal effects of galactic cosmic rays on interannual variation in global temperature', Proc. of the National Academy of Science of the USA, Vol 112, No. 34, 2015.

15. Lockwood et al; 'Near Earth solar wind speed, IMF and open solar flux', Ann. Geophys., 32, p383-399, 2014.

16. Shaviv N & Veizer J; 'Celestial driver of Phanerozoic climate', GSA Today (Geophy. Soc.), July 2003.

17. Randall L; 'Dark Matter and the Dinosaurs', Bodley Head, 2016.

18. Rial J A; 'Pacemaking the ice ages by frequency modulation of the Earth's orbital eccentricity', Science Vol. 285, 23rd July 1999.

19. Sharma M; 'Variations in solar magnetic activity during the last 200,000 years : is there a Sun-climate connection?', Earth & Planetary Science Letters, Vol. 190, issues 3-4, 2002.

20. Abreu J A et al; 'Is there a planetary influence on solar activity?', Astronomy & Astrophysics, A88, (2012).

21. Wood K D; 'Sunspots & planets', Nature 240, (1973) 91.

22. Daily Telegraph; 'Royal Society bows to climate change sceptics', 29.05.10, Page 6.

23. www.newscientist.com/article/dn11659-climate-myths-ice-cores-show-co2-increases-lag-behind-temperature-rises-disproving-the-link-to-global-warming

24. Piran & Jimenez; 'Does the anthropic principle save us from gamma ray bursts?', Physical Review letters, 23rd February 2016.

25. Edmonds I R; arxiv.org/ ftp/arxiv/papers/1404/1404.3326.pdf

26. de Wit M J & Furnes H; '3.5 Gy hydrothermal fields in the Barberton Greenstone Belt', Science Advances, 26th February 2016.

27. Svensmark H; 'The response of clouds and aerosols to cosmic ray decreases', Jo. Of Geophysical Research – Space Physics, 2016, DOI: 10.2002/2016JA022689.

Section 2 : Planetary Orbital Harmonics
Phi in the Sky?

'The fault, dear Brutus, lies not in our stars, but in ourselves'

Mark Anthony

2.1 The Solar Planets

Later we will examine the detailed structural evidence for the influence of planetary orbital dynamics on the Sun, the dynamics of the Earth – Moon system and on our geophysical and climate cycles. It is a story of non-linear systems responding to complex forcing signals.
Those signals are complex because they too are the result of non-linear interactions and resonance effects over long periods of time: the orbital periods of our planets are not random but related, sometimes obviously as in the 2 : 5 ratio of Jupiter's year to that of Saturn, next door and the 8 : 13 ratio of the Venus and Earth orbital periods and the 2 : 3 ratio between Neptune and Pluto. Already we see hints of ratios based on the Fibonacci series, 1,2,3,5,8,13,21,34... These ratios in the limit tend to Phi, the Golden Section, 1.618034....the most irrational of numbers. Hence my sub-title 'Phi in the sky?' We will see later the importance of Phi in defining the behaviour of the dynamics of non-linear systems at the 'edge of chaos' (section 3.7). We will derive in this section a model which accounts for the 'allowable' orbital period ratios which seems to work for the solar system but also for several other star systems with multiple planets.

Looking at planetary orbits is a good introduction to the kinds of processes in play throughout our story but the reader, if he or she wishes, can skip this section except for the summary 2.3, and 2.4 where we examine the evidence for other multi-planet solar systems with stellar activity cycles. Sometimes the solar system orbital relationships are more subtle, involving three or more planets, but they are there. For example several triple conjunctions of the four gas giants Jupiter, Saturn, Uranus and Neptune produce cycles of 171 – 181 years and we will see these reflected in solar activity and Earth climate spectra.

The search for order in the scaling of planetary orbits, 'the harmony of the spheres', has a long history dating back to classical Greece. More recently Johannes Kepler applied nested Platonic solids with some success to the same task. In the 18th century the Titius-Bode law seemed to confirm regular orbital scaling since orbital radii were given by

$R = (4 + 2^{(n-1)} \times 3) / 10$ for n =1 to 8, yielding in astronomical units,

0.4, 0.7, 1.0, 1.6, 2.8, 5.2, 9.6, 19.6   versus the observed

0.387, 0.723, 1.0, 1.524, 2.8, 5.203, 9.54, 19.2 for Mercury to Uranus.

The fit is good overall but fails for Neptune. Is this pattern a fluke or does it hint at some underlying resonant harmonic principle? Looking in more detail at planetary distances from the Sun and their orbital periods tells us more.

Planet	Orbital Radius	OR ratio	Orbital Period	OP ratio
Mecrury	0.387 AU		0.244 yrs	
		1.87		2.52
Venus	0.723		0.615	
		1.383		1.626
Earth	1.0		1.0	
		1.524		1.88
Mars	1.524		1.88	
		1.82		2.45
Ceres	2.77		4.606	
		1.88		2.575
Jupiter	5.203		11.86	
		1.833		2.49
Saturn	9.54		29.5	
		2.01		2.85
Uranus	19.2		84	
		1.56		1.963
Neptune	29.96		164.8	
		1.313		1.505
Pluto	39.49 mean		248.0	

We note that in four cases planetary radii ratios are on average ~1.85 and orbital period ratios on average 2.508 or close to 5 / 2. Also Earth OP / Venus OP = 13 / 8

Mars OP / Earth OP = 1.88 ~ (3 / 2) $^{3/2}$ and JupOP / CeresOP = 2.57 ~ 8 / 5 x 8 / 5 ; PlutoOP / NeptOP = 1.505 ~ 3 / 2.

Are these ratios connected to the Fibonacci series? It seems that each planetary pair occupies one of several orbital ratio slots which are members of a common harmonic family. After some experimentation, and working in terms of period ratios, for the solar planets we have

Power n	calculated OP ratio = F^n F= 1.15	Observed OP ratio	CalcORr	ObsORr
6.5	2.48	2.52	1.82	1.867
3.5	1.63	1.626	1.38	1.383
4.5	1.876	1.88	1.514	1.524
6.5	2.48	2.45	1.82	1.82
7	2.66	2.58	1.9	1.88
6.5	2.48	2.49	1.82	1.834
7.5	2.852	2.85	2.0	2.01
5	2.01	1.963	1.58	1.56
3	1.52	1.505	1.32	1.314

OP Mean absolute error 0.29%   OP Mean error 0.78%

The fit is good but more trials suggested that F = 1.149 reduced bias, mean error, to ~0. The half powers suggest the real base factor is ~1.072. The model accurately predicts all the observed orbital period ratios but not the order in which they will occur. This harmonic family model with its Fibonacci links is intriguing but like Bode's Law could be a coincidence. To be sure we need more examples of planetary systems. Fortunately the solar planets with multiple moons provide more data since they too formed from proto-planetary disks in most cases. Also data is now becoming available for other star systems.

The Jupiter System
The four primary moons of Jupiter range in size from that of Earth's moon to that of Mercury. The Jupiter system is like a miniature solar system. Does it follow the orbital harmonic model proposed?

Moon	O radius Km	OR ratio	O Period	OP ratio
Io	421.6x10³		1.769 days	
		1.593		2.008
Europa	670.9		3.551	
		1.595		2.015
Ganymede	1070		7.155	
		1.76		2.333
Calisto	1883		16.69	

A little investigation yields

Actual OPratio	ModelOPr	power	Actual ORratio	Model ORratio
2.008	2.011	5	1.592	1.593
2.015	2.011	5	1.5925	1.593
2.333	2.313	6	1.76	1.75

Mean absolute error 0.36%   Mean error -0.35%.
The fit is again excellent using F = 1.15. An F value of 1.1505 is slightly better. The model radii (scaled from Ganymede, the largest moon) would be
422, 672, 1070, 1875 x 10^3 Km, versus observed
421, 671, 1070, 1883. Similarly for the orbital periods
1.77, 3.56, 7.16, 16.6 from the model, versus actual periods
1.77, 3.55, 7.16, 16.69 days.

It is interesting that $(13/8) / \sqrt{2} = 1.149$ versus F = 1.15. 13 / 8 is of course a Fibonacci ratio and ~ phi. Root 2 is another classic irrational number.

The Saturnian System

Saturn has several medium sized moons and Titan which is larger than Mercury. The mass range is more extreme than at Jupiter. Is there a similar resonant structure here?

Moon	ORadius Km ^{3}	OR ratio	O Period	OP ratio
Tethys	294.7x10		1.888 days	
		1.28		1.45
Dione	377.4		2.737	
		1.396		1.65
Rhea	527		4.52	
		2.32		3.53
Titan	1222		15.95	
		2.914		4.98
Iapetus	3561		79.33	

We look at Titan's neighbours first since Titan is dominant in mass. The base factor for our model in this case is 1.1505.

Power n	Actual OPratio	ModelOPr	Actual ORratio	Model ORratio
2.5	1.45	1.42	1.28	1.264
3.5	1.65	1.63	1.396	1.388
9	3.53	3.52	2.32	2.325
11.5	4.98	4.99	2.92	2.937

Mean absolute error 0.7%   Mean error -0.21%.

The predicted radii (scaled from Titan, the largest moon) are
299, 378.6, 525.6, 1222, 3586 compared with actual

294.7, 377.4, 527, 1222, 3561 x 10^3 Km.

Similarly for the model periods
1.92, 2.78, 4.53, 15.95, 79.5 days, versus actual
1.89, 2.74, 4.52, 15.95, 79.33 days.

The Uranian System

Uranus has five moons of comparable mass in highly circular orbits. All are close to the equatorial plane of Uranus but that plane is almost at right angles to the ecliptic since the planet was tipped over in a major early collision. Will the model hold?

Moon	O Radius Km	OR ratio	O Period	OP ratio
Miranda	129.4x10^3		1.414 days	
		1.476		1.782
Ariel	191		2.52	
		1.393		1.644
Umbriel	266.3		4.144	
		1.64		2.102
Titania	435.9		8.706	
		1.339		1.546
Oberon	583.5		13.46	

A little work suggests a base factor of 1.151 as a good fit.

Power	Actual OP ratio	Model OP ratio	Actual ORratio	Model ORratio
4	1.782	1.756	1.476	1.46
3.5	1.644	1.635	1.394	1.388
5.5	2.102	2.16	1.637	1.67
3.0	1.546	1.526	1.339	1.325

Mean absolute error 1.53%   Mean error -0.1%

The fit is not as good as for the other systems studied so far. We now have F values of 1.149, 1.1505, 1.1505, 1.151 and a mean of 1.1502. The mix of integer n and n / 2 powers suggests the real model is 1.0724^m where m is integer in the range 5 – 23 for the systems examined so far.

Neptune and Mars

Neptune has several small moons and Triton which is 1670 miles in diameter. However Triton's orbit is inclined at 157 degrees to both Neptune's equator and the ecliptic. Triton's motion is also retrograde. This suggests it was captured after Neptune formed. The inner moon Proteus is close to the ecliptic plane as is the distant moon Nereid. Even so Triton may have affected the position of Proteus. Let us see. ORadius Triton / ORadius Proteus = 3.006; OPTriton / OP Proteus = 5.237.

Using our average base factor 1.072 we note that 1.072^{24} = 5.3 and for the distance ratio 3.04. The fit is worse than for the inner systems. F = 1.0714 gives PR = 5.234 and distance ratio 3.014.

Mars has two very small moons which are believed to be captured asteroids. They orbit very close to the planet. ORadius Deimos / ORadius Phobos = 2.501, OPDeimos / OP Phobos = 3.96. Using our average factor we have 1.072^{20} = 4.017. F= 1.0712 gives us 3.958.

Between Saturn and Uranus orbit a cluster of asteroids in eccentric orbits known as the Centaurs. The largest centaur is Chiron which is in resonance with both Saturn and Uranus. Could the harmonic model work in this extreme case? Mean ORadius Chiron / ORadius Saturn = 1.4306 and mean O Period Chiron / O Period Saturn = 1.72. For Uranus we have OR Uranus / ORadius Chiron = 1.4, OPeriod Uranus / OPeriod Chiron = 1.658. We find that 1.072^7 = 1.63 and 1.072^8 = 1.74 giving distance ratios of 1.385 and 1.44. The fit is not as good as for the planetary moon systems.

Before we examine exoplanets we will consider two additional distributed features of the solar system where resonance effects are in play : the Asteroid Belt and the Kuiper Belt.

The **Asteroid Belt** lies in a period ratio range of ~ 3 / 11 to 1 / 2 relative to Jupiter. The famous Kirkwood gaps where asteroids are ejected lie at rational fractions of Jupiter's period : 1:3, 2:5, 3:7. Large numbers of asteroids cluster in broad peaks between the gaps (1). We use F = 1.15.

No. Objects at peak	Orbital distance AU	Dist J / A dist	J OP / A OP	Model	power

200	2.36	2.204	3.274	3.28	8.5
260	2.58	2.012	2.854	2.85	7.5
190	2.74	1.9	2.62	2.66	7.0
120	2.98	1.75	2.31	2.31	6.0
160	3.12	1.668	2.154	2.16	5.5

Mean absolute error 0.42%   Mean error 0.36%.

The fit is good using F = 1.15 and even better with F = 1.148. Note that the model powers are all numerical neighbours in this case but not in order.

The **Kuiper Belt** lies beyond Neptune (2). Two peaks in object frequency are evident: a sharp peak at 40 AU and an equally intense but broader peak at 44 – 45 AU with some spread of objects from 43 – 46 AU. We use F = 1.15 again.

No. objects	Dist.	KP dist / J dist	KP period / J period	Model PR	power
200	40	7.69	21.32	21.6	22
240	44	8.46	24.6	24.9	23
190	45	8.96	26.8	26.7	23.5
or	mean 44.5	8.55	25.0	24.9	23

We might also expect gravitational interference between the peak clusters.

No. objects	Dist.	Distance / 40 AU	Period / 40 AU period	model	power
100	43	1.075	1.115	1.111	3 / 4
240	44	1.1	1.154	1.15	1.0
105	46	1.15	1.233	1.233	3 / 2
		45 / 43	1.071	1.072	1 / 2

The model appears to hold remarkably well. The appearance of low model powers 1, 1.5, 2 and 3 based on F = 1.072 is notable.

An Overall Harmonic Model

The value of F varies from 1.149 to 1.151 for the main systems. This suggests a common model is possible providing we further partition the harmonic structure. This makes it easier to match period ratios by chance so a very close match must be demanded from the analysis and / or some compelling internal patterns in the harmonic structure.
The table below compares calculated period ratios with those observed for the four systems with additional inputs for Mars, Chiron and Neptune which must be considered less certain. F = 1.15.

Power ratio	Observed OPR	Model OPR	% Error	MOP ratio / OOP
½	1.071	1.072	0.093	1.001
1	1.154	1.15	-0.34	0.996
3/2	1.233	1.233	0.0	1.0
5/2	1.4	1.418	1.29	1.01
	1.43		-0.84	0.992
	1.45		-1.38	0.98
3	1.505	1.52	1.0	1.01
	1.545		-0.65	0.983
7/2	1.626	1.631	0.25	1.024
	1.644		-0.79	0.992
	1.65		-1.15	0.988
4	1.782	1.75	-1.8	0.982
9/2	1.88	1.876	-0.21	0.998
5	2.008	2.011	0.001	1.001
	2.015		-0.2	0.998
11/2	2.102	2.15	2.6	1.026
6	2.333	2.313	-0.64	0.991
13/2	2.45	2.48	1.22	1.012
	2.49		-0.4	0.996
	2.52		-1.58	0.984
7	2.58	2.66	3.1	1.03
7.5	2.85	2.85	0	1.0
9	3.53	3.52	- 0.27	0.997
10	3.96	4.04	2.16	1.02
23/2	4.98	4.99	0.2	1.002
12	5.237	5.35	2.15	1.021

Mean absolute error 0.95%   Mean error 0.11%

The fit is remarkably good across several systems. Multiple instances occur mainly between powers 5/2 and 7. The table becomes more sparse as the power increases. A plot of the model OP ratio / Observed OP ratio suggests a slightly higher F at low powers and a slightly lower F at high powers.

A rough fit suggests F = (1.1537 – 0.000555P). It is also fascinating that computed ratios are simply related to phi – Fibonacci nos. and even e. For example

Power   period ratio   interpretation

1	1.15	$2/\sqrt{3}$; pi / e
3/2	1.233	$2/(13/8)$
2	1.323	4 / 3
5/2	1.418	$\sqrt{2}$
3	1.52	3 / 2
7/2	1.649	$5/3; \sqrt{e}$
6	2.31	$(3/2)^2$
7	2.66	$(13/8)^2$

We also noted earlier that $(13/8)/\sqrt{2} = 1.1492$.

In several cases we have multiple instances of the same period ratio across systems. It is interesting that multiple instances seem to prefer powers which involve primes: 5/2, 7/2, 5, 13/2. A second question is how well the computed ratios cluster around the closest fitting power. Neighbouring ratios are related as 0.933 : 1.0 : 1.072. Beginning with multiple observed ratio instances we have

Observed ratios	0.933	1.0	1.072
3 x 1.427		1.01	
		0.992	
		0.98	
3 x 1.64		1.002	
		0.992	
		0.988	
3 x 2.487		1.012	
		0.996	
		0.984	
2 x 1.525		1.01	
		0.983	
2 x 2.011		1.001	
		0.998	

average 0.996; range 0.98 – 1.012

We see that the computed model / actual ratios cluster tightly around 1.0 mirroring the chosen F powers accurately. For the single ratio instances we have

Observed ratios	0.933	1.0	1.072
1.071		1.001	
1.154		0.996	
1.233		1.0	
1.71		1.001	
1.78		0.982	
1.88		0.998	
2.333		0.991	
2.58		1.03	
2.85		1.0	
3.53		0.997	
3.96		1.02	
4.98		1.002	
5.24		1.021	

average 1.005; full range 0.982 – 1.03

In all cases the ratios cluster quite closely around the best fit power but with larger individual errors. If the fits were due to chance we would expect a wide spread across the range 0.933 to 1.072. So far as we know there is nothing physically special about the solar system. If so the same harmonic rules should apply to other systems

2.2 Extra-Solar Planetary Systems

In recent years several hundred planets have been identified in other solar systems (4). In some instances multiple planetary systems have been confirmed and in some cases orbital period estimates are available although subject to some uncertainty as yet. Can we apply our harmonic model to these systems? Let us try.

Kepler 62

Kepler 62 hosts at least five planets with diameter from 0.54 to 1.95 times that of Earth. The system is unusually dense in that all orbits lay within 1 AU from the star. It has been suggested that the system is almost unstable. We try F = 1.0735.

Period	Period Ratio	Power	Computed ratio Best fit
5.7 days			
	2.175	11	2.182
12.4			

Period	Period ratio	Power	Computed Ratio
18.2	1.468	5	1.43
122.4	6.725	27	6.78
267.3	2.184	11	2.182

Mean absolute error 0.82%   Mean error -0.25%

As before we look at neighbouring ratios and the error spread.

Observed Period Ratio	0.932	1.0	1.073
	2.173	1.003	
	1.468	0.976	
	6.725	1.008	
	2.184	0.999	

The fit is good.

HD 160691

We have four planets including a hot Neptune close in and three Jupiter class planets of 0.5 to 1.8 Jm. Using the solar average base factor again

Period	Period ratio	Power	Computed Ratio
9.64 days			
	32.21	49	33.2
310.5+/-0.8			
	2.071	10	2.044
643.2+/-0.9			
	6.54	26	6.42
4206+/-759			

The error spread is

Observed period ratio	0.931	1.0	1.074
	32.21	0.97	
	2.071	1.013	
	6.54	1.019	

The errors are larger here and positive suggesting the base factor is too low in this case. In fact using 1.075 gives

Observed period ratio	power	period	0.93	1.0	1.075
32.21	48	32.18		0.999	
2.071	10	2.061		0.995	
6.54	26	6.556		1.002	

The harmonic model holds very accurately. Note that $1.075^2 = 1.156 =$ Pi / e. Mean absolute error 0.26%.

47 Ursa Majoris

We have three planets with masses between 0.54 and 2.5 Jupiter. Applying base factor 1.0741

Period	Period ratio	power	computed period ratio
1078 +/- 2			
	2.218	11	2.195
239+/-80			
	5.86	25	5.97
14002+/-4500			

The fit is quite good but the longest orbital period is rather uncertain.

Period ratio	0.93	1.0	1.0741
2.21		1.01	
5.86		0.98	

Mean absolute error 0.35%   Mean error 0.165%

Upsilon Andromedae

We have four planets with 0.6 to 3.8 Jupiter masses in relatively eccentric orbits since the middle pair has e=0.26 and e=0.3. For F = 1.072 we get

Period	Period ratio	power	computed period ratio
4.62+/-0.2			
	52.23	57	52.6

241.3+/-0.6
              5.288   24   5.18
1276+/-0.6
              3.016   16   3.04
3849+/-0.7

Mean absolute error 0.6% The error spread is

Period ratio   0.933   1.0   1.072

  3.016   1.007
  5.29   1.003
  3.016   1.008

The fit is still very good. It is interesting that
$(Pi/e) = 2/\sqrt{3} = 1.156$ and $1.156 \times (21/13)^2 = 3.017$ and $2 \times (13/8)^2 = 5.281$ and
$(2 \times 2 \times 5) \times (21/13)^2 = 52.19$. We still have Fibonacci with us.

HD 10180

This is a system of 7 closely packed planets in the Neptune mass range circling a G1 sun-like star. Five of the planets are within a ½ AU of the star with periods less than 1/3 year. We should expect strong interactions between the inner planets. Will the harmonic model hold? The two outer planets have periods of 1.6 and 6.1 years. We try $F = 1.075$.

Period	Period ratio	computed period ratio	error ratio
1.777 days			
	3.241	3.138	0.968
5.76			
	2.84	2.92	1.028
16.36			
	3.041	2.92	0.96
49.75			
	2.468	2.533	1.06
122.8			
	4.897	4.82	0.984
601.2+/-8.1	(4.83-4.96)		
	3.7	3.62	0.978

(3.5-3.894)
2222+/-91

The errors are large. This suggests that the base factor is lower in this system. Using the solar system 1.072 we get

period	period ratio	computed ratio	power	error ratio
1.777				
	3.241	3.261	17	1.006
5.76				
	2.84	2.837	15	0.999
16.36				
	3.041	3.042	16	1.000
49.75				
	2.468	2.469	13	1.000
122.8				
	4.83-4.96	4.947	23	1.009
601.2				
	3.5-3.9	3.747	19	1.004
2222				

Mean absolute error 0.13% Mean error 0.09%.
The fit to all seven planets is now essentially perfect.

Mu Arae   Best fit F = 1.0723

Period days. eccentricity	Period ratio	computed ratio	power	error ratios
0.736 e=0.17				
	19.905	20.11	43	1.01
14.65				
	3.028	3.055	16	1.009
44.36				
	5.86	5.73	25	0.978
259.8 e=0.3				
	19.9	20.11	43	1.01
5169				

Mean absolute error 1.29%   Mean error 0.19%
The fit is worst for the highly eccentric orbit of the fourth planet.

Gliese 892

This star has at least seven planets with two very close to it and five in total within 0.4 AUs. Two Jupiter mass planets circle further out with periods of five and six years. The sixth planet is in a very eccentric orbit with e = 0.34.

Period days	Period ratios	Model period ratios	power	error
3.09				
	2.177	2.171	11	- 0.28%
6.735				
	3.372	3.31	17	-1.84
22.81				
	2.048	2.023	10	0.93
46.71				
	2.017	2.023	10	0.3
94.2				
	19.55	19.28	42	-1.38
1842				
	1.22	1.235	3	1.23
2247				

Mean absolute error 0.99%   Mean error -0.17%

This model uses F = 1.073 compared with our SS 1.072. The fit is good over period ratios from 1.2 to 19.

Kepler 90

Kepler 90 has seven confirmed planets including two super Earths, three mini-Neptunes and two gas giants. The whole system is contained within 1 AU of the star and with periods less than a year. There must be strong gravitational interactions between the planets although the orbits are circular. Will the model hold? The overall best F is 1.0775, some 0.4% higher than in other systems.

Period days	Period ratios	Model period ratios	power	error
7.008				
	1.244	1.251	3	0.56%
8.72				
	6.85	6.94	26	1.61
59.73				
	1.573	1.565	6	-0.44

91.93				
	1.356	1.348	4	-0.74
124.9				
	1.686	1.686	7	0
210.6				
	1.574	1.565	6	-0.57

Mean absolute error 0.66%   Mean error 0.07%

The two inner worlds are better modelled with F= 1.077 and the five outer by F= 1.078 but the compromise fit is good and impressive because it mainly involves low model powers, making chance fits lass likely.

In the last ten years data on hundreds of worlds has become available. It is of interest to see if the bulk orbital data for many exo-systems conforms to the model, as we explored for the solar system. Reference 3 provides a graph of numbers of planets at various distances from the primary stars. Figure 1 provides the following information on four clear peaks in frequency for 152 planets (3).

No. of planets	Distance in AU	Dist. Ratio	Period ratio	Model PR	power
18	0.04				
		7.9	22.2	22.42	43
9+11	0.316				
		3.165	5.63	5.67	24
24	1.0				
		2.513	3.98	3.95	19
21	2.512				
		2.512 / 0.316 (7.95)	22.4	22.4	43

Mean absolute error 0.63%   Mean error 0.28%   F = 1.075.

The fit is good but the powers involved are high and the peaks broad. Fortunately more recent data has a higher resolution. Reference (4) provides the following data for the hundreds of planets known by 2012.

Orbital period (days)	Orbital P ratio	Model OP ratio	power	error
4				
	112.5	112.3	67	-0.18%
450				
	2.333	2.329	12	-0.17
1050				
	2.0	2.02	10	1.0
2100				

Mean absolute error 0.34%   Mean error 0.26%   F = 1.073

The distribution of periods is apparently still not understood although the many planets with periods of a few days, the 'Hot Jupiters', arrived at their current locations by inward migration.
It is surprising the model 'fits' these although the period ratios to the outer clusters beyond 200 days are high and require high model powers. Perhaps the '4 day' fit is chance. We are still dealing with just a few peaks.
Reference (4) also provides a high resolution distribution of orbital radii (semi-major axes) which shows many more sharp peaks with closer separations. The main peaks are

SM Axis AU	Distance ratio	Period ratio	Model period ratio	power
0.05				
	2.2	2.83	2.82	15
0.11				
	7.27	19.6	19.48	43
0.8				
	1.513	1.86	1.862	9
1.21				
	1.653	2.13	2.138	11
2.0				
	1.15	1.233	1.23	3
2.3				
	1.44	1.728	1.735	8
3.3				
	1.515	1.865	1.862	9
5.0				

The very good fit involving low model powers and using F = 1.0715 is now more convincing. The data now shows lesser peaks at distances between the Hot Jupiters and the 1AU + cluster. These are worth a look.

Distance	Distance ratio	Period ratio	Model ratio	power
0.11				
	1.636	2.09	2.1	10
0.18				
	1.5	1.84	1.87	9
0.27				
	1.37	1.603	1.606	7
0.37				
	1.32	1.516	1.501	6

0.49
        1.633   2.086   2.1   10
0.8

The fit is good. It is also interesting that the ~4 day peak is tightly bound between 0.033 and 0.05 AU giving a period ratio of 1.865^9 or 1.072.

All these planets circle sun-like stars. Although all star systems differ in actual layout we seem to be seeing preferred orbital distance and period ratios for the planets and possibly actual repeated spatial patterns in some, but not all, cases. The Hot Jupiters are certainly spatially confined by some mechanism (3). However the observed distribution of planetary periods beyond several years is still distorted by the observation times so far available and by the sensitivity of the detection methods. Nevertheless it is worth looking a little more closely at the possible source of the regularities observed so far. Suppose we generate low order rational fractions and compare these with observed period ratios for all the systems described earlier and with our proposed harmonic model with F = 1.0725.

TABLE A

Observed Period ratios	Rational Fractions		Model PR	Power
1.071	15 / 14	1.0714	1.0725	1
	14 / 13	1.077		
	11 / 10			
	10 / 9			
	9 / 8			
	8 / 7	1.143		
1.154	15 /13	1.153	1.15	2
	7 / 6	1.166		
	6 / 5	1.2		
1.22, 1.233, 1.233			1.233	3
1.244			2 (phi-1) = 1.236	
	5 / 4	1.25		
	9 / 7	1.286		
	4 / 3	1.333	1.323	4
1.36	11/ 8	1.375		
1.4	**7 / 5**	1.4	1.419	5
1.43	10 / 7	1.429	~ √2	
1.45, 1.47	13 / 9	1.444		
	3 / 2	1.5		

1.505, 1.52, 1.524			1.52	**6**
			1.233 x 1.233	
1.545, 1.57	11 / 7	1.571		
1.6	**8 / 5**	1.6		
1.626	13 / 8	1.625	1.632	**7**
1.644, 1.65				
	5 / 3	1.666		
1.694, 1.73	19 / 11	1.727		
	7 / 4	1.75	1.751	**8**
1.782				
	9 / 5	1.8		
	11 / 6	1.833		
1.84, 1.86, 1.865	13 / 7	1.857	1.877	**9**
1.88	17 / 9	1.888		
	2 / 1	2.0		
2.015, 2.02, 2.02, 2.07, 2.08, 2.09, 2.1			2.014	**10**
	17 / 8	2.125		
2.15	15 / 7	2.143		
2.17, 2.18	13 / 6	2.166	2.16	**11**
	11 / 5	2.2		
2.22	**9 / 4**	2.25		
2.31, 2.33, 2.33			2.316	**12**
2.33	**7 / 3**	2.333		
2.45, 2.47, 2.49, 2.52	**5 / 2**	2.5	2.484	**13**
2.58	13/ 5	2.6		
2.62	**8 / 3**	2.666	2.66	**14**
	19 / 7	2.714	1.63 x 1.63	
	11 / 4	2.75		
2.83, 2.84, 2.85	17 / 6	2.833	2 x √2	
2.854			2.86	**15**
	20 / 7	2.875		
	3 / 1	3.0		
3.02, 3.03, 3.04			3.04	**16**
	22 / 7	3.142		
	19 / 6	3.167		
3.24	13 / 4	3.25		
			3.28	17
	10 / 3	3.333		
3.53	**7 / 2**	3.5	3.52	**18**
	11 / 3	3.666		
3.7, 3.7	15 / 4	3.75		

	19 / 5	3.8	3.78	19
	23 / 6	3. 833		
3.97	**4 / 1**	4.0	4.05	20
	29 / 7	4.14		
			3	
			phi = 4.235	
?	17 / 4	4.25	4.35	21
			3 √2 = 4.243	
	23 / 5	4.6		
?	14 / 3	4.666	4.66	22
4.9, 4.9, 4.93	**5 / 1**	5.0	5.0	**23**
	21 / 4	5.25		
5.273				
5.3	**16 / 3**	5.33	5.36	**24**
5.63, 5.86, 5.83	23 / 4	5.75	5.75	**25**
			(7/3)x(5/2)	
			4 √2	
	37 / 6	6.166		
?			6.17	26
	31 / 5	6.2	4	
	33 / 5	6.6	1.6	
6.54, 6.72, 6.85			6.62	**27**
			4	
	34 / 5	6.8	(21/13)	

There are clear and enlightening patterns here which are discussed below.

2.3 Discussion

The orbital radii and orbital period ratios of the solar planets and the moons of Jupiter, Saturn, Uranus and Mars appear to be members of a common harmonic family. Chiron and Neptune may also fit the model. Perhaps more surprisingly the model accounts well for the peak positions of the main Asteroid Belt clusters and for the two intense peaks in the Kuiper Belt objects. Period ratios computed from the simple model have errors typically less than 1%. There is no obvious pattern in which ratio slots are occupied in a given system. The estimated common base factor for the harmonic period model in the solar system cases is ~ 1.072 and this curiously leads to ratios which are often related to Fibonacci series ratios and simple phi functions.

For example, 3/2, 5/3, $(3/2)^2$, $(13/8)^2$, $2 / (13/8)$.

We also noted that curiously Pi / e = 1.156 and √ 1.156 = 1.075. We cannot say if this appearance of Pi, e and 'phi' relates to some underlying physical, mechanism controlling orbital spacing and periods but it is intriguing. The model appears to hold for bodies known to have formed by accretion from proto-planetary disks but also for captured bodies like Phobos and Deimos and disturbed systems like that of Neptune although the data is less strong here. The same harmonic model appears to apply to the eight extra-solar planetary systems examined, all rather different from ours, in three cases with multiple Jupiter mass planets.

The other systems have super-Earth class and Neptune class planets packed in orbits within 1 AU of their star. In three cases the same solar system period ratio base factor of 1.072 fits well. The other systems use 1.073, 1.073, 1.074 1.075 and 1.077. But the observed range for F is <0.4% across all systems.

The period ratios and model power ranges differ markedly between systems but the model holds. We have

	Period Ratio Range	Power range (F = 1.072 – 1.077)
Solar system	1.4 – 5.3	5 – 24
Kepler 62	1.5 – 6.7	5 – 27
HD 160691	2 – 32	10 – 48
47 Ursa maj.	2.2 – 6	11 – 25
Up. Androm.	3 – 52	16 – 57
HD 10180	2.5 – 5	13 – 23
Mu Arae	3 – 20	16 – 43
Gliese 892	2.2 - 19.6	3 – 42
Kepler 90	1.2 – 6.9	3 – 26

Given the differences in planetary mass ratios and density of orbital packing it is surprising that one model with an almost fixed, single parameter, holds. When the model fits well, as it does, the errors are small and usually the 'correct' power to apply is obvious. There is rarely ambiguity as shown by the model / observed period ratios usually being very close to unity. Is this an indication of an underlying resonant mechanism which only 'allows' certain radii and period ratios to be stable and expels objects from period ratios which do not conform? Such is the case in the Kirkwood gaps of the asteroid belt. There are some curious numerical coincidences which may or may not have physical significance. We noted that pi / e = 1.15573 and its root is 1.075 close to the maximum base factor observed. Also 2 /√3 = 1.15473 with root 1.0746, close to several factors observed.

It is also strange that the lowest, quite common, factor, 1.072, gives us 1.0474 for the orbital radii ratio base factor and 1.0474 x 3 = 3.1422 , within 0.02% of pi. Could it be that the orbit ratios we observe are those which are 'trapped' between local, major rational harmonic resonances? Resonances may stabilise or destabilise orbits. In the Asteroid Belt for example the Kirkwood gaps exist at periods which are 1 / 3, 2 / 5, 3 / 7, 1 / 2 that of Jupiter. Ceres, the largest asteroid, sits at 1 / 2.575 the period of Jupiter, just avoiding the 2 / 5 resonance gap, and we noted a large cluster of asteroids centred at 1 / 2.62 Jupiter's period. Note that 2.62 is 1.6186 x 1.6186 or ~ phi x phi. The observed cluster ratio 3.274 is very close to (13 / 8) x 2 = 3.25 and 2.154 is identical to (21/13) x 4 / 3. Also 2.854 is close to (13 / 8) x 7 / 4 = 2.845 (but also 2 $\sqrt{2}$ = 2.83).

TABLE A expands these 'hints'. We compared all observed orbital period ratios, low and medium order rational fractions and our proposed model with F = 1.0725. The following patterns are clearly seen. There are clusters of period ratios at or near model powers of 3, 6, 9, 12, 15, 18, - , 24, 27. These are harmonics of a base ratio of 1.233. We note that 2 x (phi − 1) = 1.236 and using the nearest Fibonacci ratio 2 x (21 / 13 − 1) = 1.231. Continuing the phi connection, there are major clusters near model power 7, 'trapped' between the 8 / 5, 13 / 8, 5 / 3 Fibonacci ratios. The harmonic mean here is 1.63, while the model gives 1.632, and the observed mean cluster period ratio is 1.63. A related cluster is 'trapped' between 5 / 2 (model power 13) and 8 / 3 (model power 14). This cluster is 'near' ratio phi x phi.

There are also clusters at period ratios 2 / 1, 3 / 1, 4 / 1, 5 / 1. However we also have overall, good clusters at $\sqrt{2}$, 2, 2 $\sqrt{2}$, 4, 4 $\sqrt{2}$ at powers 5, 10, 15, 20, 25. The key seems to be the spacing patterns of lower order rational fractions which presumably represent possible orbital resonances. The model seems to capture all these locations quite accurately. How is this possible? Are the various patterns in the period ratio sequence we have found really related? Consider the two classical irrational numbers phi and $\sqrt{2}$. Let's take the Fibonacci approximation of phi = 13 / 8 = 1.625 and the classical, architectural $\sqrt{2}$ approximation, 24 / 17 = 1.412. Note that 13 x 17 / (8 x 24) = 1.151. Note that $\sqrt{1.151}$ = 1. 0728, very close to our overall model factor, F = 1.0725. Note also that $\sqrt{3}$ / phi = 1.0705 or in nearest Fibonacci terms, $\sqrt{3}$ / (21 / 13) = 1.0724. Also note that $\sqrt{3}$ / $\sqrt{2}$ = 1.225, close to the key model ratio, 1.233 discussed earlier. Period ratios are observed at all model powers except 19, 21, 22. Why? It may be down to system sampling limits and chance but power 19 gives a ratio of 3.78 ~19 / 5 and close by is the low ratio 4 / 1 above and 7 / 2 (power 18) below. 4 / 1 has one observed ratio and 15 / 4, next to 7 / 2, 2 ratios.

Also power 20 gives 4 / 1 and 1 observation and power 23 gives 5 / 1 with 3 observations. Power 21 yields 17 / 4 and power 22, 14 / 3, less sharp harmonics.

Do strong neighbours entrain objects from nearby weaker ratios?

We can at least say that phi related ratios play a role in determining local orbital period ratios and the process involved leads to harmonically related orbital periods. In the solar system this leads to a potential gravitational forcing pattern with a complex harmonic structure which appears to act on other physical systems (such as the Sun and climate) which then add their own dose of non-linearity, sometimes also involving Phi harmonics. It is remarkable that anything can be dissected out of this mess and it explains why proving a path of causation is so difficult.

The model represents well the system of natural harmonics which appear to define orbital period ratios. What the model does not do is predict which combination of ratios will be observed in a given system. Is this a matter of chance related to the chaotic collapse of proto-planetary disks and perhaps later planet migrations, or are the period ratios observed in specific systems relatable to, for example, the range of planetary masses formed or the scale of the star system? To answer that would require a detailed analysis of the hundreds of systems now known.

2.4 Stellar Activity Cycles

In the coming century we will not only find many more planetary systems but spectral studies will build up long time series of variations in stellar activity for a large sample of stars similar to our own. Here are a few current examples.

In a few cases records of Ca 11 H+K lines in stellar spectra (a good measure of solar activity as confirmed for our Sun) have been measured for some decades. HD81809, a K0V class star, has been monitored since 1965. A long activity cycle ranging from 8 to 9 years has been established. This compares with our own primary ~11 year activity cycle. Similarly HD10476 has a cycle ranging from 10 to 12 years. Xi Bootes has two interacting cycles of 3.5 and 5.3 years. Interestingly these cycles are related in length by a ratio of 1.514 or closely as 3 : 2, a familiar Fibonacci number ratio. HD 78366 also has two cycles of 5.9 and 12.2 years a ratio of 2.06 or closely 2 : 1, another Fibonacci ratio. HD 149661 has cycles of 12.4 and 4.4 years, a ratio of 2.82, not Fibonacci but still highly irrational since ($\sqrt{2}$) cubed = 2.83. In our Sun we noted various harmonic relationships including 22.33 years and the third harmonic, 7.43 years for the Hale magnetic time series.

Notice in the image below, 136202 has a 23 year cycle but of small amplitude compared to our Sun imaged next to it. The other examples are of similar activity amplitude to the Sun. The range covered here is from K5 to F8 class stars.

The prospect of exploring and comparing stellar activity cycles in future is very exciting, especially in those with planetary systems. In terms of climate modulation by stellar variations we may soon have answers. One of the most contentious issues in the current climate debate is the impact or not of 'grand minima' in solar activity when the indicators of activity, such as the surface sunspot cycle turn off.

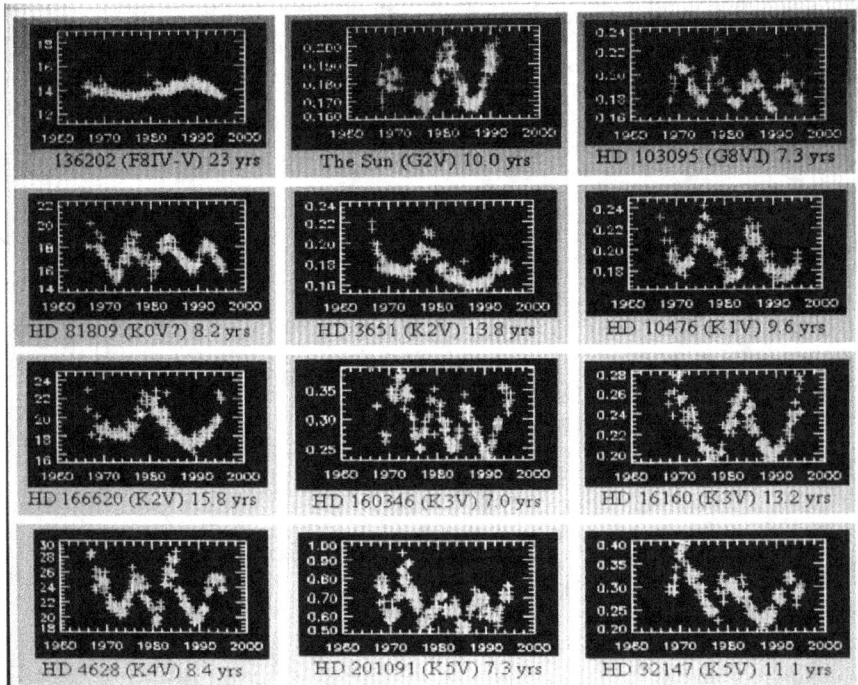

Notoriously the last grand (Maunder) minimum from ~1640 to ~1715 occurred at the same time as the 'Little Ice Age', a period of very low global temperatures. Figure 2 illustrates the Maunder Minimum in terms of sunspot activity and the Beryllium 10 solar activity proxy series. Since 1990 solar activity has begun to fall sharply and some astrophysicists believe that a new grand minimum is on the way. Notice the unprecedented high levels of activity developing over the last three centuries after the Sporer and Maunder minima. Some appear to deny even the existence of these solar minima.

Fortunately astronomy comes to the rescue with its Ca 11 H+K measurements (5). We can look at the proportion of the, now large, sample of Sun like stars which show zero or low activity. The raw number is 5 – 30%. As usual nothing is simple. This number caused great consternation among astrophysicists and the climate folk. Could these stars really be in temporary or permanent Maunder Minimum states? It is said that many of the stars are brighter than the Sun and therefore older and intrinsically less active. One claim is that such stars go through activity minima only 3% of the time.

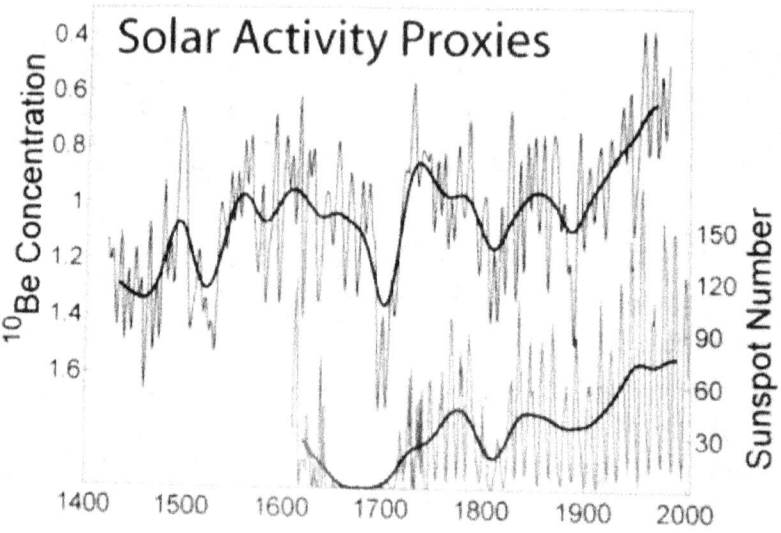

However a survey of coeval stars in the open cluster M67, which is slightly younger than the Sun, also shows about 25% in a quiet state. The question is said to 'remain open' by some, settled by others.

What is the situation with the history of the Sun? Well the evidence for the Maunder Minimum is good but there were other minima judging by solar activity proxy time series (such as tree ring C14) and in between, activity maxima, over many millennia. In the last few thousand years several have been claimed. Consider the period from the beginning of the Maunder Minimum until today. The minimum lasted ~71 years so we have an upper limit of 71 / (2016 – 1645) = 0.15 or 15% for the proportion of minima. If we also accept the Sporer Minimum of 1450 – 1550 we get (71 + 101) / (2016 – 1450) = 0.302 or 30.2%. So if distant observers had looked at the last 600 years of solar activity via Ca 11H+K line intensity they would estimate a 'grand minima time fraction' of ~22 %, similar to the M67 survey result. Clearly evidence will continue to build up and any corrections for star age and slow rotation will become clearer. The author would still take a bet that our middle aged Sun is not that unusual and that long term solar activity grand minima are real and climate significant.

2.5 Self Organised Physical Structures : Planetary Spacing

There seems little doubt that planetary spacing in stellar systems is far from random and that the patterns of orbital period ratios exhibit a preference for or avoidance of certain values. The author has taken this observation into the possible fine detail of such patterns but the existence of broader ratio patterns is now accepted thanks to the many example systems available beyond the solar system. The figure below shows the orbital period ratios for 932 pairs of exo-planets collected during the KEPLER mission (6) with 310 pairs having 'gap free' sequences in the ratio pattern proposed by the authors. There are frequency peaks at the proposed locations but also others which correspond to this author's 'fine detail'. We see peaks at 5:4, 4:3, 3:2, 5:3, 2:1, 5:2, 3:1. We note familiar Fibonacci ratios such as 2:1, 3:2, 5:3, 5:2. The authors look into the theory of multiple resonances and propose a dual origin mechanism compatible with the facts.

Planets form as condensations in proto-planetary disks around young stars. The process is believed to be essentially chaotic, forming planetesimals at arbitrary distances around the star by accretion and collision. Gravitational interactions can lead to nominally circular orbits becoming eccentric and some orbits will become unstable. Other secular changes can lead to orbital resonances which may be stable or unstable. The authors propose that these processes act to 'self organise' the planets into a system which is mutually stable over long time periods, while involving low order harmonic resonances. Even so we know from our solar system that such systems may hover on the edge of stability and bodies may still be ejected from time to time. The classic case involves the Kirkwood Gaps in the main Asteroid Belt where asteroid orbits at certain harmonic distances from Jupiter are unstable e.g. 2:1, 5:2, 3:1, 7:2, 4:1. Study of the 3:1 resonance orbit using very long term computer simulation of the solar system shows that a body orbiting here follows a very irregular path subject to sudden large, brief increases in orbital eccentricity on time scales of hundreds of thousands of years. If the mean eccentricity is 0.05 the peak eccentricity can suddenly increase to 0.35 in a few hundred years. This makes a body orbit overlap Mars' orbit and the body is ejected. By contrast the Hilda group of asteroids at a 3:2 Jupiter resonance is stable because the simulation proves the overall pattern of planetary forces here does *not* create a chaotic zone.

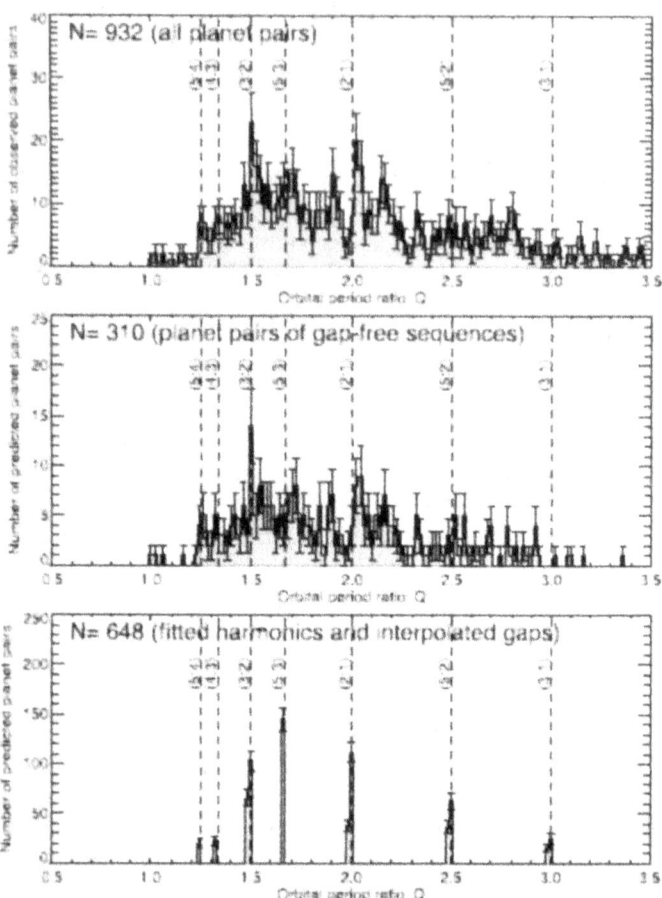

Each harmonic orbit has unique properties but overall we still find non-random patterns again and again in planetary systems. The self organization of planetary orbits through natural processes of interaction teaches many lessons for other complex natural systems with non-linear behavior…which are the majority. Is there direct evidence for a chaotic solar system? A paper in nature in 2017 reported a study of limestone and shale layers covering several million years during the Cretaceous period around 87 million years ago in Colorado in the bed of the shallow North American Seaway (7). Variations in the rock layers of clay and calcium carbonate proportions could be read as a response to a changing climate. The authors looked at several driving forces but remarkably evidence of a clear 'resonance transition' between the orbits of Mars and Earth affecting the angle of the Earth's axis to the ecliptic (and presumably more so that of Mars) and the time of year of the closest approach to the Sun by the Earth.

These changes (as in the case of the Croll-Milkanovitch cycles in the present ice epoch) change the pattern of solar radiation received over a given area and thence the climate. This is the first direct geological proof of chaotic behavior in the planetary system affecting climate. It seems likely that it is only a matter of time before many other climate / planetary effects are recognized.

We will see that recent research provides evidence for the dynamics of other systems relevant to climate change being generated by near Chaotic and Self Organized Critical processes including earthquakes, volcanic eruptions and many climate features over a wide range of times scales from days to hundreds of thousands of years. We have a pathway of causation here and each stage in the path involves similar non-linear processes. The key property of SOC systems is that internal drivers continually move the components towards a critical state which means that small periodic external forces can precipitate large effects on longer time scales. That is remarkable but apparently the case. In the next section we will move on to the fingerprints of solar activity variation and show their relationship to the fingerprints of planetary orbital dynamics. We will also see the evidence pointing to both Chaos and Self Organized Critical processes in the Sun.

Section 2 References

1. Chamberlain A; http: //ssd.jpl.nasa.gov/images/ast-histo.ps
2. 'Stability of the Solar System; Wikipedia.
3. Papaloizou J, Terquem C ; 'Planetary Formation & Migration'; Rep.Prog.Phys 69 (2006), 119-180.
4. 'Extrasolar Planets Encyclopedia'; http: //exoplanet.eu, Paris Observatory.
   5. Mount Wilson HK Project; Extra-Cyclical Activity in Solar-like Stars.
   6. 'Order out of Randomness: Self Organised processes in Astrophysics'; M J Aschwanden et al, Space Science Reviews 214, article.55 (2018); 06.03.18.
   7. 'Theory of Orbital variations confirmed by Cretaceous geological evidence'; Chao Ma et al; Nature 542 (7642): 468 (23 Feb 2017).

Note

Examining the model powers observed also shows an excess of odd integers (63%) in the significant model powers and of the 18 rational fractions 'confining' large period ratio clusters, 17 involve primes : 1, 2, 3, 5, 7.

Early Fibonacci numbers and their ratios frequently feature primes: 1, 2, 3, 5, 13… This is interesting. Could there be a link with prime sequences here? For example in 1772 Euler was searching for formulae to generate prime numbers.

By chance he found that the formula $X^2 + X + Q$ with $Q = 41$ and $X = 0$ to $Q-2$ generated many primes including 41, 43, 47, 53, 61, 71, 83, 97, 113…This sequence seems to generate the range of base factors we have see.

$43 / 41 = 1.0488$ and $1.074^{2/3}$ $47 / 41 = 1.147 = 1.071^2$ $53 / 43 = 1.232 = 1.072^3$

$61 / 53 = 1.151 = 1.073^2$ $61 / 43 = 1.419 = 1.0725^5$ $71 / 53 = 1.3396 = 1.075^4$

$83 / 53 = 1.565 = 1.076^6$ $97 / 71 = 1.3663 = 1.046^7$ and 1.078^4

Euler found that Q= 2, 3, 5, 11, 17 also generated primes exclusively and some of these values may also give ratios related to our problem. For example $Q = 17$ gives

17, 19, 23, 29, 37, 47…Note that $29 / 19 = 1.526 = 1.073^6$; $37 / 23 = 1.609 = 1.0703^7$ ~phi ; $47 / 29 = 1.62 = 1.0713^7$ ~phi ; $59 / 47 = 1.255 = 1.077^3$; $23 / 17 = 1.351 = 1.076^4$. Using $Q = 11$ gives us

11, 13, 17, 23, 31…We note that $31 / 23 = 1.347 = 1.076^4$; $23 / 13 = 1.77 = 1.074^8$; $17 / 11 = 1.545 = 1.075^6$; $23 / 17 = 1.35 = 1.076^4$. Using $Q = 5$ gives us 5, 7, 11, 17…

and $17 / 11 = 1.544 = 1.075^6$; $17 / 7 = 2.43 = 1.076^{12}$; $11 / 7 = 1.077^6$; $11 / 5 = 2.2 = 1.075^{11}$; $7 / 5 = 1.4 = 1.07^5$.

The connection to certain prime sequences is perhaps worth considering further.

Section 3: Planetary Forcing of Solar Activity
More Phi in the Sky?

'When you have eliminated the impossible whatever remains, however improbable, must be true'
                                                        Sherlock Holmes
3.1 Introduction

It has long been suggested that planetary dynamics modulate solar activity either via tidal effects or torque effects as the spinning sun moves around the barycentre of the solar system. Yet the physical forces involved seem 'too small' to produce such effects. However some classes of non-linear dynamical systems exist where small inputs produce large outputs. Is it possible that the Sun hosts such a system? It has also long been argued that solar activity affects terrestrial weather but support for this idea ebbs and flows. Both the solar dynamo and the weather system are complex and non-linear, exhibiting quasi-periodic and perhaps chaotic behaviour. It would be remarkable if a simple link was consistently observable. Nevertheless the author became convinced of a link many years ago via the work of Professor R H Dicke (1). Dicke examined a long data series of deuterium / hydrogen ratios found in ocean sediments. The D / H ratio is an accepted indicator of historical temperature. Dicke found a very sharp spectral peak at 22.33 years which is identical to the primary solar, Hale 'magnetic' activity cycle. In fact Dicke used the *stability* of this terrestrial 'weather' signal to suggest that the Sun contained a 'hidden chronometer' controlling the Hale cycle (1, 2). More recently several studies using MESA, maximum entropy spectral analysis, have confirmed the stability of the Hale magnetic cycle (3). We will explore the probable planetary origin of this solar chronometer.

In all that follows we will see the importance of non-linear behaviour of dynamic systems in the phenomena we explore. The lay reader might wish to scan through Appendix 1 for a briefing on these issues. Until recently many sciences dealt with linear mathematical models of the systems they studied or used locally linearised, non-linear models subject only to small perturbations, where ever they could. Nonlinearity means trouble and was to be avoided whenever possible. However with this attitude remarkable richness of behaviour is excluded from the start and complex, non-linear links between natural phenomena are not recognised. We will see how this ignorance has happened, damagingly, in the case of the solar activity long term time series itself.

Only in recent decades has non-linear behaviour become moderately familiar to professional scientists as opposed to specialist mathematicians. In a linear oscillatory system we might expect to see at most a small response to an external forcing signal of arbitrary frequency. In a non-linear system a much wider range of behaviours may occur with large responses related to harmonics (integer multiples or rational fractions) of the forcing signal. If the forcing signal is itself complex the problems multiply!

The tool for studying dynamic systems forced by external, possibly periodic, signals is spectral analysis. This in essence attempts to reduce a complex signal to a set of component sine waves of different periods and to estimate the amplitude and phase of each component (see Appendix 1). In practice data record length and analysis choices can affect observed spectral frequencies, complicating interpretation. It may also be that the signal components change in relative importance over time. With older spectral analysis techniques spectral peak resolution was also limited. Newer techniques such as MESA and wavelet analysis reduce these problems and in recent years high resolution spectra for solar activity and weather related variables have become more widely available. These are more suitable for testing the planetary 'influence' hypothesis by direct comparison of supposed 'forcing' and target system 'response' frequencies.

We can now in effect compare fingerprints in detail. We will calculate hypothetical 'forcing' frequencies by looking at harmonics of planetary orbital periods, at their conjunction periods and at the frequencies that could then be generated by simple non-linear dynamic systems. We will find clear links between the planetary periods, solar activity cycles and some weather cycles through this simple process of allowing the possibility of non-linear dynamic behaviour within and between their generating systems. Along the way we will also show how the planets influence the Earth – Moon orbital system, the Earth's spin and earthquake energy release. This provides a valuable lesson to those who glibly deny a 'cosmic' link to our weather. Planetary effects on the Earth are far from trivial. Here we go! Hang on tight!

3.2 Solar Activity : The Wolf Sunspot Number Series

This analysis is based on high resolution MESA spectra derived from Wolf annual sunspot numbers for the years 1700 to 1990 reported in (3). The spectrum for the sunspot activity series contains ~30 sharp frequency peaks of various magnitudes down to periods of four years. The main peaks are

Large Amplitude	Medium Amplitude	Small Amplitude	
~100 years	179 years	~63 years	12.8
11.1	52	43.9	9.4
10	11.8	29.3	8.43
	10.5	22.2	8.11
	8.43	16.8	7.44
		15.4	5.75
			5.5
			5.26

Large Peaks in the Sunspot Spectrum

We note the frequently reported largest spectral peak at 11.1 years. This is half the Hale magnetic series main cycle at ~22.3 years. Testing the planetary link involves consideration of the orbital and conjunction periods of the gas giant planets, Jupiter, Saturn, Uranus and Neptune, namely $P_j = 11.86$, $P_s = 29.46$, $P_u = 84$, $P_n = 165$ years. We have two near resonant pairs here since $J/S = \sim5/2$ and $N/U = \sim2$.

C_{js}, the JS conjunction period is given by $1/11.86 - 1/29.46 = 1/19.852$

C_{jsn} is ~178.8 years. C_{jsu} is ~180 years.

C_{jsun} on average is ~179.4 years but JSN and JSU shift out of phase over time.

Simply we have $1/P_j - 1/C_{js} = 1/19.852 - 1/178.8 = 1/22.33$ which is identical to the period found by Professor Dicke for his solar 'chronometer'.

In the sunspot intensity cycle the key period would therefore be 11.167 years compared with, in this observed spectrum case, ~11.1 years.

The observed 10 year period compares with $C_{js}/2 = 19.852/2 = 9.926$ years. The large peak near 100 years is related to both the other large peaks since $1/10 - 1/11.1 = 1/100.9$.

Medium Amplitude Peaks.

We note immediately the peak at 11.8 years close to Jupiter's orbital period of 11.86 years. This shows a direct linear response by the Sun but the non-linear effects are much bigger. The medium peak at 179 years is close to the Cjsn and Cjsu conjunction periods at 178.8 and 180 years. The peaks at 52 and 8.43 years are also related since

$1 / 8.43 - 1 / 10 = 1 / 52.5$ but $Cjs / 2 = 9.93$ years and we noted that
$1 / Pj + 1 / Ps = 1 / 11.86 + 1 / 29.46 = 1 / 8.45$. The relationships could not be clearer.

The strong peaks at 10.5 years and 9.4 years can be similarly explained. Half the Cjs conjunction period is 9.93 years and so

$2 / Cjs - 1 / Cjsun = 1 / 9.93 - 1 / 179 = 1 / 10.51$ while $1 / 9.93 + 1 / 179 = 9.41$
and we see that $\sqrt{(10.5 \times 9.4)} = 9.934$ so the 10.5 and 9.4 peaks can be interpreted as side bands of the second harmonic of Cjs. Triplets of main peak and side bands will be found to be common in these analyses.

Note that the Uranus orbital period is also apparent since for the medium amplitude peaks 10.5, 8.43 and 9.4 we have

$Pu / 8 = 84 / 8 = 10.5$ $Pu / 9 = 84 / 9 = 9.4$ $Pu / 10 = 84 / 10 = 8.4$.

We have several simple harmonics of the Uranus orbital period in the sunspot spectrum. All the medium amplitude peaks are simply related to the orbital periods of the four gas giants with J and S dominant.

Small Peaks

Note the small peak at 29.3 years which is very close to the Saturn orbital period of 29.46 years. Here is more evidence of a direct linear link although the non-linear effects are again much bigger.

There is also a small peak at 22.3 years or 11.15 x 2, which is Ph, the Hale magnetic cycle. The ~63 year peak can be seen as $2 / Cjs - 1 / Pj = 1 / 10 - 1 / 11.86 = 1 / 63.7$ but also for Uranus $84 \times 3 / 4 = 63$ and for Neptune $165 / phi^2 = 63.02$. For Neptune we see yet another Fibonacci series ratio. The 43.9 year peak is obtained from medium amplitude peak interactions $1 / 8.45 - 1 / 10.5 = 1 / 43.3$. But also Csu, the conjunction period of Saturn and Uranus is 45.3 years.

16.8 years is a simple harmonic of $Ph / 2 = 11.16$ as $11.16 \times 3 / 2 = 16.74$.

3 / 2 is another Fibonacci ratio.
Medium amplitude peak interactions define other small peaks.

1 / Pj − 1/ 52 = 1 / 11.86 − 1 / 52 = 1 / 15.36 versus 15.4 years observed.
1 / Pj − 1 / Cjsn = 1 / 11.86 − 1 / 179 = 1 / 12.7 versus 12.8 years.

Note that √ (11.1 x 12.8) = 11.9 We can see the 11.1 and 12.8 year periods as 'sidebands' of the Pj = 11.86 year orbital period.
Cjn, the Jupiter − Neptune conjunction period is also 11.07 years. The sum of these orbital frequencies gives us the 12.78 year peak. Similarly for 8.78 and 8.11 years note that √ (8.78 x 8.11) = 8.44, the observed medium amplitude peak. So 8.78 and 8.11 are 'sidebands' of 8.43.

The 7.44 year peak is the third harmonic of Ph = 22.33 years and 22.33 / 3 = 11.16 x 2 / 3 = 9.93 x 3 / 4 = Cjs x 3 / 8 = 19.852 x 3 / 8 = 7.44. The link to the JS conjunction period is clear. Note for later the presence again of factors 3 / 2 and 8 /3, which are Fibonacci series ratios. The last three peaks also form a family: 5.75, 5.5, 5.26 since √ (5.75 x 5.26) = 5.5 and 5.5 = 11.1 / 2 = Ph / 4 = 22.2 / 4. These outer peaks are 'sidebands' of the second harmonic of the main, 11.1 year sunspot intensity cycle.

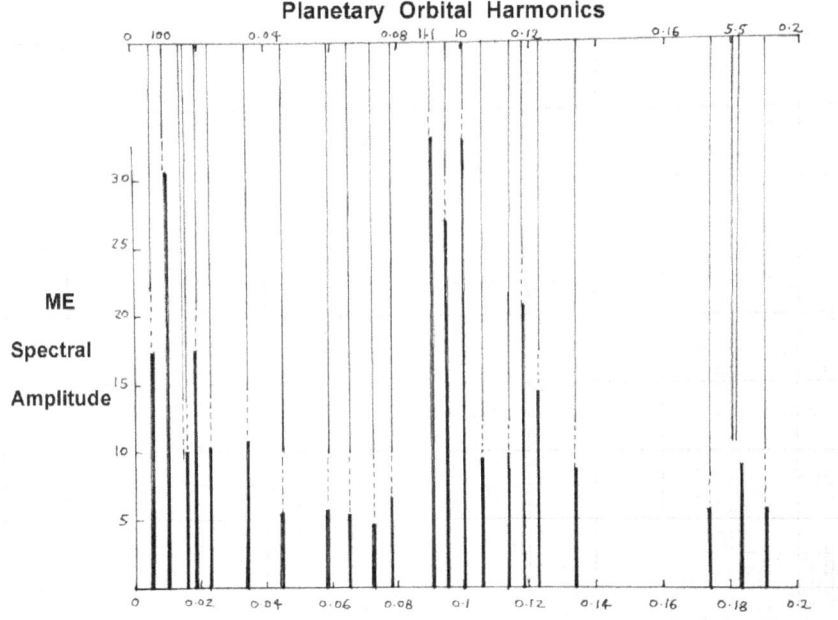

Planetary Orbital Harmonics

ME Spectral Amplitude

Sunspot Intensity Cycle Frequency (cycles per year)

97

The resolving power of MESA is sufficient to track changes in the sunspot spectra over time. Berger et al (3) do this for Wolf sunspot numbers over a period of 1700 – 1980 by computing spectra for overlapping data windows of 70 years duration. This provides 20 snapshots of the evolving spectrum. Although for long periods the ~11 year cycle is dominant, deviant periods are notable. The 4 most recent spectra show a shift in the ~11 year peak towards ~10 years. From the second data window (1710 – 1780) to the eighth (1770 – 1840) the ~11 year cycle is disrupted with power dispersing downwards towards 10, 9 and 8 year peaks and upwards towards ~14 years. The ~50 to ~100 year peaks also evolve over time.

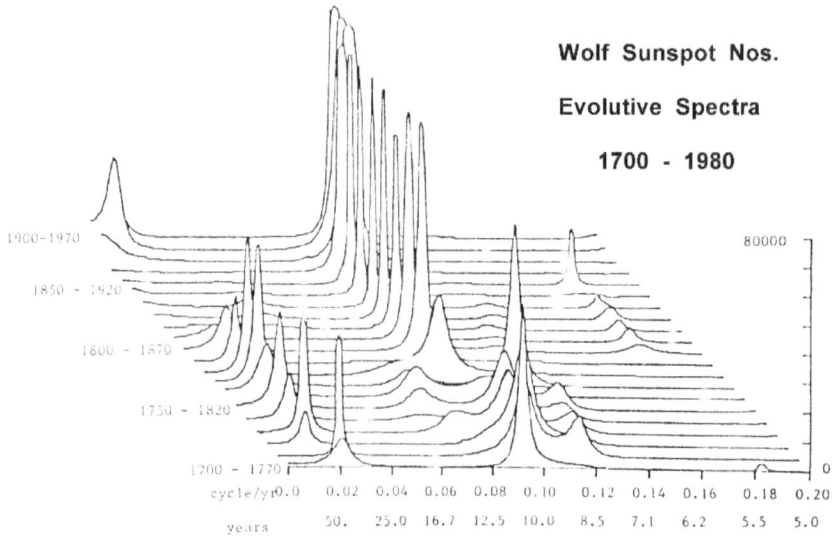

This instability in the solar activity spectrum is sometimes attributed to 'bad data' in the late 18th and early 19th centuries. However the shifts in spectra power over time are not random from data window to window but follow clear trends. The author pointed this out to Berger who included this analysis in his 1984 paper (3). By taking Berger's main spectral peaks we can demonstrate this. Take $f = 1 / 11.11$ years; $f2 = 1 / 58$; $f3 = 1 / 98$. Then :

1 / (f1+/-f2) gives us   9.3 and 13.7 years

1 / (f1 +/-f3)   "   "   10 and 12.5 years

1 / (f2 +/-f3)   "   "   8.4 and 11.9 years

The temporary loss of coherence in the primary 11 year cycle leads to an organised redistribution of power which gives further clues to causation.

Notice that the new periods 10 and 11.9 are familiar since Cjs / 2 = 19.85 / 2 = 9.93 and Pj = 11.86 years. We are seeing the signatures of the gas giants more directly during this interlude. Note also that the periods 8.4, 9.3, 12.5 and 13.7 are very close to other planetary 'conjunction' periods, namely:

fj + fs = 1 / 8.45 ; 2 fjs + fjsn = 1 / 9.4 ; fj - fn = 1 / 12.8 ; fj – fu = 1 / 13.8.

All these peaks are directly related to the gas giant orbital periods. The question of 'bad' sunspot data is important since its supposed existence has been used to deny a solar–weather link and recently to justify homogenising the sunspot number and sunspot area time series to eliminate the non-existent 'great solar anomaly'. In fact later we will see that although sunspot number and area are obviously related they have somewhat different dynamic properties as oscillators (4). They should not be homogenised. That is mathematically naïve. This homogenisation has been done in such away that the upward trend in sunspot peak amplitude has been reduced over the period from 1700 to today. That in turn reduces any estimated impact on climate if sunspots are used in a crude way in a sunspot – climate model. That is deeply troubling but not unusual in the climate story (nor in other areas of science: see App.3). However we have seen that the Sun's behaviour changed for some decades during the 'great anomaly' not in a random way, indicative of 'bad data', but by organised shifting of spectral power between harmonically related sunspot periods. The changes were real. Fortunately such unwanted data manipulation cannot negate the clear message of the long term spectral fingerprint of solar activity. In discussing the Hale magnetic time series we will look at Professor Dicke's later 1988 analysis (2) which confirms the long term phase stability of the sunspot cycle, eliminates the 'bad data' myth and gives a convincing physical interpretation of how the solar cycle may operate. His work and much else appears to have been forgotten…or worse. When we 'know' that climate change is man made why look at evidence of other, natural factors?

Summary

All the peaks in this high resolution sunspot intensity spectrum can be simply related back to the orbital periods of Jupiter, Saturn, Uranus and Neptune, their harmonics and conjunction periods and interactions between the main frequency peaks so generated. Nonlinearity is manifest as is the presence of phi and Fibonacci series ratios.

The periods when the primary ~11 year cycle is disrupted show clearly the movement of power towards Pj, the Jupiter orbital period, and planetary conjunction periods. This close spectral correspondence between orbital periods and solar activity strongly suggests a physical link between the Sun and the orbital dynamics of the gas giant planets.

3.3 The Hale Magnetic Solar Activity Cycle

It has been noted many times that the Hale, 22.33 year cycle, allowing for the reversal of sunspot magnetic polarity every 11.16 years, yields a more stable and simpler spectrum whichever analysis technique is applied. This is true for the MESA spectrum for this data which displays ~15 clear peaks compared with ~30 for the 11.1 year sunspot intensity series. Berger's (3) 'evolutive' MESA studies of the Hale cycle over time from 1770 to 1980, demonstrate the remarkable stability of the cycle. However we can see some changes in the 'shoulders' of the 22.3 year peak between ~12 and ~26 years.

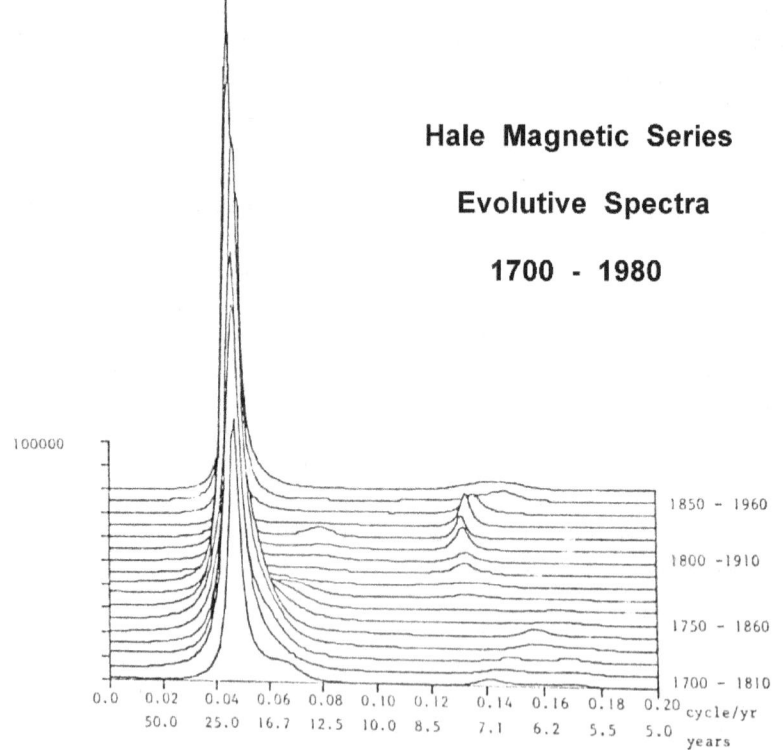

This movement reflects the bigger shifts apparent in the Wolf series between ~7 and ~14 years we looked at above. The only other feature in the evolutive spectra is the growth in power of the 22.3 year 3rd harmonic at 7.4 years in the ninth to the fourteenth data windows.

By plotting the Hale series with a pure 22.3 year sine wave we can see that the phase of the Hale cycle is preserved over the 1700 – 1980 period but with occasional, temporary loss of phase due to changes in cycle shape. E.g. from 1770 to 1810 and less extremely, from 1955 to 1980. These excursions correspond to the breakdown observed in the primary 11 year cycle of the Wolf spectra treated earlier. Note in phase SS maxima are usually higher; out of phase maxima are lower, except for ~1790.

We noted in the last section how spectral power was redistributed in an organised way between related spectral components during the 'great solar anomaly'. In 1988 Professor Dicke took the question further and answered it definitively in the journal Solar Physics (2). Dicke used data **after** the alleged anomaly (1817 – 1986) to construct a semi-physical model of the Hale cycle. This assumed that the average transit velocity of magnetic flux from the deep Sun to the surface is proportional to some power of the amplitude of the solar cycle (averaged over the transit period). The best fit model used a transit time of 12 years and an exponent, s, of 0.71.

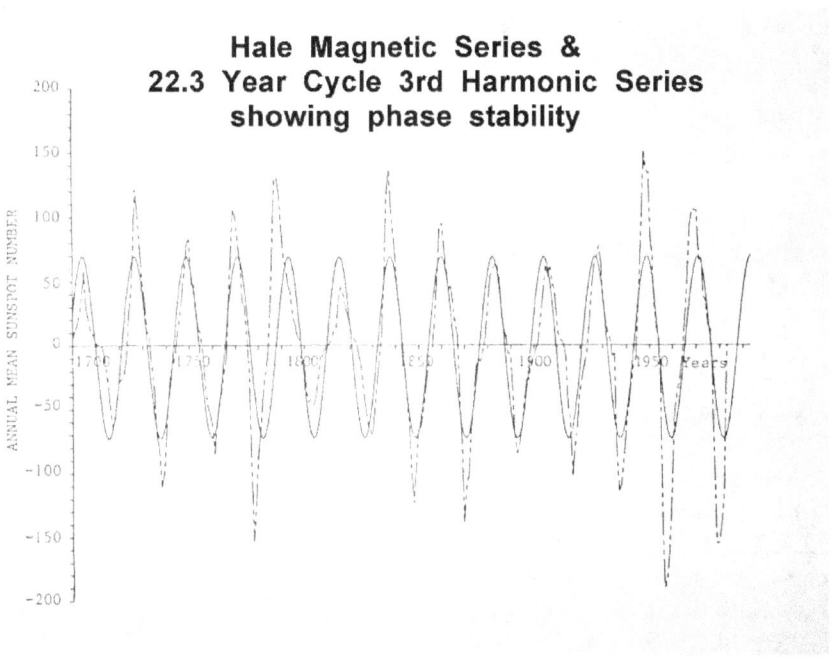

Hale Magnetic Series & 22.3 Year Cycle 3rd Harmonic Series showing phase stability

That is

$$V_m = C + At^{0.71}$$

The model gave a Hale period of 22.22 years, very close to the period 22.33 years he found in sea bed D / H data and to our evolutive spectra. The fit of the model is remarkably good over the data fitting period. However the fit was **just as good** for the historical Hale series data **before the 'great solar anomaly'**. Remember this model is based on physical assumptions set firmly in solar physics theory. During the 'anomaly' the model does not capture the observed phase shift. This result is very clear. It shows us that the Hale cycle phase is stable over centuries and indeed the 'chronometer' oscillator in the Sun looks to be phase locked in some way. Rarely, phase locking is lost. During the 18th century anomaly, phase was temporarily lost but then recovered perfectly. Even though phase was lost spectral energy was redistributed among other periods in an **organised way** which we can relate simply to planetary periods. The myth of 'bad data' was a product of linear thinking, or worse, and can be dismissed. (See App.3 for more on wishful thinking and data fiddling).

Let us now look at the full history Hale spectrum in detail with some confidence that what we are seeing is real. The dominant spectral peak is at 22.2 years with other large peaks at 25.5, 20 and 17.9 years. Smaller peaks are at ~111, 51, 39.6, 29.6, 16.7, 15.5, 14.6, 13.66. There are also clusters of peaks at 8 to 6 years around the third harmonic at Ph / 3 = 7.4 years.

As noted above the main Ph peak is derived from the Cjs given by 1 / 11.862 – 1 / 29.46 = 1 / 19.859 and the Cjsu and Cjsn conjunctions near 179 years. For example

1 / 19.859 – 1 / 178.8 = 1 / 22.34.   Also 1 / 19.859 + 1 / 178.8 = 1 / 17.874.

Note that 22.34 = 9 x 19.859 / 8 and 17.874 = 9 x 19.859 / 10 versus observed 17.9. Also 22.34 / 17.874 = 5 / 4 or (5 / 2) / 2 and 17.874 / 11.167 = 8 / 5, Fibonacci ratios again.

The three main peaks at 22.2, 20 and 17.9 are thus readily explained in terms of planetary orbital dynamics. Also for the small peaks, 111/ 5 = 22.2; 39.6 / 2 = 19.8 but Cjs = 19.85. As before there is clear evidence of non-linear behaviour. Third in magnitude is the peak at 20 years which is very close to the JS conjunction period at 19.85 years. Next in magnitude is the peak at 25.5 years.

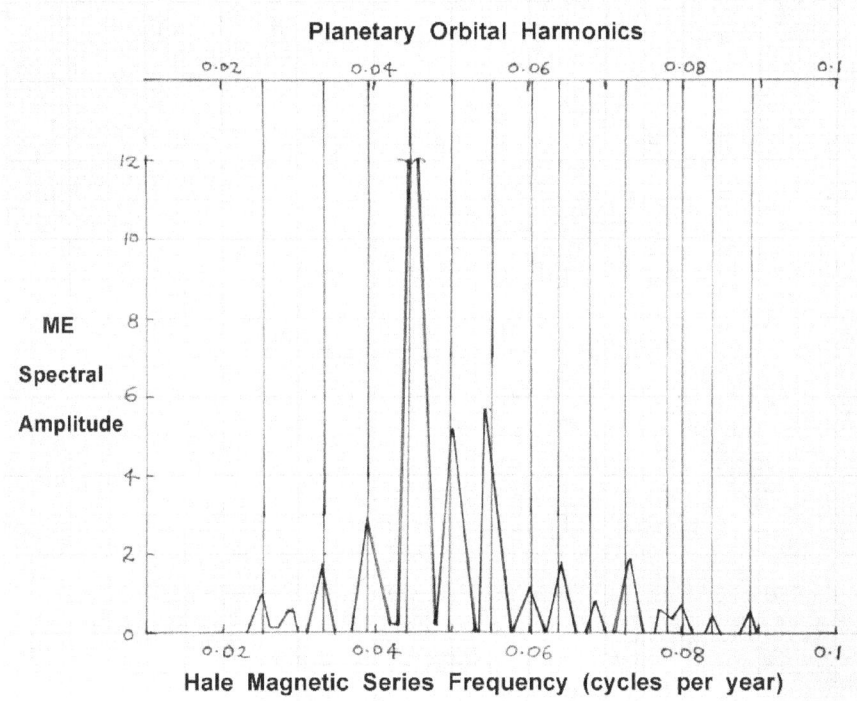

Note that √ (22.1 x 29.46) = 25.51 so we could see 22.1 and 25.5 as simply related to the Saturn orbital period which also occurs directly in the spectrum at ~29.6 years. We also note that summing the orbital frequencies for Saturn and Neptune yields a period of : 1 / 29.46 + 1 / 165 = 1 / 25 exactly. Note also for now that $22.32 / 2 = (8 / 13)^2$ x 29.46, another Fibonacci relation.

The next peak in size is at 13.66 years and 11.86 x 8 / 7 = 13.6 ; 22.2 x 8 / 13 =13.66.

Recall that the Cju period is also 13.8 years. Buck and Macaulay also noted the presence of Fibonacci ratios among the small peaks between ~17 and ~10 years. Some of these also appear to be phi 'echoes' of the four main peaks.

For 16.7 we have 22.2 x (3 / 2) / 2 = 16.65.

For 15.5   (8 / 13) x 25.5 = 15.7

103

For 14.6   (2 / 3) x 22.2 = 14.8

For   13.66   (8 / 13) x 22.2 = 13.66

But 13.8 years is also the Uranus / Jupiter conjunction period.

For   12.9   25.5 / 2 = 12.8

But 12.78 is the Jupiter / Neptune conjunction period.

For   12.5   (5 / 8) x 20 = 12.5   20 is equivalent to Cjs= 19.85

For   11.93   (3 / 5) x 20 = 12.

But the Jupiter orbital period is 11.86 years

For   11.3   22.33 / 2 = 11.17 and (5 / 8) x 17.9 = 11.2.

As noted earlier the sum of orbital frequencies for Jupiter and Neptune gives us 11.07 years. Buck and McAuly (5) also examined the fine detail of the Hale magnetic spectrum in the vicinity of the third harmonic Ph / 3, and noted some near phi relationships.

Further detailed inspection shows that every small peak between 8.1 and 5.1 years are phi related either to the main Hale peaks or to planetary periods or both.

7.73 years observed. $Cjs / (8/5)^2 = 7.75$ years where as before Cjs = 19.85.

7.35 years   "   Ph / 3 = Pj / phi = 11.86 / 1.618 = 7.33.

7.143   "   17.9 / (5 / 2) = 7.16 $Cjs / (5/3)^2 = 7.15$

6.895   " $17.9 / phi^2 = 6.84$. (Ph / 2) / phi =11.16 / 1.618 = 6.897
            Cju / 2 = 6.9 $Ps / (13/8)^3 = 6.87$ $4 \times Pj / phi^4 = 6.92$

6.67   "   Cjs / 3 = 6.62   (Ph / 2)/ (5 / 3) = 6.665

6.38   "   Cjn / 2 = 6.39   25.5 / 4 = 6.38 $16.7 / phi^2 = 6.38$

6.14   "   (Cjs / 2) / phi = 6.135

5.95   "   17.9 / 3 = 5.97   (Cjs / 2) / (5 / 3) = 5.98   Pj / 2 = 5.93

            Csn / 5 = 6.04 $25.5 / phi^3 = 6.02$ Ps / 5 = 5.9

5.55	"	Ph / 4 = 5.55	(17.9 / 2) / (21 / 13) = 5.54
5.33	"	Ph / (21 / 13) = 5.27	
5.13	"	7.73 / (3/2) = 5.15 25.5 / 5 = 5.1 Ph / (13/8)^3 = 5.17	

Ph / 3³ line: Ph / 4 = 5.55 with exponent 3 above, and Ph/(13/8)³ = 5.17

12.8 / (5 / 2) = 5.12   and Cjn = 12.78

In low resolution spectra this detail is absorbed into a broader peak spread around Ph / 3 = 7.4 years. At high resolution we have a central peak at 6.895 = (Ph / 2) / phi = 11.16 / phi, directly related to the main sunspot intensity period, which is flanked by pairs of sidebands making the close relationships of all these peaks starkly obvious. That is :

√ (7.143 x 6.67) = 6.9   √ (7.35 x 6.37) = 6.85   √ (7.73 x 6.14) = 6.89

√ (8.12 x 5.95) = 6.95   The mean peak value here is 6.897 years.

As we have seen Fibonacci series ratios are widespread throughout the spectrum. We have 2/1, 3/2, 5/3, 8/5, 13/8, 21/13 and phi factors.

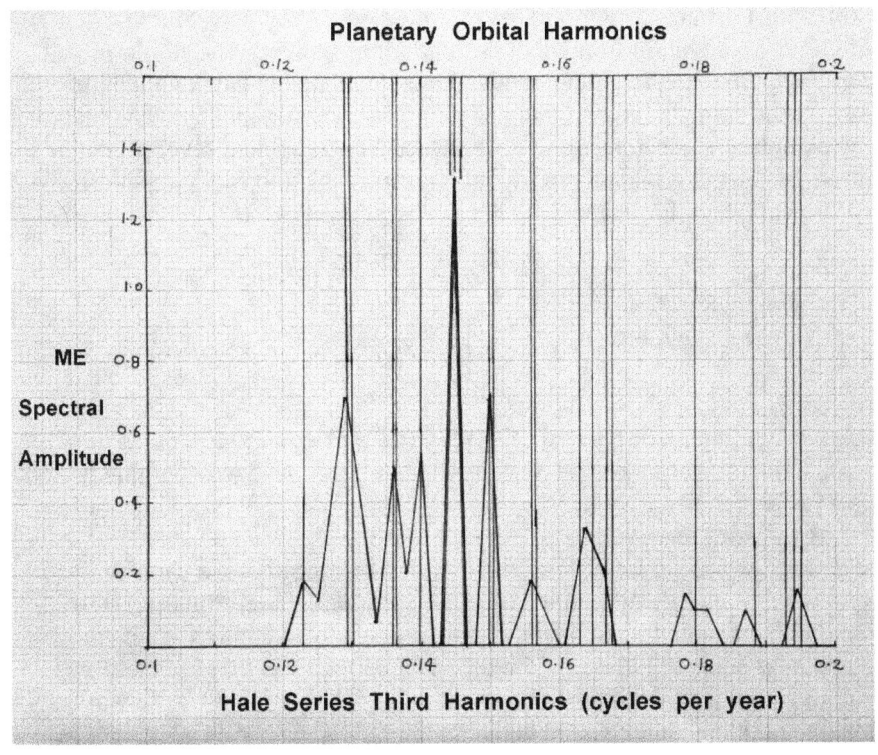

105

For example

11.86 x 5 / 8 = 7.41 and 19.859 x 3 / 8 = 7.44 versus 7.44; 29.46 x 3 / 5 = 17.7 or 29.46 x 8 / 13 = 18.1 versus 17.9 observed; 17.9 x 5 / 8 = 11.19 versus 11.17; 22.2 x 8 / 13 = 13.66; 29.46 / phi^2 = 11.25; 84 / phi^3 = 19.83

84 / $(8 / 5)^4$ = 12.82.

For the sunspot activity spectrum we also had 11.16 x 3 / 2 = 16.75 ; 11.16 x 2 / 3 = 7.44; 9.4 x 13 / 8 = 15.3 and 9.4 x 8 / 13 = 5.78; 10.5 x 8 / 5 = 16.8 ; 84 x 3 / 4 = 63 ; 84 / phi = 52 ; 84 / phi^3 = 19.83; 165 / phi^2 = 63; 22.33 x $(8 / 13)^2$ = 8.46; 19.852 x phi^2 = 52; 29.46 x $(8 / 13)^2$ = 11.16.

The origin of these Fibonacci harmonics will be discussed later. Overall the Hale MESA spectrum seems to be clearly related to the gas giant orbital periods and conjunctions via some non-linear mechanism which often produces phi or Fibonacci ratio harmonics. We saw a similar result for the Wolf SS series. If the reader finds the phi story difficult we can look at the same harmonic data a different way which, if anything, makes the planetary-solar links more transparent. We noted that several integer multiples of the gas giant conjunction periods seem to focus on a period of ~179 years. Cun for example is 171.2 years. But also

9 x Cjs = 14 x Cjn = 178.8   5 x Csn = 13 x ju = 179.44

4 x Csu = 14 x Cjn = 180.1 and of course 8 x 22.33 = 9 x 19.85 = 10 x 17.87 = 178.65.

We will see later that periods of ~179 years and its sub-harmonics turn up regularly in long range climate series spectra. Suppose we treat this period as a fundamental and calculate its integer harmonics. A surprising pattern emerges which is shown overleaf. Considering the Wolf SS and Hale spectra together we see that all the identified spectral peaks can be generated as simple harmonics of 178.65 years. That is, from the 1st to the 35th harmonic. In just a few cases the peaks we see are half integer harmonics such as 178.65 / 3.5 = 51 ; 178.65 / 4.5 = 39.6 (and of course 2 x 19.85 = 39.7); 178.65 / 11.5 = 15.5 ; 178.65 / 33.5 = 5.33. A few other integer harmonics fail to 'lock on' but the failures are informative.

So there is a small Hale peak at ~111 years but 178.65 / phi = 110.4 (and of course 5 x 22.3 = 111.5). Also the small Hale peak at ~ 68 years is 178.65 / (phi x phi) = 68.2 (and of course 3 x 22.33 = 67).

The 178.65 detailed harmonic match is remarkable. Look for example at the phi echoes of the Hale third harmonic near 7.43 years. The relationship of solar activity to planetary orbital periods and conjunctions is surely beyond doubt.

Harmonics of the ~179 Year 'Conjunction' Period

		Wolf SS Periods	Hale SS Periods	
1	178.7	179	---	
2	89.3	(~100 = 5 x 19.85)	(~111 = 5 x 22.3 = 178.7 / phi)	
3	59.55	(52, 63)	(51, 68) 178.7 / 3.5 = 51	
			178.7 / phi^2 = 68.2	

(Ironically the 89 and 59 year cycles show up strongly in many climate time series !)

4	44.67	44.6	(39.6)	
			178.7 / 4.5 = 39.71	
5	35.74	---	---	
6	**29.7**	29.3	29.6	Ps = 29.5
7	25.52	---	25.5	
8	**22.33**	22.2	22.2	
9	**19.85**	---	20	Cjs = 19.85
10	**17.87**	---	17.9	
11	16.24	(16.8, 22.2 x 3 / 4 = 16.75)	16.7	
12	14.89	(15.4, 178.7 / 11.5 = 15.5)	14.6	
13	**13.74**	---	13.66	Cju = 13.8
14	**12.76**	12.8	12.9	Cjn = 12.77
15	**11.91**	11.8	11.93	Pj = 11.86
16	11.17	11.1	11.3	
17	10.51	10.5	---	
18	**9.93**	10.0	---	Cjs / 2 = 9.93
19	9.4	9.4	---	
20	8.93	---	---	
21	8.5	8.43	---	
22	8.12	8.11	8.1	
23	7.77	---	7.73	
24	**7.44**	7.44	7.35	
25	7.15	---	7.143	
26	6.87	---	6.895	

27	6.62	---	6.67	Cjs / 3 = 6.62
28	6.38	---	6.38	
29	6.16	---	6.14	
30	5.95	---	5.95	
31	5.76	5.75	---	
32	5.58	5.5	5.55	11.16 / 2 = 5.58
33	5.41	---	---	
			5.33,	178.7 / 33.5 = 5.33
34	5.25	5.26	---	
35	5.1	---	5.1	

3.4 Hale Magnetic Cycle Wavelet Analysis

Wavelet analysis also produces high resolution spectra but also allows the detailed tracking of the variation of power at particular frequencies over time. Reference (6) examines the wavelet spectra of the magnetic sunspot series from 1700 to 2007 and Central England Temperature, a classic weather series, from 1659 to 2007.

The mean power of the enhanced wavelet transform of the Hale series shows a prominent peak at ~22.4 years and at the 3rd and 9th harmonics along with several other familiar peaks. In terms of planetary periods we have

		J 11.86	S 29.46	JSconj. 19.85
3rd H	7.46	7.42 5 / 8	7.4 1 / 4	7.44 3 / 8
9th H	2.49	2.4 1 / 5	2.46 1 / 12	2.48 1 / 8

The harmonic relationship to the Cjs period is particularly clear. The enhanced instant wavelet power is also provided for 1850, the mid point of the time series, yielding very sharp isolated peaks at

Large amplitude   22.4 years

Medium   39, 11.9, 8.1, 5, 3

Smaller   136, 90, 60

These peaks are simply related harmonically. For example 136:90; 90:60; 60:39; 11.9: 8.1 are all as ~3 : 2. Also 8.1: 5 is 8 / 5 and 5:3 is 5 / 3.

These are Fibonacci series ratios yet again as were the 3rd harmonic – Cjs and the 3rd harmonic – Pj ratios.

We also see several periods familiar from the Hale series MESA study including 60, 39, 22.4, 11.9, 8.1, 5 years. Note the 11.9 year peak …the Jupiter orbital period directly present. Several longer period peaks are simply related to the main 22.4 year cycle since $22.4 \times 6 = 134.4$ versus 136 ; $22.4 \times 4 = 89.6$ versus 90; $22.4 \times 8 / 3 = 59.7$ versus 60. Also $3 \times Csu = 136.1$, $2 \times Csu = 90.7$, $3 \times Csn = 90.6$ and $2 \times Csn = 60.4$. For Neptune,

$Pn / phi^3 = 38.95$ versus 39.

Note also that $22.4 / (5/3)^2 = 8.07$ versus observed 8.1 and $22.4 / (5/3)^3 = 4.9$ versus 5. Harmonics of the planetary orbital periods of Jupiter and Saturn also occur frequently.

Spectral peak	J 11.86	S 29.46	JSconj. 19.85
136	$(13/8)^5$	$(5/3)^3$	phi^4
90	$(8/5)^4$	3	$(5/3)^3$
60	5	2	3
39	$2 \times 5/3$	$(2/3) \times 2$	2
11.9	1	2/5	3/5
8.1	2/3	$(5/3)/6$	2/5
5	$(5/3)/4$	1/6	1/4
3	1/4	1/10	$(3/5)/4$

We have several simple integer harmonics but Fibonacci series ratios are again common. The wavelet spectra are very similar but not identical to the MESA solar spectra and emphasise that power moves across the frequency range over time. However it is clear that we are dealing with closely related families of frequencies and that each is harmonically related to the planetary orbital periods of Jupiter and Saturn and often via Fibonacci series ratios. We will now look at the X Ray Sun before seeking to explain these correlations.

3.5 Solar X Ray Flare Cycles

Our examination of solar activity has so far centred on longer term variations in sunspots and magnetic cycles. Other energetic processes operate on shorter time scales and do not obviously correlate with the main solar activity cycles. X ray flares for example may occur at any stage of the sunspot cycle. Do X ray phenomena on time scales of a year or months relate to the planets? Landscheidt (7) provides a high resolution MESA spectrum for solar X ray flare activity for >/ = class X1 based on the period 1970 – 1982. This is a short data window of 312 monthly records but most of the spectral power observed is on the order of several months. Large peaks occur at ~150, 4.84, 2.81 and 1.117 months. There are moderate peaks at 14.8, 6.3 – 7.2, 2.26, 1.34, 1.22 months. We will first relate the X Ray flares to general solar activity.

The large ~150 month peak is at 12.5 years, close to the 12.78 years in the Wolf sunspot spectrum. 12.5 is also related to the third harmonic of the Hale cycle since (22.3 / 3) x 5 / 3 = 12.4 years or 149 months. Note the Fibonacci ratio. The Wolf activity spectrum has a prominent peak at 8.4 years but 8.4 x 3 / 2 = 12.6 years or 151 months. The Hale series spectrum has a major peak at 19.85 years and 19.85 / (8 / 5) = 12.4 years or 149 months. The Fibonacci harmonic links are clear.

The large X Ray peak at 14.8 is also ~150 / 10 months. The primary Wolf spectrum peak is at 11.165 years and 11.1 / 9 = 1.233 years or 14.8 months. Alternatively we consider the third harmonic of the Hale series, 7.4 years which gives 7.4 / 6 = 1.233 and 14.8 months exactly. The 19.85 year peak is also represented since 19.85 / 16 = 1.24 years or 14.9 months. The Wolf peak at 8.4 years is also present since

$$^{4}$$

8.4 / (21 / 13) = 1.233 years and 14.8 months.

The broad X Ray peak 6.3 - 7.3, centred on 6.73 months hides at least two other peaks. We note that 1 / 6.73 +/- 1 / 88.8 gives periods of 6.2 and 7.3 months. 88.8 months is 7.4 years and 22.3 / 3. Also 22.3 / 4 = 5.57 years and 66.9 months but 66.9 / 10 = 6.7 months, the observed peak mean. The 8.4 year Wolf period gives us 8.4 x 12 / 15 = 6.72 months. We seem to have higher order harmonics of the longer solar activity cycles.

For the 4.84 month X Ray period note that 14.8 / 3 = 4.92 and considering the next peak and the previous peak, 1 / 2.81 – 1 / 6.73 = 1 / 4.82. The spectral peaks are simply related. This 4.84 peak is also related to the Hale cycle since 22.3 / 3 = 7.4

years and 88.8 months but $88.8 / (13 / 8)^3 = 4.83$ or equivalently $7.4 / 18 = 0.41$ years or 4.92 months.

2.81 months is also $14.8 / (3 / 2)^2 = 2.91$ $1 / 4.81 + 1 / 6.73 = 1 / 2.81$

2.26 months is $6.73 / 3$ and $14.8 / (8 / 5)^4 = 2.26$

1.34 months $14.8 / 11 = 1.34$ $6.73 / 5 = 1.34$

1.22 months is $4.83 / 4 = 1.21$ $14.8 / 12 = 1.23$

The last large peak at 1.12 months is $6.73 / 6 = 1.122$ and $2.26 / 2 = 1.13$

The longer periods of the X Ray flare spectrum are clearly related to the harmonics of the Hale solar activity cycle such as 7.4 and 5.57 years along with harmonics of the Wolf sunspot series such as 8.4 years. The natural period of the X Ray generator in the Sun is an order of magnitude shorter than the periods of the sunspot activity cycles yet the response is still clear. The shorter X Ray spectrum periods are simply harmonics of the longer periods. Also Fibonacci ratio harmonics are widely present and all these results again point to the responses of a strongly non-linear system. Now we examine a long term data record before turning to how these phenomena arise. Then we will consider the non-linear behaviour of Earth's weather systems and their response to complex forcing from the Sun and planets. We have non-linearity piled upon non-linearity.

3.6 Long Term Beryllium 10 & Cosmic Ray History

There are a number of measures of solar activity which may be of interest when we come to consider climatic effects on Earth. Variations in surface sunspots and in total or say, UV range, radiation from the Sun are merely three indicators of activity. Some climatologists believe that aspects of the open solar magnetic field and the solar wind which are likely to be more important in climate forcing as we will see later. It is accepted that as the solar magnetic field varies over the solar cycle, galactic cosmic rays reaching the Earth's upper atmosphere will also change. A strong solar field reduces the cosmic ray intensity on Earth and some believe this reduces cloud formation at certain levels in the atmosphere, reducing the albedo of the Earth and promoting warming: the Earth is darker and absorbs more solar energy. Cosmic rays also produce radio-isotopes in the atmosphere through energetic collisions. One such isotope is beryllium 10 for which we now have very long, high resolution records from ice cores collected in Greenland and Antarctica (9).

It follows we can track the variations in the strength of the Sun's magnetic field and possibly climate forcing through the Beryllium 10 record. McCracken has produced a high resolution amplitude spectrum of Beryllium 10 variation based on a 9,400 year ice core record. This provides an opportunity to look for very long, 100+ year, signals in the Sun. Here are the highest intensity, significant peaks from that spectrum where the amplitude of peaks increases with cycle period. Beginning with the largest peaks we have:

2300, 970, 705, 515, 350, 208, 150, 130, 104, 87 years.

First we should ask if and how these peaks are harmonically related, suspecting that such relationships might be informative about non-linearities and so on. There is no obvious pattern of regular harmonics in successive peaks but taking pairs of peaks:

1 / 87 − 1 / 104 = 1 / 532 compared with the spectral peak at 515 years

1 / 104 − 1 / 130 = 1 / 520 compared with 515.

1 / 130 − 1 / 150 = 1 / 975 compared with 970.

1 / 150 + 1 / 208 = 1 / 87.1 versus 87.

1 / 150 - 1 / 208 = 1 / 535 versus 515.

1 / 208 − 1 / 350 = 1 / 513   versus 515.

1 / 208 + 1 / 350 = 1 / 130.4 versus 130.

Several peaks are closely interrelated. We may be seeing here evidence of frequency modulation. We also note cycles related to other familiar periods in the Hale sunspot 'magnetic' series. The primary cycle we found therein is ~22.3 Years. Note that 87 / 4 = 21.8 years. The 87 year peak may be the 4th sub-harmonic of the Hale cycle but also the positive side band for Saturn −Uranus period interaction is 21.8 years. There is more. Note that

6 x 21.8 = 130.8 but the observed peak is at 130 years.

7 x 21.8 = 152.6 versus observed 150.

9 x 21.8 = 196 and 9 x 22.3 = 200.7 versus observed 208.

16 x 21.8 = 348.8 and 16 x 22.3 = 356.7 versus 350.

Even for the longer cycles we may have integer harmonics.

24 x 21.8 = 523.2 and 23 x 22.3 = 513.6 versus 515.

32 x 21.8 = 697.6 and 32 x 22.3 = 713.5 versus 705.

106 x 21.8 = 2311 and 103 x 22.33 = 2300.3 versus 2300.

It is highly likely that we have 22.33 year cycle harmonics in the Beryllium 10 record. The Hale 3rd Harmonic also seems to be very precisely present as higher harmonics even in the very long Beryllium 10 spectral cycles, indicating more non-linear forcing behaviour.

309 x 7.443 = 2300 years ; 130 x 7.443 = 967.3 ; 95 x 7.443 = 707.1 ; 69 x 7.443 = 513.6 ; 47 x 7.443 = 349.8 ; 28 x 7.44 = 208.4 ; 20 x 7.443 = 148.9 ; 14 x 7.443 = 104.2.

Only the 130 year cycle is not accurately related to the Hale 3rd harmonic or the base period but note that 130 / 11 = 11.82 while Pj = 11.86;
the Cjs = 19.85 shows up in the Wolf sunspot series as a strong 9.925 year cycle but here 130 / 13 = 10 years. In fact Cjs harmonics are everywhere.

$$87 / 19.85 = 4.38 = 1.63^3 \sim (13/8)^3 \quad 104 / 19.85 = 5.23 = 1.51^4 = (3/2)^4$$

$$130 / 19.85 = 6.55 = 1.6^4 = (8/5)^4 \quad 350 / 19.85 = 17.63 = 1.613^6$$

$$515 / 19.85 = 25.94 = 1.5^8 = (3/2)^8 \quad 970/19.85 = 48.87 = 1.626^8 = (13/8)^8$$

The Cjs cycle is the strongest in determining the planetary torque effect and here it is as a remarkably consistent and accurate series of Fibonacci harmonics. The Cjn conjunction period is also visible.

$$87 / 12.78 = 6.81 = 1.615^4 \quad 104 / 12.78 = 8.1 = 2^3 \quad 130 / 12.78 = 10.1$$
$$208 / 12.7 = 16.2 = 2^4 \quad 350 / 12.78 = 27.38 = 1.51^8 = (3/2)^8$$
$$515 / 12.78 = 40.3 = 1.59^8 = (8/5)^8$$

Cju may also be present. $87 / 13.8 = 6.4 = 1.593^4 = (8/5)^4$

$$104 / 13.8 = 7.54 = 1.657 \overset{4}{=} (5/3) \quad 350 / 13.8 = 25.34 = 1.498 \overset{8}{=} (3/2)$$
$$515 / 13.8 = 37.29 = 1.495 \overset{9}{=} (3/2) \quad 970 / 13.8 = 70.2 = 1.603 \overset{9}{=} (8/5)$$

The Hale series contains other large spectral peaks near the main 22.33 year cycle such as that at 25-25.5 years. Note that the Saturn-Neptune, Csn conjunction gives us 25 and 35.87 years. Several harmonics of this period are also present. $4 \times 25.53 = 102$ versus observed 104; $5 \times 25.5 = 127.7$ versus 130 ; $6 \times 25.5 = 153$ versus 150 ; $8 \times 25.5 = 204$ versus 208 ; $20 \times 25.5 = 510.7$ versus 515 ; $90 \times 25.5 = 2298$ versus observed 2300 years.

There are no consistent patterns of harmonics related *directly* to planetary periods, just the odd isolated correlation such as $14 \times Pn = 14 \times 165 = 2310$; $3 \times 165 = 495$; $4 \times Pu = 4 \times 84 = 336$; $6 \times 84 = 504$; $7 \times Ps = 7 \times 29.46 = 206.2$; $12 \times 29.46 = 353.2$; $30 \times Pj = 30 \times 11.86 = 355.5$. This could simply be chance. However recall that pair wise and triple conjunctions between the four gas giants also have long 'conjunction' periods between ~171 and ~180 years. For example the Uranus – Neptune conjunction time is $1 / 84 – 1 / 165 = 1 / 171.14$. Perfect conjunctions take on the order of 1100 years and 2200 years but near perfect conjunctions (in terms of the forces generated) have periods clustering around 178 years and the second harmonic. We have:

3 planets	Cjs	19.8585 years	$9 \times 19.8585 = 178.72$
	Cjn	12.782	$14 \times 12.782 = 178.95$
4 planets	Cju	13.81	$13 \times 13.81 = 179.53$
	Csn	35.87	$5 \times 35.87 = 179.35$
	Cjn	12.781	$7 \times 12.781 = 89.47$
	Csu	45.36	$2 \times 45.36 = 90.6$

Is there any sign of gas giant triple or quad conjunctions in the Beryllium 10 record? There is.

$88 \times 2 = 176$ and $350 / 2 = 175$ years compared with the range of $171 – 181$ years in the conjunction multiples.

$515 / 3 = 171.7$; $705 / 4 = 176.3$; $970 / (11 / 2) = 176.3$; $2300 / 13 = 176.9$ years.

This seems clear enough but also, considering Be10 spectral peak interactions :

1 / 178 + 1 / 515 = 1 / 132 compared with 130.

1 / 178 − 1 / 970 = 1 / 216 versus 208 and 1 / 178 + 1 / 970 = 1 / 149.7 versus 150.

The gas giant planetary conjunction periods seem to be present.

Since Beryllium 10 is definitely produced by the, solar wind modulated, galactic cosmic ray flux, the Sun is in the frame. The solar wind and the open solar magnetic field, based on the Beryllium 10 record, must be modulated in turn…somehow… by the orbital dynamics of the gas giant planets.

The author has mainly concentrated on exploring cycles in the 10 to 100 year range with a view to understanding 'medium' term solar and climate interactions. While researching the Beryllium question for this book it was extremely satisfying to discover that, at least in terms of longer cycles in the 100 to ~1000 year range, a direct comparison has recently been made between Beryllium variation as a solar activity proxy and the kind of planetary torque effect on the Sun suggested by Landscheidt and discussed by this author in the 1980s (9,10).

This new work (11) is based on a plausible physical model of the boundary layer, the tachocline, between the inner, rigidly rotating Sun and the upper layers subject to differential rotation and convection. The tachocline is ellipsoidal and the rotation axis tilt of the Sun leads to torques. The authors made detailed comparisons of the 'solar modulation potential' derived from Beryllium 10 and Carbon 14 time series going back 9,400 years. The Be10 based spectrum was very similar to that which we have analysed already. Solar potential and the planetary torque spectra for periods up to ~500 years in duration were then generated and analysed.

The idea of torque effects modulating the solar dynamo sounds a little crazy but in section 4 (1, 2) we will look at new work which supports the same process in relation to the Earth-Moon system. Varying lunar tides appear to generate the core turbulence that drives the Earth's geomagnetic field and perhaps modulates it cyclically.

The solar and torque cycle match is remarkably good for the most prominent spectral peaks except for the ~360 year peak in the solar modulation potential spectrum which is very small in the torque spectrum. However we noted earlier that 360 / 2 = 180 matching some of the pair wise and triplet conjunctions of the gas giants. It is worth looking further at this pair of spectra to see what is hidden in them. There is much more evidence in the spectra, taking into account non-linearities, than the authors appreciated, which we will now examine.

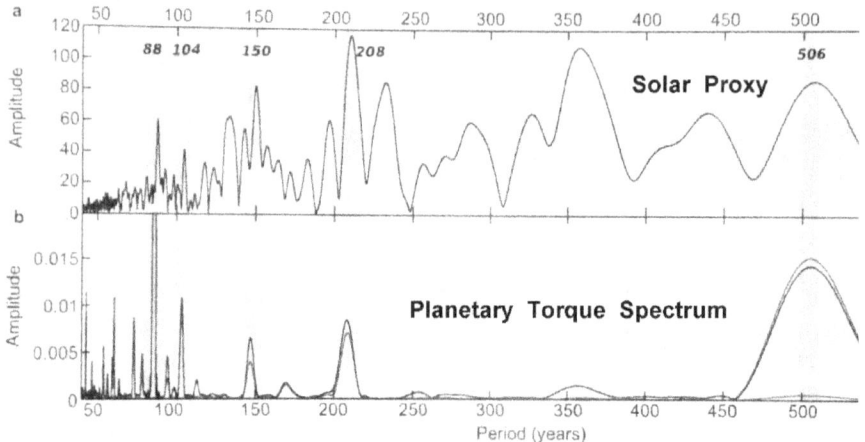

It is interesting that both ~360 and ~170 - 180 are not strong in the torque spectrum. However 176 / 2 = 88 year period is **very** strong in the torque spectrum as is the large, broad peak centred on ~515 years which is 3 x 171.2, the Uranus – Neptune conjunction period. But note also that P_n = 165 and 165 x 3 = 495 years.

Also P_u = 84 and 6 x 84 = 504 years while 84 x 5 / 2 = 210 compared to the main peak at 208 years. Also P_n = 165 and 165 x 5 / 4 = 206.3 years. We see simple harmonics of the actual planetary periods here. It is periods of 1/2, 2 and 3 times the 170 – 180 year planetary conjunctions which show up in the solar proxy spectrum, plus direct orbital period harmonics.

A large peak with the same amplitude as the 515 peak appears at ~235 years in the solar proxy spectrum but not in the torque spectrum. However 235 / 88 = 2.67 and assuming 88 is actually 4 x 22.33, we note that 235 / 89.2

$$= 2.635^2 \text{ or } 1.624 \text{ . This is of course the familiar Fibonacci ratio } 13 / 8.$$

Similarly there is a strong solar proxy peak at ~132 years but 89.2 x 3 / 2 = 133.8, another Fibonacci ratio harmonic.

The moderate peak at ~285 is close to 89.2 x 2 x 8 / 5 or 178.4 x 8 / 5 = 285.4. The moderate, broad peak at ~440 is close to 89.2 x 5 = 446 and 88 x 5 = 440: more evidence of non-linear responses to torque effects in the Sun.

This spectrum is very close to the earlier Be10 spectrum we compared with planetary conjunction periods in detail and those results hold here also. Planetary conjunctions in pairs, triplets and the quad are obvious via Fibonacci harmonics.

The authors used a simulation approach to try to estimate the probability of the spectral matches of their five main peaks occurring by chance. They claim that the chance of a coincidence is about 5×10^{-7} or less. This is impressive but their calculation has been challenged on their 'wrong' statistical assumptions. The new analysis claimed a 1 in 4.5 probability that the results would occur by chance for a white noise model and a 1 in 13.3 probability of coincidence for red noise. Strictly such levels are normally considered to be insufficiently significant to be accepted as indicating a real effect.
Aliasing is also claimed to have generated phantom peaks.
However as we noted these cycles are not randomly distributed across the spectra but are harmonically related. Also there are ~12 cycles in total with large or moderate amplitudes which we also related in our earlier analysis harmonically within the solar potential spectrum **and** back to planetary period harmonics and conjunctions either as simple sub-harmonics or Fibonacci ratio harmonics. Perhaps the authors' original data processing fortuitously acted as a non-linear filter mimicking the actual torque-sun forcing mechanism. Their original estimate of chance coincidence may be in error but the detailed, structured, planetary links cannot be talked away, taking into account the harmonic and physical coherence between the two spectra *and* by anticipating strong non-linear, rather than merely linear, forcing.

The authors (11) also discovered long term phase locking between torque and the solar potential for the periods 208 and 506 years in particular. When forcing torque and solar potential are in anti-phase overall, the authors point out that this condition occurs more often in periods of 'grand minima' in solar activity. They rightly suggest that these phenomena also speak for a real physical coupling between the planets and the Sun.

More recently Scarfetta (15) has compared the MEM spectrum of the HadCRUT3 global temperature record with prominent 'astronomical' periods and harmonics (solar-lunar and planetary orbital periods) and solar activity proxies. Prominent periods include: 5.2, 5.95, 6.54, 7.5, 8.25, 9.1, 10.4, 14.5, 20.7, 30-34, 61...We can immediately read 5.95 as $11.86 / 2 = 5.93$ or $17.9 / 3 = 5.96$; 6.54 as $19.85 / 3 = 6.62$; 7.5 as $22.3 / 3 = 7.43$; 8.25 as 8.1 or 8.4 from the Wolf or Hale sunspot series; 9.1 as $17.9 / 2 = 8.95$ or the 8.85 lunar apsides cycle; 10.4 from the Wolf ss series; 14.5 as $22.3 \times 3 / 5 = 14.8$ or $19.85 \times 3 / 4 = 14.8$ or $29.46 / 2 = 14.73$; $19.5 - 21.9$ as Cjs = 19.85 and Hale ss 19.85 and/or Hale 22.3; 61 as $3 \times 19.85 = 59.6$ or $5 \times 11.86 = 59.4$. Scarfetta does not address some highly significant shorter periods including : 4.8 years or $19.85 \times (5/8)^3 = 4.84$ and $22.33 \times (3/5)^3 = 4.82$; 3.7 years or $22.33 / 6 = 3.72$ or $19.85 / 5 = 3.96$ or $18.6 / 5 = 3.72$; 2.9 years or $22.33 \times (3/5)^4 = 2.89$ or $19.85 / phi^4 = 2.9$ or $8.85 / 3 = 2.94$ or $18.6 \times (5/8)^4 = 2.84$. Note : 18.6 is the lunar nodal cycle and 8.85 is the apsides cycle ; see section 4.

Scarfetta then displays MEM spectra for global northern and southern hemisphere ocean surface, global, northern and southern land surface series, etc, 9 in all, and shows the close periodic similarities. The prominent periods are again easily linked to the Sun and / or lunar-planetary forcing periods. Scarfetta points to significant chi squared tests of spectral coherence for 10 of the prominent temperature periods and the 'astronomical' periods (or typically 10 out of 12 total peaks in each spectrum). This is good evidence for such influences.

Under certain conditions the planets appear to disrupt the production of sunspots and significantly weaken the solar magnetic field. This new work and the author's work on shorter solar and planetary dynamics time scales, cannot be explained away as coincidence, particularly when we explore in more detail the unfamiliar but enlightening theory of non-linear dynamic systems in section 3.7 (and Appendices 1 and 2). The case for planetary modulation of solar activity is strong but not formally proven.

Solar Activity & Planetary Tidal Forcing

Certain groups of planets also have direct tidal effects on the Sun which are 'small' but may be significant. The main group consists of Venus, Earth and Jupiter, the latter although much further from the sun is so massive it still has a major contribution to tides. Mercury being so close to the sun, in a fast orbit has a large, high frequency tidal impact. Dr. Hung of NASA was tasked to investigate possible effects because of the impact of 'space weather', such as solar flares, on communication satellites in the hope of improving storm forecasts (16).

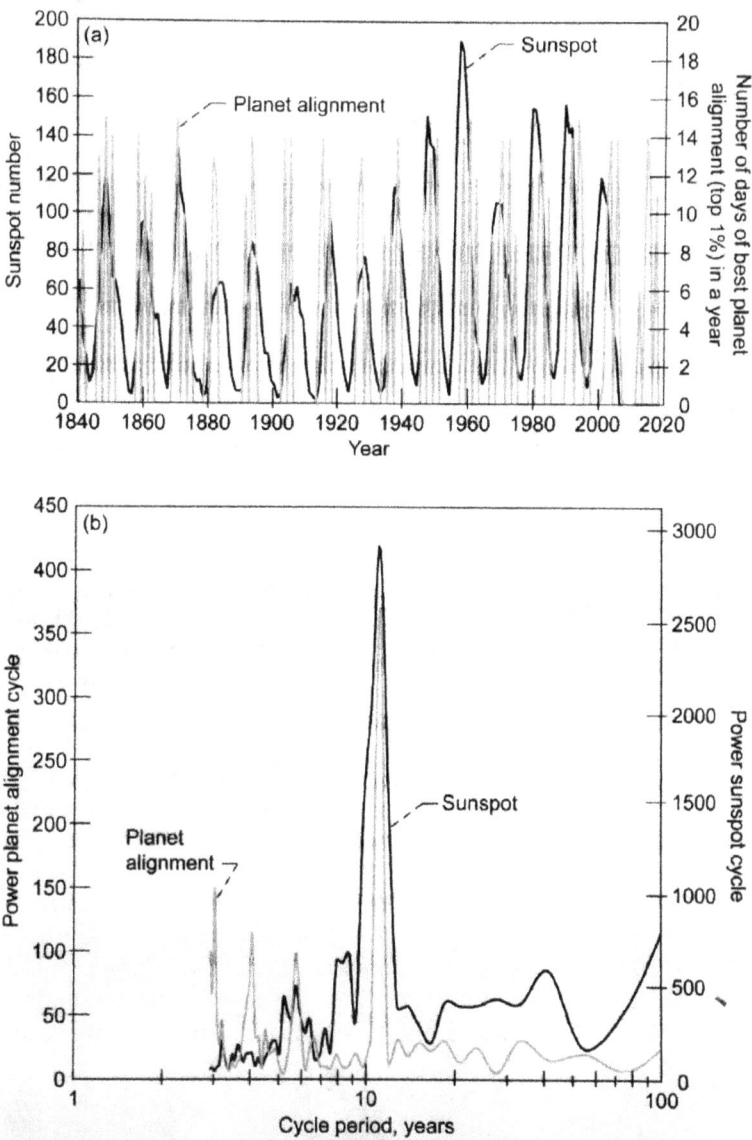

Figure 3.—Average number of sunspots every calendar year from 1840 to 2019 and number of most-aligned days (top 1 percent) each year for three-planet system Venus, Earth, and Jupiter. (a) Number of sunspots and most-aligned days. (b) Fast Fourier transforms of number of sunspots and number of most-aligned days.

Hung looked at a long sunspot series from 1840 to 2019 and at more resent solar flare times series. He compared the Fast Fourier Transforms for sunspots and a tidal measure. This was the number (top 1%) of planetary aligned days representing maxima in tidal forcing. We look first at the V, E, J effect which is easier to interpret.

The three planets periods are in fact harmonically related. For example $Pv / Pe = 224.7 / 365.25 = 0.6152$ or very closely, the Fibonacci convergent to phi of $8 / 13 = 0.161538$. In fact the three planets are related in a way which generates an interesting and familiar solar activity period.

$$1 / ((3 / Pv) - (5 / Pe) + (2 / Pj)) = 22.16 \text{ years}$$

We will see it is the first harmonic which shows up in tidal forcing, $22.16 / 2 = 11.08$. The figure below shows the sunspot time series with the tidal maxima superimposed. It is interesting that the tides and solar activity are in phase when solar activity is high and out of phase in low or falling activity periods. i.e. the peaks between 1880 and 1910 and the post 2000 peaks.

Our most recent cycle 24 peak was very low and cycle 25 is predicted to be even lower by many analysts, some expecting a decay into a grand solar minimum for some decades. The spectrum shows a major peak for sunspots at ~ 11.07 years, closely matched by the tidal measure. We also have several lesser sunspot and tidal peaks worth consideration.

We note the 11.07 peak has a shoulder at ~ 10 years which is the familiar $19.85 / 2 = 9.925$, JS conjunction. There is a smaller double peak at ~8 – 8.9 years.

We recognize $17.9 / 2 = 8.95$, from the Hale cycle sideband but also 8.85 is the lunar apsides cycle. The next SS peak coinciding with a tidal peak is at ~ 5.8 or $11.6 / 2$, a Jupiter orbital period harmonic. The peak at ~5.3 in SS is probably $22.16 / 4 = 5.5$. The last significant tidal peaks are at 3 and 4 years but are not matched in this SS spectrum. These are not random.

2
4.0 $19.86 / 5 = 3.97$; $11.08 \times (3 / 5) = 3.99$; $11.86 / 3 = 3.95$

4 is a harmonic of the main tidal period but also picks up the Jupiter period directly and a Hale cycle spectral sideband we have seen high resolution spectra.

3.0 $11.86 / 4 = 2.96$; $17.9 / 6 = 2.98$; $8.85 / 3 = 2.95$, $18.6 / 6 = 3.1$

We pick up Jupiter again, a Hale cycle sideband and interestingly, harmonics of the lunar nodal and apsides cycle.

The latter tells us that the lunar orbital cycles are perhaps generated by the V, E, J resonances in addition to the links to the gas giant interactions. We can conclude that the main V,E,J tide is reflected in SS activity along with some of its harmonics.

Now we consider the messier spectrum of the M, V, E, J 'tidal' series where the high speed Mercury orbit dramatically shifts the spectrum. The largest 'tidal' peak is now at ~ 7.3 years with other peaks at 4.9, 4, 3 with low broader peaks at ~18 and ~33 years. These new tidal peaks seem unconnected with the 11 year SS cycle at first sight although 3 and 4 are simple harmonics as above. Let's look more closely considering also the Hale magnetic SS spectrum, already familiar.

7.3   22.16 / 3 = 7.38, i.e. the 2nd harmonic of the main Hale SS cycle; 11.08 x (2 / 3) = 7.38; 11.86 / phi = 7.33; 29.46 / 4 = 7.36

4.9   19.85 / 4 = 4.96, Hale sideband at 19.85; 29.46 / 6 = 4.91, Saturn orbital period

4   19.85 / 5 = 3.97; 11.86 / 3 = 3.95, see VEJ analysis above.

3.1   11.86 / 4 = 2.97; 22.16 / 7 = 3.16, see VEJ analysis above.

Mercury has shifted the 'tidal' power to a large peak at 22.16 / 3 = ~ 7.38 years or 1 / 3 of the Hale magnetic solar cycle we will see many times in weather phenomena. It is interesting that the MVEJ tidal quartet can generate periods also generated by interactions of the four gas giant planets. This is because of the various harmonic relationships between the orbital periods of sub-sets of the planets creating either tidal or torque effects in the Sun or both.

The boarder, longer peaks in the MVEJ 'tidal' series also paint a similar picture. The peak at ~18 years is 11.08 x phi = 17.93, but this is also a Hale solar sideband at 17.9 years. There is a wide plateau in the SS spectrum from ~18 years rising to a larger peak at ~44 years. This plateau contains unresolved peaks at 19.85, 22.3, 25, 29.46, 44.5 years which we see in higher resolution MESA solar spectra.

The last tidal peak is at ~ 33 years which is just 3 x 11.08 = 33.24. Tidal energy falls away strongly beyond ~35years. The solar signal generating properties of the MVEJ (tidal) and the JSUN (torque) planetary quartets are mathematically quite similar and probably complimentary.

On the same basis of comparing the spectral 'fingerprints' of planetary dynamics, solar activity measures and several terrestrial weather phenomena, acknowledged to be of great importance by the climatologists, we will show that the evidence supports the extreme heresy that **the planets modulate our weather and climate, probably via the Sun.**

3.7 Review : Non-linear Oscillators and Phi

The general properties of dynamic systems driven by quasi-periodic signals are explored in Appendix 1 and readers unfamiliar with such issues may wish to start there. The generation of so many spectral peaks from planetary periods and conjunctions and from interactions between peaks, with frequent Fibonacci ratios is striking. Is this some weird fluke? Buck and Macaulay (5) point out that this kind of spectrum identifies a non-linear system capable of complex behaviour. In fact one of the simplest non-linear models, the iterated circle map, can produce a family of frequencies related by Fibonacci ratios and, in the limit, phi.

A single oscillator of this type is

Yn+1 = (Yn + O − (K / 2 *pi) x sine (2*pi*Yn))   mod 1

K is the coupling constant linking the oscillator and forcing signal. O controls the frequency of the basic oscillator. The winding number is W = lim ((Yn − Y0)/n), the limiting angle of rotation per step as n increases. We can say generally that Wn = p / q. If p and q are integer the oscillation is periodic. Rational p / q ratios could be, for example, the ratios of successive Fibonacci numbers. (See 12 for an introduction to non-linear oscillators).

If multiple oscillators of this form are coupled rich dynamic behaviour is possible including periodic, quasi-periodic and chaotic behaviour. Why in the case of the Sun and the planets do we find quasi-periodic behaviour along with both integer and Fibonacci ratio harmonics? Well in the circle map at K =1 and with O = 0.60666, experiments show a critical transition to a regime that exhibits phi related harmonics. Note that 1 / phi = 0.618 and 3 / 5 = 0.6. The question of the stability of oscillatory dynamic systems is the subject of Kalmagorov-Arnold-Moser or KAM theory. The trajectories of dynamic systems in a phase space describing all possible states of that system are confined to the surface of a doughnut shaped torus. Various sets of initial conditions can define a nested, concentric family of such tori (see the next figure below). KAM theory is concerned with what happens if the system is perturbed by a small external forcing signal.

In a non-linear system resonance can occur for a wide set of rational harmonics of the frequencies in the signal. With sufficient perturbation some of the phase space tori may be destroyed and some merely distorted. The most vulnerable tori are those where the system can best respond to rational harmonics of the forcing signal frequencies. Tori break up into rings of small islands of stability surrounded by areas of unstable fixed points. These stable islands in turn have rings of fine structure within them including locally unstable fixed points. The breakdown is self similar on all scales. The surviving tori have the most 'irrational' properties. Since Phi is the most irrational number, it and its rational approximations are often seen as survivors in forced non-linear systems. Quasi-periodic motion can survive where strict periodic motion cannot. As the complexity of such systems, in terms of the number of dimensions, increases the volume of phase space occupied by stable tori decreases. Surviving tori become disconnected, Cantor sets or Cantori in phase space. Nevertheless quasi-periodic motion can persist and remain stable. We can say that Phi enables multiple oscillators to coexist without mutual destruction. Interestingly the surviving tori can continue to interact at 'the edge of chaos'.

The above behaviours have close relevance to our analysis of planetary orbits and solar activity cycles. We noted phi related near resonances in the solar system. The orbital periods of Jupiter and Saturn are close to 2 : 5 so $2 \times f_j - 5 \times f_s \sim 0$ leading to the so-called 'small divisor' condition in the mathematical description of the joint motion. Uranus and Neptune periods are close to 1 : 2 giving $f_u - 2 \times f_n \sim 0$ again. We are not exempt since the orbital period of the Earth to that of Venus is as 13 : 8, very closely a Fibonacci ratio harmonic. It is also interesting that Jupiter, Mars and Earth share a near resonance since $3 \times f_j - 8 \times f_m + 4 \times f_e \sim 0$. Mars and Earth are Jupiter's closest neighbours. We will see later that Venus, Earth and Jupiter are the main long period creators of solar tides.

The solar system dynamics appear to lie on the 'edge of chaos' and planetary orbits are not predictable on times scales of hundreds of millions of years. The shapes of some orbits are also unstable with large excursions in eccentricity which in the limit could lead to the crossing of orbits and mayhem. In the early lives of solar systems planetary migrations and even ejections may be commonplace until the unstable orbital period ratios are destroyed, leaving the quasi-stable harmonic patterns we saw earlier.

Solar activity is also clearly quasi-periodic and the solar dynamo has the aspect of a complex, non-linear system. The author developed a non-linear model for a solar oscillator and a linear model for the modulation envelope of sunspot cycle peaks which was published in Solar Physics (13).

A KAM TORUS

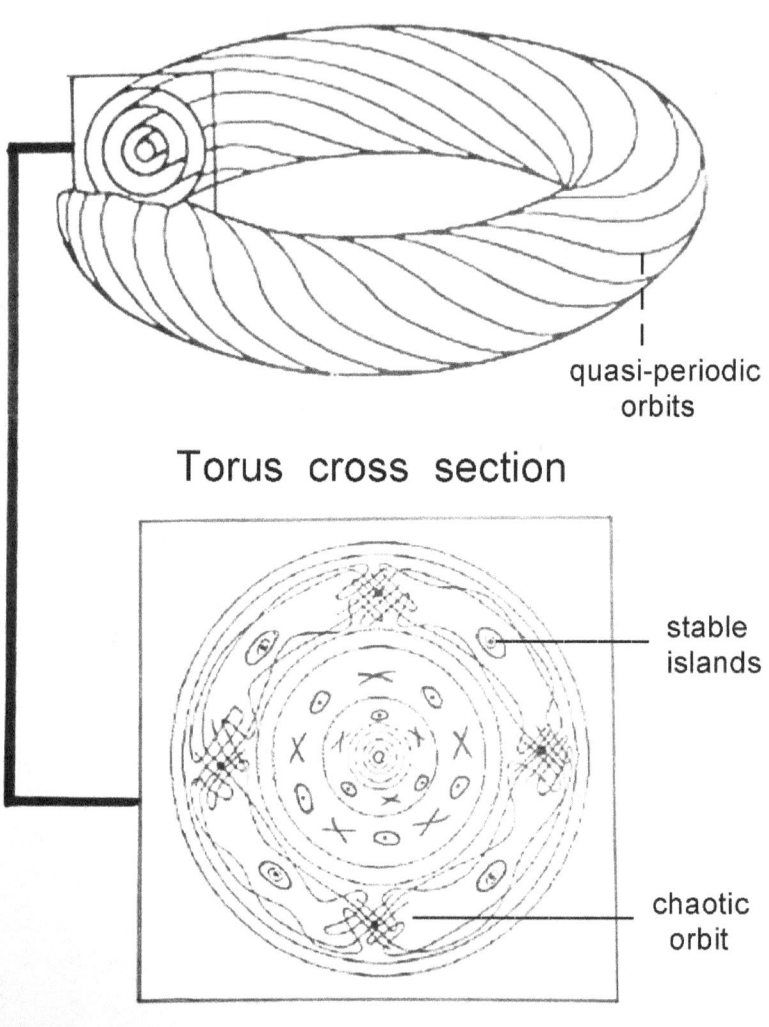

quasi-periodic orbits

Torus cross section

stable islands

chaotic orbit

The models described the long term history of the sunspot cycle well, back to 1700. The ten years forecasts from the models were less successful although they predicted the general decay in peak solar activity which is now apparent. Within a year the author had concluded that the Sun was more non-linear than he had thought and began to suspect the presence of chaotic behaviour.

A second published paper described the argument including the presence of odd and even harmonics in the solar activity spectra of very long solar time series and proxi-time series (carbon 14) and the clear relationship to planetary conjunction periods (10 and Appendix 2). In addition to the major spectral peaks we explored earlier even the fine detail could be explained by non-linear planetary 'forcing'. For example taking $1 / Cjs = f1$ and $1 / Cjsn = f2$ we get

Solar activity period	Combination frequencies	F(f1, f2) Period
22.33	f1 – f2	22.35
19.85	f1	19.85
17.86	f1 + f2	17.74
25.5	f1 – 2 f2	26
16.4	f1 + 2 f2	16.3
10.5	2f1 – f2	10.4
9.92	f1 / 2	9.92
9.4	2f1+ f2	9.4
7.43	3 (f1 – f2)	7.44

The patterns are very clear. The question has often been asked as to whether the solar dynamo is simply stochastically disturbed or truly chaotic. Recent work has confirmed chaos (4). Shuang Zhou et al show that sunspot numbers and sunspot area monthly time series for the period 1874 to 2012 are best described by strange attractors set in a 3D phase space. The attractor was reconstructed by using a smoothed monthly number or area and two time delayed values : $X(t)$, $X(t + d)$, $X(t +2 d)$. The attractors have a dimension of ~1.2 and the characteristic predictability times for sunspot numbers and areas are 3.5 years and 2.5 years (see Appendix 1).

This result explains the failure of the many attempts to forecast one cycle (11 years) or more ahead, including the author's attempts. As suspected the solar dynamo is indeed non-linear and chaotic. However the good news is that these results explain in general terms how the orbital periods and harmonics of the gas giant planets and their conjunctions, turn up consistently in various solar activity spectra. Small perturbing signals somehow generated by planetary movements are being captured and magnified by the 'chaotic' solar dynamo in the rotating Sun.

If some joint magnetic field effect is discounted only two candidates remain: direct planetary tides and the shifting centre of mass of the solar system, thanks to the gas giants, relative to the centre of the Sun. Both tides and torque effects are said to be 'too small' in scale to affect a 880,000 mile diameter ball of gas and plasma.

But could small forces affect the deep boundary zone where the dynamo supposedly generates the Sun's magnetic field? Could small forces modulate the localised zones of differential rotation in the Sun where the fields of sunspots are supposedly generated? We saw in the last section impressive evidence for this. Whatever the mechanism the 'fingerprints' of the gas giant planets are in the solar activity spectra. In the final sub-section we will examine a mechanism which can magnify small driving forces in some classes of highly non-linear system. However we would expect a real physical process to scale in some way to the properties of the planets involved. The obvious parameters are mass and solar distance. For the gas giants we have

	J	S	U	N
Mass ratios	1.0	0.32	0.04	0.05
Radius AU	5.2	9.54	19.2	30
Mass x radius	1.0	0.585	0.16	0.29

Based on mass and distance (tides) and mass x distance (torque), Jupiter and Saturn would dominate the hypothetical forces. In the solar activity spectra we found mainly Jupiter and Saturn harmonics but also conjunction periods involving Uranus and Neptune, Cjs, Cjn, Cjsn, Cjsun played a detectable role. This Jupiter – Saturn dominance is suggestive of a real physical mechanism.

Next we will look at possible links between solar activity and Earth's weather and climate. We face the further complication that the weather system itself is non-linear and in some aspects, quasi-periodic and perhaps chaotic. We possibly have here a complex set of planetary orbital cycles modulating a non-linear, chaotic (or SOC) dynamo in the Sun which in turn somehow modulates the non-linear weather system.

We should not expect simple, clear or consistent relationships describable by simple models. However, despite all this complexity the 'fingerprints' of the planets do show up in solar spectra and in several weather and climate time series. The planetary periods show up as simple harmonics and phi related harmonics in the solar activity spectrum. Curiously we can also see the direct gravitational effect of the planets on some orbital properties of the Earth - Moon system. All these systems show significant cyclical commonalities to the extent that taken in isolation we could claim that the Moon's orbit generated sunspots!

Equally, in isolation we could claim the planets affect the weather! In fact the planets modulate the Sun's and the Moon's orbits and solar activity which in turn modulate Earth's weather systems. Again we will see evidence of non-linear behaviour with climate spectra containing simple harmonics and phi related harmonics of solar activity periods and planetary orbital periods. In some non-linear systems involving heterodyne phase conjugation, such as in soliton (standing wave) formation, the sequence of frequency spacings in a spectrum may follow a 1, phi,1, phi, 1 pattern (14). However other favoured irrational ratios may also occur such as 1.466, 1.38, 1.325 (the 'plastic ratio') and 1.839 (the Tribonacci constant) along with $\sqrt{2} = 1.414$, $2^{1/3} = 1.26$, $\sqrt{3} = 1.732$, $3^{1/3} = 1.442$. Although phi is dominant we will also see these period - frequency ratios in our analyses of spectral fingerprints.

3.8 Self Organized Criticality in Solar Activity Processes ?

We have seen that planetary orbital dynamics and solar activity measures share similar non-linear features and common frequencies in their spectra. We can see clear signs of the Jupiter-Saturn conjunction period of 19.85 years in solar activity along with triple conjunction periods for combinations of the four gas giant planets and their simple and Fibonacci related harmonics and interaction frequencies. However we have not yet answered the question of how it is that the planetary movements with orbital periods from ~0.24 to 165 years can generate tidal or torque forces large enough to significantly modulate solar activity. The computed forces are said to be too small to significantly affect solar variation. The mean rotation period of the Sun is ~27 days and turbulent convection zone has high frequency properties. As we saw flare activity is also a high frequency phenomenon as is individual sun spot generation. The answer may lie in the occurrence of self organized criticality in some of the physical processes of the Sun.

Self organization is a process in which patterned behavior at the global level of a system emerges solely from numerous interactions among the spatially distributed lower level components of the system. The interactions are executed locally using only local information without reference to the global pattern. The key property is that there is a slow and *continuous* energy input rate across the system. The second property is a local instability threshold in some phenomenon which if exceeded results in a release of energy, a so-called avalanche. This process leads to a critical state where small increases in input energy cause avalanches. Avalanches come in all sizes so with a similar driving force it follows that the number of large avalanches is lower than the number of small avalanches. The result is a power law distribution where the log of occurrence frequency is linearly (and negatively) related to the log of avalanche size (e.g energy release). Such power laws may hold over scales of several orders of magnitude. It follows that there will always be large avalanches available to release if small external forces (energy increments) are applied. Such systems act like non-linear amplifiers where small inputs can result in highly magnified outputs at low frequencies. This is how small external forces can create large effects.

How can this concept be applied to the Sun? The sun is a spinning ball of gas and plasma with a core fusion reactor producing energy, a layer dominated by radiation based energy transfer and a convection dominated layer. The convection layer is turbulent. The spinning sun also generates a powerful magnetic field and this is distorted by the fact that different layers and latitude zones) in the sun rotate at different velocities.

It is traditionally said that there is a single dynamo in operation but from time to time evidence suggests there may be two stacked dynamos in effect (see section 6). Convection creates areas of turbulence on various scales which interact with the variable magnetic field. The result is the formation of magnetic 'knots' which reach a point of critical energy which is then released. These are the 'solar flares' which in their larger sizes can play havoc on Earth. The continuous generation of sun spots may be something similar and we noted that they are harmonically related to flares although the times scales are different. So is there direct observational evidence for the origin of sunspots and flares in SOC processes? There is.

Figure 25 shows the spectrum of flare frequency against flare intensity in the X-ray zone from recent solar satellite records. It is remarkable that a power law holds for three orders of magnitude in flare intensity. This is a strong indicator of SOC behavior (17). Other recent work looks for the role of SOC and chaos in the shaping of sunspot activity and finds both (18).

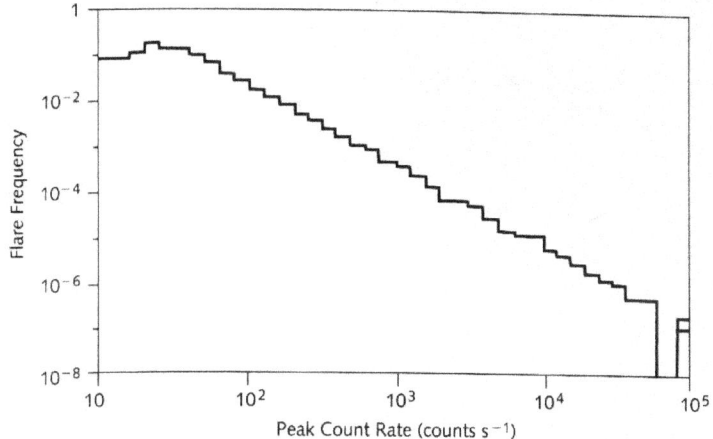

Figure 25. Histogram of x-ray intensity from solar flares, as measured by the NASA satellite ISEE 3/ICE (Dennis, 1985). The diagram shows the relative amount of flares with a given energy, as represented by the "counting rate." The data fit a straight line over four orders of magnitude. The statistics is poor for the few large events.

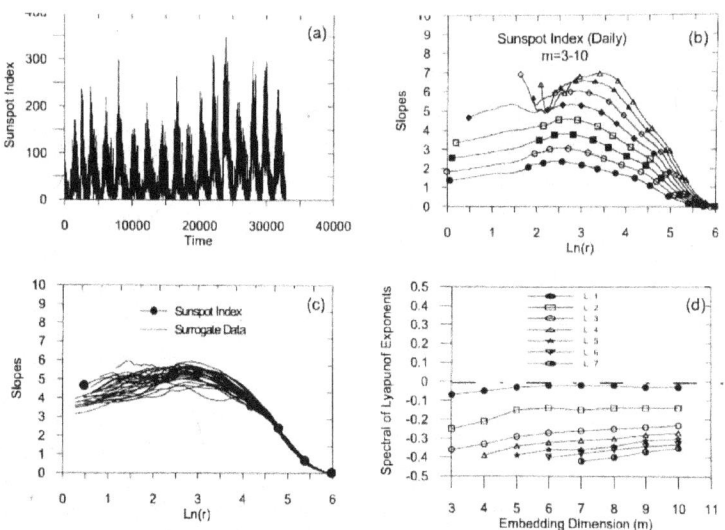

FIG. 1: (a) Time series of Sunspot Index concerning the period of 184 years. (b) Slopes of the correlation integrals of the Sunspot Index time series estimated for delay time $\tau=200$ and for embedding dimensions $m=3$-10, as a function of $Ln(r)$. (c) Slopes of the correlation integrals estimated for the Sunspot Index time series and its thirty (30) surrogates series estimated for embedding dimension $m=7$ and delay time $\tau=200$, as a function of $Ln(r)$. (d) The spectrum of the first seven (7) Lyapunov exponents estimated for the Sunspot Index as a function of embedding dimension.

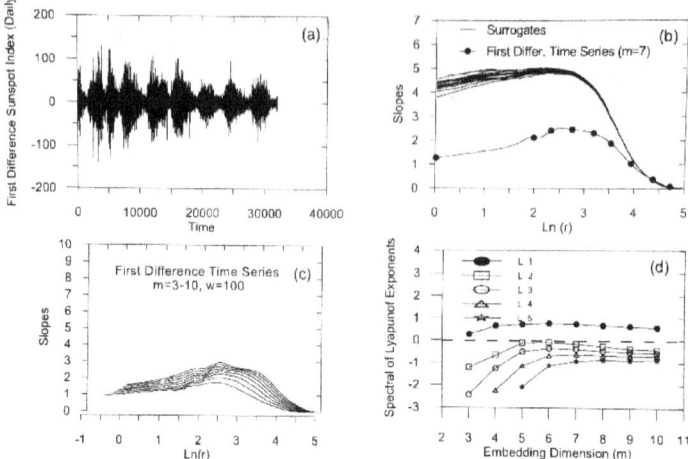

FIG. 2: (a)Time series corresponding to the First Differences of the Sunspot Index time series. (b) Slopes of the correlation integrals of the First Difference time series and its thirty (30) surrogate estimated for delay time $\tau=15$ and for embedding dimension $m=7$, as a function of $Ln(r)$.(c) Slopes of the correlation integrals for the First Difference time series estimated for delay time $\tau=15$, Theiler parameter $w=200$ and for embedding dimensions $m=3\text{-}10$, as a function of $Ln(r)$. (d)The spectrum of the first seven (7) Lyapunov exponents estimated for the Sunspot Index as a function of embedding dimension.

Karakatsanis & Pavlos analysed the underlying dynamic properties of 184 years of daily sunspot data in various ways looking for indicators of a low dimensional attractor (Figure 1). They looked at a range of possible attractor embedding dimensions and calculated the Lyapunov exponents as a function of dimension. All the LEs were less than unity. For a strange attractor at least one LE should be positive. They concluded there was no evidence for a chaotic attractor using daily data. They note something else of interest. For the daily data they estimated the main periodic component as 4300 days and identify this as the primary solar cycle. However 4300 days is not 11 years but 4300 / 365 = 11.78 years. The reader may compare this with the Jupiter orbital period of 11.86 years.

The authors speculate that the strong ~11 year periodicity may be distorting underlying nonlinearities in the daily time series. They therefore also study the first difference of the data which acts as a high pass filter. The differenced series is shown in Figure 2. They describe this as non-periodic which is a simplification. In fact we can see a moderate amplitude modulation where the envelope has a short period for the first two cycles but a longer constant period for the remaining cycles. The first two cycles are ~4.8 and 7.4 years long and the rest, on average ~11.9 years long. So even looking at the rates of change of daily sunspot numbers we still see the presence of the long solar cycle. Again we appear to see the Jupiter orbital period in the differenced daily data. Repeating the same analyses as for the original series they find the first LE is positive and approaching unity, the second is near zero and the rest are negative.

They conclude the overall analysis indicates a low dimensional strange attractor with an embedding dimension of 5-6 and a dynamical fractal dimension of 2-3. The observed LEmax was found to be significantly higher than the calculated LEmax for a stochastically driven system at 99% probability. The authors conclude that the two dynamical models describe firstly a self organized critical process in the turbulent convection zone of the sun driving a process in the photosphere which is well described by a low dimensional chaotic process. Our proposed planetary forcing of solar activity (itself possibly the product of self organized solar system structure) may act through the deep SOC process in the convection layer which then drives the chaotic upper process. That non-linear system would be capable of generating the rich harmonic content of the spectra of the various solar activity measures we have examined. It is remarkable that the complex causal pathway preserves the signals of the planets so clearly.

We began with flares and continued with sunspots. Just to reassure the reader on our speculations there is now direct evidence for interaction between flare modulation and the long sunspot cycle. Satellite records of soft X ray peak flux and flare rise time are now available from the mid 1970s (19).

Fig. 10

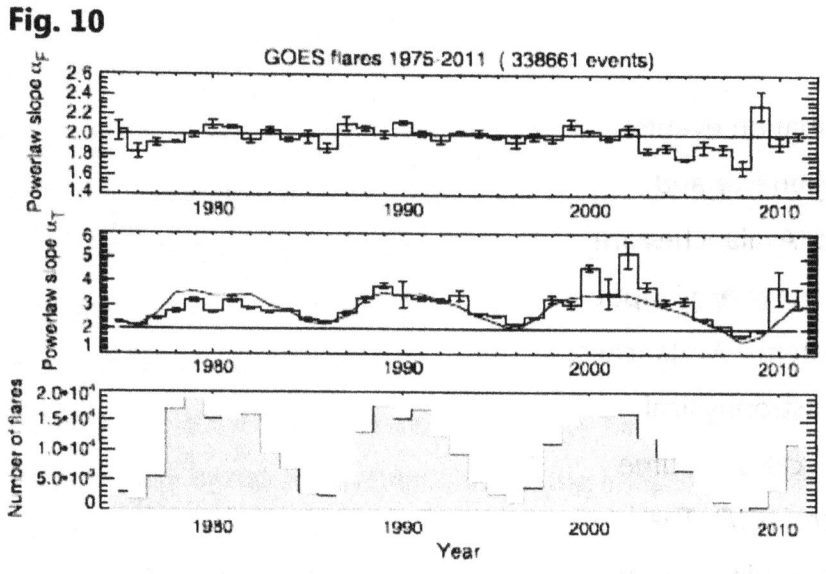

If we look at the power law slopes for the (self organized) X ray frequency versus flare rise time relationship of Figure 10 we see the slope variation through time closely correlates with the long ~11 year sunspot cycle.

The flare generation process and the longer term measures of solar activity like sunspots are physically linked as suggested by the earlier results.

Finally in the last few decades satellite studies of the solar wind have discovered more power law size distributions for energy fluctuations, disturbance durations and waiting times between bursts of a given energy (19). This is said to indicate more self organized behavior. It is unclear where this structuring process takes place.

Readers will recall this result is of some interest in climate terms since the solar wind influences the levels of cosmic rays (over a particular energy range) reaching the Earth's upper atmosphere which according to Svensmark et al, modulates cloud formation over the oceans, the Earth's albedo, energy balance and ultimately temperature.

Section 3 References

1. Dicke R H; 'Is there a chronometer hidden deep in the Sun?', Nature, 276, (1978), pp676-680.

2. Dicke R H; 'The phase variation of the solar cycle', Solar Physics, Vol. 115, pp171-181.

3. Berger et al; 'Predictability of the Wolf sunspot numbers', Solar- Terrestrial Predictions: proceedings of a workshop at Meudon (Paris Observatory), France; June 18-22, 1984. Publisher : US Air Force Geophysics Laboratory, Bedford, Massachusetts.

8. Shuang Zhou; 'Low-dimensional chaos and fractal properties of long term sunspot activity', Research in Astronomy & Astrophysics, Vol. 14, No. 1, pp104-112; 2014.

9. Buck B & Macauly V; 'Fine structure in the sunspot record', Theoretical Physics Dept., University of Oxford.

10. Johnson R W; 'Enhanced wavelet analysis of solar magnetic activity with comparison to global temperature', J Geophys. Res. , Vol 114, Issue A5, May 2009.

11. Landscheidt T; 'Long-range forecast of energetic X-Ray bursts based on cycles of flares', Meudon S-T workshop ref. 3.

12. McCracken et al; 'Long term changes in cosmic ray intensity at Earth', Space Science Review, No 176, p59, 2013.

13. Landscheidt T; 'Long range forecast of sunspot cycles', Meudon Solar-Terrestrial workshop ref. 3.

14. Gregg D P; 'Long term solar cycle modulation', Meudon S-T workshop, see ref. 3 and Appendix 2.

15. Abreu J A; 'Is there a planetary influence on solar activity?' Astronomy & Astrophysics, 548, A88, (2012).

16. Essl G; 'Circle maps as simple oscillators for complex behaviour II; experiments', Proc. of the 9th conference on digital audio effects, Montreal Canada, September 18-20, 2006.

17. Gregg D P; 'A non-linear solar cycle model with potential for forecasting on a decadal time scale', Solar Physics, 90, pp185-194, 1984.

18. Bovenkamp F; 'Why Phi : a derivation of the Golden mean ratio based on heterodyne phase conjugation', www. science.trigunomedia.com /whyphi2007'.

19. Scarfetta N; 'Discussion on the spectral coherence between, planetary, solar and climate oscillations: a reply to some critiques', Astrophysics & Space Science 354, pp275-299, 2012; arxiv.org/pdf/1412.0250.pdf

20. Hung C C; 'Apparent Relations between Solar Activity and Solar Tides caused by Planets'; NASA Glenn research Centre, Cleveland Ohio, NASA/TH-2007-214817.

21. Dennis B. R., Solar Physics, 100 (1985), 65.

22. Karakatsanis L P & Pavlos G P; 'SOC & Chaos in Solar Activity'; Democritus University of Thrace, Dept. of Electrical & Computer Engineering, 67100 Xanthi, Greece' 31.12.2007.

23. '25 Years of Self Organised Criticality: Solar and Astrophysics'; Aschwanden et al. A major review article of ~100 pages, arXiv : 1403.6528v1 [astro-ph.IM] 25 March 2014.

24. Kauffman S; 'The Origin of Order';Oxford University Press, 1993

S4. 'Cosmic' Forcing of Earth – Moon Orbital Dynamics and Geophysical Cycles

> You sulphurous, and thought-executing fires
> Vaunt-couriers to oak cleaving thunderbolts
> Singe my white head.
> And thou, all shaking thunder
> Smite flat the thick rotundity o' the world.
>
> King Lear

4.1 Introduction

Can we relate Terrestrial weather to the solar activity cycle as many have claimed? We noted earlier the classic case of ocean sediment D / H ratios exactly recording the 22.33 year Hale cycle (section 3, 1). Can we further relate the weather to planetary dynamics? To explore this we must also ask if and how the planets physically affect Earth-Moon system dynamics and the motion of the Earth itself. If they do not then we might reasonably doubt the power of planetary dynamics to affect other phenomena such as solar activity and our weather. But if they do then…

4.2 The Earth - Moon System

The motion of the Moon is notoriously complex but two simple phenomena are of possible relevance. Firstly the plane of the Moon's orbit is inclined at ~5.18 degrees to the ecliptic. Where the Moon's orbit crosses the ecliptic we have two nodal points. The period of return of the Moon to the same node is 27.2 days compared with the orbital period of 27.32 days. The difference occurs because of the apparent motion of the Sun in the ecliptic and the regression of the nodes. If the Sun is at a node it will return to the same node in 346.62 days, the so called eclipse year. But the line of the nodes also completes a circle in 18.6 years. The long axis of the lunar ellipse also precesses giving an apsides (apogee, perigee) cycle of 8.8504 years. In fact for the nodal and apsides cycles

Number of sidereal months = number of draconic months - 1

Number of sidereal months = number of anomalous months + 1.

Can these primary orbital cycles be simply related to our planetary periods? Before we address this note that Hale primary cycle = 22.33 years and 22.3 x 5 / 6 = 18.58 while 22.3 x 2 / 5 = 8.9.

Note that we have factors $(5/3)/2$ and $2/5 = (3/5) \times (2/3)$ and Fibonacci ratios again.

Since it seems unlikely that the Moon's dynamics cause sunspots or vice versa both must be related to another cause. We know that solar activity is related to planetary forcing. Could the lunar orbit also be modulated by the planets? We have, using simple rational fractions

Jupiter $11.86 \times (3/2)/2 = 8.89$ Saturn $29.46 \times (3/5)/2 = 8.84$

Jupiter $11.86 \times (5/4) \times (5/4) = 18.53$ $29.46 \times (4/5) \times (4/5) = 18.8$

Or $(11.86 \times 29.46)^{1/2} = 18.68$ compared with the nodal cycle of 18.6 years.

Pu = 84 and $84 / (3 \times (3/2)) = 18.66$.

Cjs conjunction period $19.86 \times (2/3) \times (2/3) = 8.83$ and $19.86 \times 15/16 = (5/8) \times (3/2)$ is exactly 18.61.

Cun is 55.7 years and $55.7/3 = 18.57$. $55.7 / (2/5)^2 = 8.9$

Csn or $30.23 / (13/8) = 18.603$. Cju or $13.8 \times (3/2)/2 = 18.4$. The sum of fj + fn gives us 11.07 years but $11.07 \times 5/3 = 18.5$ and $11.07 \times 4/5 = 8.856$.

For Neptune's orbital period of 164.79 years, note that, weirdly, nodal x apsides cycle = $18.61 \times 8.85 = 164.7$.

One last observation is intriguing. We saw in section 1 that planetary orbital period ratios in the solar system and elsewhere are not random but conform to a harmonic series F^n where $F = 1.072 - 1.078$. Could something similar link planetary periods and the lunar orbital properties we have explored? Putting $F = 1.077$ we obtain a good match.

	Pj = 11.86	Ps = 29.46	Cjs = 19.85
8.85	x 1.34	x 3.327	x 2.244
	$1.077^4 = 1.345$	$1.077^{16} = 3.28$	$1.077^{11} = 2.26$
18.6	x 1.568	x 1.58	x 1.068

135

$$1.077^6 = 1.561 \qquad 1.077^6 = 1.561 \qquad 1.077^1$$

Also note the model gives Nodal / Apsides = 1.077^{10} = 2.0997 versus actual 2.1015. The conjunction period of Earth and Jupiter is 398.88 days but as noted earlier the 'eclipse' year is 346.62 days. We have 398.88 / 346.62 = 1.151 or 1.073^2. Curiously the Earth – Saturn conjunction period is 378.09 days giving us 378.09 / 365.255 = 1.0352 and √ 1.072.

The nodal and apsides cycles of the Moon do seem to be closely related to the gas giant orbital periods and conjunctions. We seem to have harmonic links which consistently involve Fibonacci ratios again. We can expect that the interpretation of any terrestrial weather spectrum may be difficult given all these harmonic links and this numerical confounding.

4.3 Global Earthquake Energy and Polar Nutation.

In areas of the planet subject to relatively frequent earthquake activity we sometimes find the old folk concept of 'earthquake weather'. More modern myths of course have linked earthquakes, even the end of the world, to planetary alignments: most recently in 2012. Such ideas are easily dismissed as nonsense but can we demonstrate more modest links between the shaking Earth and 'cosmic' forces?
Thanks to high resolution spectral analysis it turns out that we can. Landscheidt (section 3, ref. 9) provides a MESA spectrum for annual global earthquake energy release for the period 1904 – 1965.

The spectrum shows several well defined peaks of interest.

Period (yrs)	Description	Ratio to 11.05 yr period
~11.05	small amplitude, broad peak	1.0
6.7	largest, sharp peak	$1 / (5/3) = 6.64$
3.9	sharp, medium amplitude peak	$1 / (5/3)^2 = 3.96$
2.78	" " " "	$1 / 4 = 2.76$
2.37	" " " "	$1 / (5/3)^3 = 2.39$

136

We already see clear hints of a solar activity signal in the 11.05 year peak versus the Wolf peak at 11.1 and the Hale second harmonic at 22.33 / 2 = 11.16 years. Note also that the strong peak at ~ 2.78 is 22.3 / 8 = 2.787. We also see the signature of phi and Fibonacci ratio harmonics of 11.05 in the biggest peaks at 6.7, 3.9 and 2.37 years suggesting we have yet another near chaotic, non-linear system. We can say that **all** of the earthquake energy spectral power is transparently related to the main solar activity cycle and the Hale magnetic cycle. This clear picture is quite surprising.

The broad peak at 11.05 may hide other solar activity frequencies. Let us see.

6.7 year peak :   11.16 / (5 / 3) = 6.69   8.43 / (5 / 4) = 6.72   19.85 / 3 = 6.63

$$2$$
17.9 / (13 / 8) = 6.76   7.4 (11 / 10) = 6.73   **mean : 6.694 years**

$$23$$
3.9 years :   19.85 / 5 = 3.96   11.15 / (5 / 3) = 3.98   17.9 / (5 / 3) = 3.87

$$42$$
25.5 / (8 / 5) = 3.89   8.43 / (3 / 2) = 3.75   **mean : 3.89 years**

2.78 years :   22.3 / 8 = 2.79   19.85 / 7 = 2.83   25.5 / 9 = 2.83

$$42$$
8.43 / 3 = 2.81   17.9 / (8 / 5) = 2.74   7.4 / (13 / 8) = 2.8

mean : 2.8 years

$$34$$
2.37 years :   11.16 / (5 / 3) = 2.42   17.9 / (5 / 3) = 2.33

$$3$$
(19.85 / 2) / (8 / 5) = 2.42   **mean 2.39 years**

There is a surprisingly strong correlation between earthquake energy release and solar activity far beyond that expected by chance. On this basis, since the solid crust of the planet itself shows a solar link, perhaps we need to be ready to entertain a significant solar link to the weather itself which manifests merely though the movements of low density fluids. As usual the path of possible causation is confounded. Is earthquake energy released indirectly via solar effects on terrestrial weather or directly by physical planetary effects on the spinning Earth? We may have other clues in other Earth dynamic phenomena. The Earth is subject to polar wandering on many time scales. The shortest recognised cycle is the Chandler Wobble with a period of 433 days, 14.22 months and 1.1855 years.

The wobble represents a divergence between the crust and the axis of rotation. This wobble interacts with the orbital year, presumably though seasonal weather changes, to produce a longer cycle of 6.39 years. These polar nutation cycles have been linked to changes in ocean and atmospheric circulation with the oceans dominant. If so we have a direct link between spin and seasonal weather.

Looking at the earthquake energy spectrum and the medium peak at 2.37 years we note that $1.1855 \times 2 = 2.371$. Our shortest earthquake cycle is a sub-harmonic of the Chandler Wobble. But of course the earthquake spectral peaks are related so

$(1.1855 \times 2) \times (5/3) = 3.95$ years versus the observed peak at 3.9 years

$(1.1855 \times 2) \times (5/3)^2 = 6.6$ versus 6.7.

$(1.1855 \times 2) \times (5/3)^3 = 10.98$ versus 11.05.

All are related to the Chandler Wobble via Fibonacci ratio harmonics. The peak at 2.78 years is not part of this sequence but is $22.24 / 8 = 11.12 / 4 = 2.78$. So earthquake energy release is related to the Chandler Wobble but also directly to a harmonic of the primary solar activity cycle. But the CW and the primary cycle are also harmonically related since $(11.05 / 2) / 1.1855 = 4.65$ and $(5/3)^3 = 4.63$. We can also link the CW to familiar Earth-Moon orbital parameters, namely the nodal and apsides cycles.

$18.6 / 2.371 = 7.844 = 1.67^4 \sim (5/3)^4$ or $18.6 / 1.1855 = 15.68 = 2.502^3$
$= (5/2)^3$. $\qquad 8.85 / 1.1855 = 7.47 = 1.655^4 \sim (5/3)^4$

The presence of lunar orbital factors suggests direct planetary forcing of nutation and energy release so let us examine this.

Looking at earthquake energy we note immediately that $1.1855 \times 10 = 11.855$ but the orbital period of Jupiter is 11.86 years.
But we also see that $11.86 / phi^3 = 2.78$ and $11.86 / 3 = 3.94$ versus the earthquake peaks at 2.78 and 3.9 years.

Also $11.86 / 5 = 2.372$. For Saturn $(29.46 / 4) / (13/8)^2 = 2.79$ versus 2.78 years and $(29.46 / 3) / (5/2) = 3.92$ versus the earthquake spectrum

3.9 years. For Neptune we have $(165 / 5) / (13 / 8)^3 = 7.68$ versus the primary peak of 7.6 years. Planetary orbital period harmonics, or these harmonics modified by low order Fibonacci ratio harmonics, are strongly present in the earthquake energy spectrum. Pairwise planetary conjunctions also show up clearly. For Cjs = 19.85 we have : 19.85 / 5 = 3.96 versus 3.9. 19.85 / 7 = 2.83 versus 2.78. (19.85 / 5) / (5 / 3) = 2.38 versus 2.37.

The other conjunctions are also visible.

$(12.8 / 2) (13 / 8) = 3.93$ versus 3.9 $12.8 / (5 / 3)^3 = 2.77$ versus 2.78

$13.8 / 2 = 6.9$ versus 6.7 $13.8 / 5 = 2.76$ versus 2.78

$30.2 / (5 / 3)^3 = 6.6$ and $(30.2 / 2) / (3 / 2)^2 = 6.71$ versus 6.7

$30.2 / (5 / 3)^5 = 2.35$ $30.2 / (5 / 3)^4 = 3.91$

$45.3 / (21 / 13)^4 = 6.65$ $(45.3 / 3) / (3 / 2)^2 = 6.71$ versus 6.7 observed

$(45.3 / 5) / (3 / 2)^2 = 4.02$ versus 3.9

$55.7 / 5 = 11.1$ $55.7 / 8 = 6.96$ and $(55.7 / 5) / (5 / 3) = 6.66$ versus 6.7

$(55.7 / 9) / (8 / 5) = 3.87$ versus 3.9

The planetary effects are clear and consistent. We can now say that global earthquake energy release is linked to the Chandler Wobble which is also related to the lunar nodal and apsides cycles and to planetary orbital periods. The relationships involve simple harmonics or Fibonacci ratio harmonics of the dominant 'forcing' periods: the signatures of quasi-periodic behaviour in non-linear systems at 'the edge of chaos'.

The patterns are remarkably self-consistent. The Earth's spin is being affected by planetary dynamics and the Chandler Wobble is said to be associated with changes in ocean and atmospheric circulation i.e. weather patterns. Earthquake energy release contains Chandler Wobble, solar activity and planetary orbital signatures. Perhaps multiple paths of causation are in play here but all lead back to the planets. Given these clear physical links we might reasonably expect to see solar activity and planetary dynamics links to specific terrestrial weather time series. In fact we note now as an appetiser that for the Chandler Wobble, 2 x 1.186 = 2.373 but the famous Quasi Biennial Oscillation, the QBO, in the equatorial stratospheric winds has a period of 2.375 years.

We will explore such links for weather time series chosen because high resolution spectra are now available for them in the literature.

These analyses are based on earthquake energy data for the period 1904 – 1965 using the MESA technique with a single choice of low model order. We have noted that any spectral analysis method will produce results dependent on the data window length and the start and end date of the window and the number of auto-regressive terms in the spectral model. In the kind of natural phenomena we are considering it may be that spectral power at particular frequencies comes and goes over time meaning that what we see depends on where we look in the time series and the length of the data window in relation to the frequency composition. We have seen this with the solar activity spectra and with analyses of weather phenomena over time. The AR order of the spectral model (the number of terms) will also affect which frequencies appear in the final spectrum. Studies which take these factors into account are therefore particularly valuable.

We will look at a second longer 'earthquake' series below. This demonstrates how data windows and model orders affect the results very clearly. It also demonstrates that *all* the frequencies found are part of a coherent family that can be easily tracked back to other coherent families in phenomena we believe to be related. Such analyses strengthen the case for causation. Liritzis et al (3) published a classic paper thoroughly examining the effects described above on a time series of global seismic energy release from 1898 to 1985. The series included all shallow earthquakes (h <= 60 km) of magnitude 7 or greater. These were converted into energy equivalents using Bath's formula

$$Log\ E = 12.24 + 1.44\ M7$$

MESA analyses were carried out for a range of data window lengths and start dates and for a range of Autoregressive order with m between 7 and 54. Stable results were found with m in the range 20 to 40. Table 1 shows the main results. The table suggests a relatively stable situation for seismic energy release over the whole data period except for the longest spectral periods. The full data set gives a period of 36.3 years. The mid 20th century short windows give a mean period of ~25.2 years. The period from 1898 to mid 20th century gives us ~44.2 years. In this area of the spectrum the periods are roughly half the full series length or 87 / 2 = 43.5. We might expect some instability. However the periods are clearly related and are relatable to periods in other phenomena already suspected earlier. So 44.2 / 2 = 22.1 years but the Hale solar cycle main period is 22.0 – 22.33 years. We also note that 22.33 x (13 / 8) = 36.3, using a familiar phi convergent.

Also the Hale side band 17.9 x 2 = 35.8, Jupiter period, 11.86 x 3 = 35.6, lunar apsides cycle is 8.85 x 4 = 35.4, lunar nodal cycle 18.6 x 2 = 37.2, with a mean of $\sqrt{(35.4 \times 37.2)} = 36.3$. The period ~25.2 is familiar from several places. The Jupiter –Neptune conjunction period is 12.8 years so 12.8 x 2 = 25.6. The Jupiter – Saturn conjunction period is 19.85 years and it and the first harmonic, 9.925 appear in the solar activity spectrum. So we note 9.925 x (8 / 5) x (8 / 5) = 25.4, more phi shadows. The other Hale cycle sideband is 17.9 years and so 17.9 x $\sqrt{2}$ = 25.3. Also for the classic 11.15 year solar peak 11.15 x (3 / 2) x (3 / 2) = 25.1, another phi shadow. Perhaps most interesting of all we note that

$$1 / 11.86 - 1 / 22.4 = 1 / 25.12$$

25.1 also can be generated by interaction between the Hale solar cycle and the orbital period of Jupiter. We should not be surprised by now since we linked solar activity to planetary orbital and conjunction periods. So even where widow length causes some instability the frequencies estimated are related and can be linked back to other candidate causal phenomena in simple, consistent ways.

TABLE I

Maximum Entropy method used to obtain the following periods for truncated records

Truncated time records in intervals	Periods in years										
1898–1940		15.3		6.7	5.6		4.4	3.6	2.9		2.3
1941–1985	24.8		9.0	6.4		4.97		3.7	2.7		2.3	2.2
1908–1985	23.4	15.3	8.84	6.4	5.4		4.47	3.72	2.9	2.7	2.3	2.1
1918–1985	26.5	14.7	9.04	6.4			4.5	3.68		2.65	2.34	2.08
1928–1985	26.5	15.9	8.65	6.2				3.68		2.7	2.37	2.08
1938–1985	24.9		9.47	6.3			4.68	3.75		2.73	2.37	2.08
1898–1985 ($m = 31$–34)	36.3	15.3	8.85	6.42	5.52		4.52	3.59	2.9	2.72	2.31	2.09
1898–1968	44.2	15.3	9.04	6.4	5.6		4.47	3.55	2.9	2.73	2.3
1898–1958	44.2	15.3		6.12			4.37	3.59	2.8		2.31
1898–1948		14.7			5.76		4.42	3.55	2.86		2.28
1898–1938		14.2		6.75	5.6		4.33	3.52	2.84		2.29

(___): Significant pronounced spectral power $S(f)$.
(---): Less significant.

Whatever the pathway, this further analysis confirms that planetary orbital dynamics, solar activity and global seismic energy release are somehow causally related. The Liritzis stable frequencies results gives further insights into how the MESA method behaves in uncovering period relationships in a time series. Starting with the longest period 44.2 years we note that the shortest stable period is 2.1 years and that 2.1 x 21 = 44.1 years. In fact most of the shorter periods are close to integer harmonics of the longest period. The same is true for the 25.2 period.

2.1	2.32	2.85	3.62	4.46	5.58	6.41	8.98	15.1	25.2	44.2
21X	19	31/2	12	10	8	7	5	3	√3	1X
12X	11	9	7	11/2	9/2	4	2√2	5/3	1X	-

Table 2 gives more insight into how the autoregressive model order, m, forming the basis of the MESA generation affects the frequencies found as m is increased from 7 to 54. We see that at various points, periods split into daughter periods which are often stable for some time until further splits occur. Changing the parameter m provides another way of probing the underlying structure of a times series, generating more information than using a single m value (and single model). The point is that all the periods generated are related and m acts like a tunable lens focusing on different frequency sets in turn.

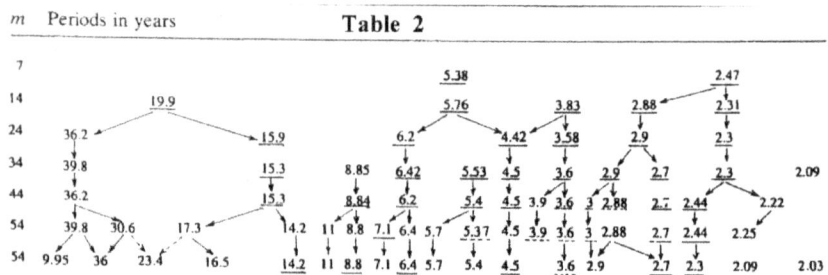

Table 2

We will take a few examples in illustration. With m = 14 the longest period is 19.9 years. We note another period at m =14 at 5.76 and 19.9 / 5.76 = 3.458

or $(1.51)^3 = (3/2)^3$. We have a Fibonacci shadow. Similarly 19.9 / 3.83 = 5.196

or $(1.51)^4 = (3/2)^4$, another phi shadow. Also 19.9 / 2.88 = 6.91

or $(1.622)^4 = (13/8)^4$ and 19.9 / 2.31 = 8.614 = 2 x (1.627)^3 =

2 x (13 / 8)^3.

Of course we know that 19.85 is a major peak in the Hale solar activity series, 19.85 / 2 is a major side band in the sunspot series, and 19.85 is also the Jupiter – Saturn conjunction period.

142

We also found 19.85 prominently in our weather time series. The analysis with m = 14 tells us that all the global seismic energy release variation is related to the Jupiter – Saturn interaction and its harmonics, an interesting result. Let us look at frequency splitting. With m = 24 the 19.9 prime period exits and we get instead 36.2 and 15.9 year peaks. Are these peaks related ? We see that

1 / 19.9 – 1 / 44. 2 = 1 / 36.2 where 44.2 is from Table 1 analyses and 2 x 22.1, the primary Hale solar cycle period. Also 3 x 11.86 = 35.6, 2 x 17.9 = 35.8, 22.33 x 8 / 13 = 36.3, 4 x 8.85 = 35.4, 2 x 18.6 = 37.2. We should expect to pick up power around 36 years under some conditions.
Also 1 / 19.9 + 1 / 79 = 1 / 15.9, and 79.6 = 7 x 11.2 = 78.4, 4 x 19.85 = 79.4, 9 x 8.85 = 79.6.

For Jupiter we have 11.86 x $(8/5)^4$ = 77.7.
Also for Saturn's orbital period 29.46 x (5 / 3) x (5 / 3) = 81.8.
A solar periodicity around 80 years has long been picked up and is known as the Gleissberg cycle but it appears to be a rough conjunction of the sub-harmonics of several shorter cycles. In Fibonacci terms we note also
15.9 x $(5/3)^2$ = 44.16. 15.9 is a simple phi harmonic of the double Hale period.
Scanning though m has lessons to teach. As m increases to 34, 19.9 returns as 2 x 19.9 = 39.8 years. The next longest period
becomes 15.3 which is 39.8 / $(13/8)^2$ = 15.26 and the next 8.85 (the apsides lunar
cycle) and 39.8 / 8.85 = $(1.651)^3$ ~ $(5/3)^3$ and 5.53 appears which is just 22.1 / 4.

Consider the next m = 14 peak splitting of 5.76 into 6.2 and 4.42. We see that 1 / 5.76 – 1 / 79 = 1 / 6.21 (see 79 origin above).

Also (19.9 / 2) / (8 / 5) = 6.22; 18.6 / 3 = 6.2; 44.2 / 7 = 6.3. We note that the 4.42 peak is just 44.2 / 10 = 22.1 / 5 and also 1 / 5.76 + 1 / 19.9 = 1 / 4.46; 8.85 / 2 = 4.41; 17.9 / 4 = 4.47. For m >24 this peak stabilizes at 4.5 years and 22.3 / 5 = 4.46. Scanning through m allows us to pick up many periods we may know from other phenomena, the physical links with which we wish to explore. Increasing m beyond 34 we begin to pick up other periods such as 8.85, the lunar apsides cycle, 11.0 which is just 22.1 / 2, the first Hale solar cycle harmonic, and 9.95 which is the first harmonic of the 19.85 Jupiter – Saturn conjunction period…and a major solar activity spectral peak.

Liritzis also the examined the effect of taking subsets of the time series at various sampling intervals which he believed provides clearer long period terms (Table 3). An SI of 2 picks up 39.8 or 2 x 19.9; 17.3; 13.5;8.7;6.4. Liritzis suggests 17.3 is the lunar nodal cycle of 18.6 years but we also noted the Hale solar sideband of 17.9. 8.7 is close to the lunar apsides cycle of 8.85 years, clearer at SI = 1, 3, 4. Also 19.9 / 3 = 6.6 compared with 6.4 and 6.8. and 18.6 / 3 = 6.2. SI = 4, 11 pick up 33.9 or 3 x 11.3 and 33 or 3 x 11, the main sunspot cycle. Even longer periods appear at SI = 8 and SI = 11. The period 72.4 = 4 x 18.1, probably reflecting the Hale sideband of 17.9. The period 61.7 is close to 3 x 19.9 = 59.7 and even closer to the apsides cycle multiple, 7 x 8.85 = 61.95 and 8 x 8.85 = 70.8. However we also can see that $\sqrt{(72.4 \times 61.7)} = 66.8 = 3 \times 22.3$, the Hale solar cycle. The longest periods are interactions between a 3 Hale cycle period and a long period of ~831 years. This is interesting but not convincing.

Evenspaced time series and periods obtained from Maximum Entropy method

Evenspacing (yrs)	Periods in years									
						Table 3				
2		39.8	17.3*	13.5	8.7	6.4				
3		30.6			8.9	6.8				
4		33.9	16.4	12.9	8.9					
8	72.4	31.5	19.9							
11	61.7	33.0								
1		39.8	15.3		8.8	6.4	5.5	4.5	3.6	2.9	2.3

For completeness we now examine the origins of all the stable periods in Tables 1 and 2 for moderate m values, from longest to shortest.

44.2 22.1 x 2 = 44.2, or 2 Hale solar cycles; 19.85 x (3 / 2)^2 = 44.6 where 19.85 is the Jupiter –Saturn conjunction period; 29.46 x (3 / 2) = 44.2 where 29.46 is the Saturn orbital period; 5 x 8.85 = 44.25 where 8.85 is the lunar apsides cycle. The links could not be clearer.

39.8 19.85 x 2 = 39.7 the JS conjunction period directly but also a Hale Solar sideband; 11.86 x (3 / 2) x (3 / 2) x (3 / 2) = 40.0.

25.22 conjunction of Saturn-Neptune is 25; 2 x 12.8 = 26.5, 2 Jupiter – Neptune conjunctions; 11.15 x (3 / 2) x (3 / 2) = 25.1, main solar cycle; $\sqrt{2} \times 17.9 = 25.3$, two Hale sideband periods; $2 \sqrt{2} \times 8.85 = 25.04$, apsides cycle is 8.85 years.

15.1 22.33 x (2 / 3) = 14.9; 9.93 (3 / 2) = 14.9, solar period, JS conjunction / 2; 29.46 / 2 = 14.75, half Saturn orbital period; 18.6 x (2 / phi) =15.05, lunar nodal period.

8.98 17.9 / 2 = 8.95, Hale solar sideband; 19.85 x (2 / 3)^2 = 8.82;

The lunar apsides cycle 8.85.

6.41 $19.85 / 3 = 6.61$; $22.1 \times (2/3)^3 = 6.53$; $17.9 \times (3/5)^2 = 6.44$; $12.8 / 2 = 6.4$, half Jupiter – Neptune conjunction period.

5.58 $22.3 / 4 = 11.15 / 2 = 5.575$; $8.85 \times (5/8) = 5.53$; $18.6 \times (2/3)^3 = 5.52$ and we have both lunar nodal and apsides cycles present.

4.46 $22.3 / 5 = 4.46$; $17.9 / 4 = 4.47$; $8.85 / 2 = 4.43$; $18.6 \times (5/8)^3 = 4.54$; $29.46 \times (5/8)^4 = 4.49$; $11.86 \times (8/13)^2 = 4.49$.

3.62 $22.3 / 6 = 3.7$; $17.9 / 5 = 3.58$; $18.6 / 5 = 3.72$; $29.46 / 8 = 3.68$; $11.86 \times (2/3)^3 = 3.52$; $19.85 / 2) \times (3/5) = 3.58$.

2.85 $22.3 / 8 = 11.15 / 4 = 2.79$; $19.85 / 7 = 2.84$; $11.86 / 4 = 2.96$; $8.85 / 3 = 2.95$; $18.6 \times (5/8)^4 = 2.84$.

2.32 Quasi-biennial Oscillation is ~ 2.37; $11.86 / 5 = 2.37$; $18.6 / 8 = 2.33$.

2.1 $12.8 / 6 = 2.13$, JN conjunction; $25 / 12 = 2.08$, SN conjunction; $11.1 \times (2/3)^4 = 2.19$; $(19.85 / 2) \times (3/5)^3 = 2.14$; $8.85 \times (13/21)^3 = 2.1$.

4.4 Power Law Models & SOC

In the 1990's the author was fortunate for some years to visit the Santa Fe Institute to explore Complex Adaptive Systems Theory on behalf of Unilever Research. SFI is very much concerned with non-linear dynamic systems in physics, economics and biology. One ambitious theme was the search for universal laws of behavior in complex systems. The author and colleagues worked on industrial optimization problems with Professor Stuart Kauffman who in his 'day job' was trying (with remarkable success) to understand molecular self-organisation and ultimately the origin of life and organization in general (6). In a limited sense the author's work on climate herein has the same objective as we seek out the common harmonic structures underlying several, physical dynamical systems.

The work of Per Bak and others at SFI centred on dynamic systems which 'self organised' into what he called a critical state. We met SOC in sections 2.5 and 3.8 in discussing the origins of resonances in the orbits of the solar planets and the problem of how small planetary forces can be magnified to modulate solar activity. If we think of the spatial and temporal distribution of earthquakes SOC is also highly relevant. Stresses steadily but slowly build up as tectonic plates move, setting up faults and points where eventually the earth is able to move. The result is earthquakes. Bak was able to show that such a process follows a power law distribution in earthquake magnitude and later he and others realized that the distribution in time at given locations should also follow power laws (4).

The figure below shows plots of P1, the 'first return time' after an earthquake in a given small area of California and P t, 'all return times' after an earthquake. The vertical axes are the frequencies of experiencing given durations of P1 and P t. This data is for 8,000 earthquakes. P t follows a power law with an exponent of 0.5 across an order of magnitude in earthquake frequency. P1 follows a power law with an exponent of 1.4 over one and a half orders of magnitude. This is a striking result. Bak concludes that earthquakes (and many other phenomena) are not periodic. This is true over part of the power law regions below.

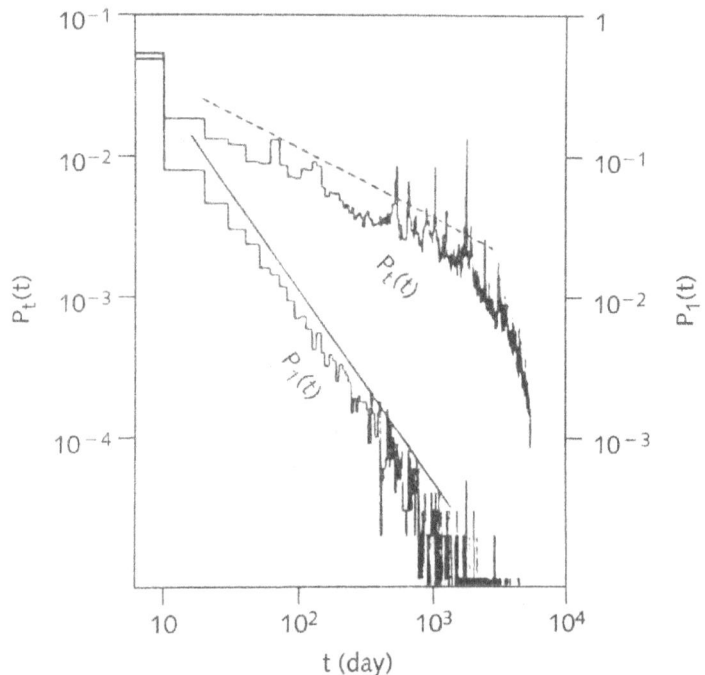

However for periods beyond ~400 days it is not true. We need to look closer at return times greater than a year. I assume Bak simply treats this region as noise although there are very clear peaks in P t and P1. Let us see if these peaks have any significance beginning with P t.

7.6 years $22.33 / 3 = 7.44$; $19.85 / (21 / 13)^2 = 7.61$; $29.46 / (8 / 5)^3 = 7.2$

6.85 $19.85 / 3 = 6.62$; $11.16 / (13 / 8)^2 = 6.86$; $17.9 / \text{phi} = 6.83$; $11.86 / \sqrt{3} = 6.85$.

5.07 $19.85 / 4 = 4.96$; $11.16 / (3 / 2)^2 = 4.96$; $17.9 / (3 / 2)^3 = 5.3$
$11.86 / (3 / 2) = 5.27$.

3.56 $11.16 / 3 = 3.71$; $11.16 / (8 / 5)^4 = 3.42$; $17.9 / 5 = 3.58$; $11.86 /(3 / 2)^3 = 3.5$

2.82 $22.33 / 8 = 11.16 / 4 = 2.79$; $19.85 / 7 = 2.84$; $11.86 / (21 / 13)^3 = 2.816$.

1.75 $22.33 /13 = 1.72$; $11.16 / 1.6^4 = 1.7$; $17.9 / 10 = 1.79$; $11.86 / (13 / 8)^4 = 1.7$

1.42 $11.86 / 8 = 1.48$; $19.85 / 14 = 9.93 / 7 = 1.42$; $22.33 / 16 = 11.16 /8 = 1.39$
$18.6 / 13 = 1.43$; $7.443 / 5 = 1.49$; $29.46 / 21 = 1.4$.

We can see simple harmonics of the Hale (main and sidebands) and the sunspot main 11 year cycle with Phi harmonics and direct signs of Jupiter and Saturn periods. The largest peaks 5.07 years and 2.82 years are 6.5 X and 2.8 X above the background distribution level. 5.07 is just 19.85 / 4, the JS conjunction period and a primary Hale solar series cycle. 2.82 is just $22.33 / 8 = 11.16 / 4$ from the main Hale period and the main sunspot series period and also 19.85 / 7 from the JS conjunction period. The period 1.75 years is also a known interaction of the QBO of period ~2.37 years with the solar year. It stands at 2.8 X the background distribution.

The P1 distribution is noisier than P t and the longer periods based on fewer observations. Even so there seem to be large peaks beyond the background.

7.66 $22.33 / 3 = 7.44$; $19.85 / \text{phi}^2 = 7.58$; $17.9 / (3 / 2)^2 = 7.9$

5 $19.85 / 4 = 9.925 / 2 = 4.96$; $11.16 / (3 / 2)^2 = 4.96$

3 $22.33 / 7 = 3.19$; $19.85 / (8 / 5)^4 = 2.03$; $11.86 / (8 / 5)^3 = 2.89$;
$17.9 / 6 = 2.98$

1.9-2.2 $22.33 / 10 = 11.16 / 5 = 2.23$; $22.33 / 11 = 2.03$; $19.85 / 10 =$

$9.925 / 5 = 1.99$; $17.9 / 9 = 1.99$; $11.86 / 6 = 1.98$

1.3 $11.86 / 9 = 1.32$; $19.85 / 15 = 1.32$; $(17.9 / 2) / \text{phi} = 1.306$

1.1 $11.16 / 10 = 1.116$; $19.85 / 18 = 9.925 / 9 = 1.1$; $17.9 / 16 = 1.19$

We should also note that $\sqrt{(1.1 \times 1.3)} = 1.195$ whereas the Quasi Biennial Oscillation is ~2.37 or 1.185 x 2 years and 1.185 years is the Chandler Wobble and 1.185 x 10 = 11.86, the orbital period of Jupiter. Between ~1.1 and 1.3 years there is a deep fall off in earthquake return frequency and a negative spike. The negative spike is ~ 5 X lower than the power law background. So first earthquake returns cluster slightly at $11.86 / 11 = 1.08$ and $11.86 / 9 = 1.32$ and avoid $11.86 / 10 = 1.186$, the exact Chandler Wobble period. We also see excess return frequency at ~1.64 and 1.86 years but another fall off in frequency between them and a negative spike. The spike is 3 X below the power law level. We see that $\sqrt{(1.64 \times 1.86)} = 1.75$ years, which as for P t is a QBO interaction with one solar year: $1 - 1 / 2.37 = 1 / 1.73$. Are these features a purely Californian phenomena or commonplace? If we believe Bak's hypothesis of universal critical behaviour they may be universal although he did not apparently notice the fine structure involving larger earthquakes recurring on multi-year time scales.

Overall we seem to be seeing the usual suspects emerging beyond Bak's power law range. The largest peak is at 5 years (as for P t) which is ~ 5.5X background. 7.6 years, which is ~ 22.3 / 3, is next at ~ 2.3X background. We could see the 7.66 year peak as 22.98 / 3 which can also be seen as 2 x 11.49 / 3. Interestingly $\sqrt{(11.16 \times 11.86)} = 11.5$ years or the harmonic mean of the sunspot cycle and the orbital period of Jupiter. Perhaps the Earth is picking up both the indirect solar / weather periods and the planetary periods.

Bak's criticality model serves us well here. The earth crust system is in a critical state and from time to time additional stresses, perhaps directly from planetary orbital dynamics cycles (or those cycles acting via lunar orbital changes, or both via weather changes such as the QBO, AMO, PDO, NAO and El Nino), exceed the crustal strength locally and we see an earthquake and a stress release. In that location stress rebuilds until another peak in the 'cosmic' forcing function releases it. *Bak's criticality model may answer the wider question* of how supposedly very small forces, such as planetary tides and torque effects acting on the Sun and on Earth's weather systems can cause large effects on Earth. These small forces repeatedly provide the straws that break the camel's back in SOC systems.

One last comment on recent work on the Earth-Moon system may have implications for our later look at planetary dynamics, the solar magnetic field and Earth's weather. Dr. Andault, of the centre national de la recherché scientifique, and his team have shown that the Moon's varying gravitational tides have always caused turbulence in the liquid iron of the Earth's core and contributed to the continued operation of the dynamo which generates our geomagnetic field (1, 2). We perhaps owe our long term protective shield to the Moon. We will also see later the importance of the magnetic field in weather issues. We saw in section 3 that the tides and torques caused by the planets on the tachocline at the base of the Sun's convection layer, similarly modulate its magnetic field and solar activity cycles with profound consequences for the Earth's climate.

4.5 Volcanic Activity Modulation ?

Given the earthquake energy result it is reasonable to look also at volcanic activity but there is little in the literature. The most detailed analysis is that of Strestik as long ago as 2003 in a symposium paper (5). He examines a long term volcanic activity index based on ice acidity in Greenland glaciers and at annual sunspot numbers. The spectra have common, large (95 – 99% confidence level) frequency peaks at ~ 628 and ~ 219 years and two other smaller common peaks at ~ 105 and ~ 79 years.

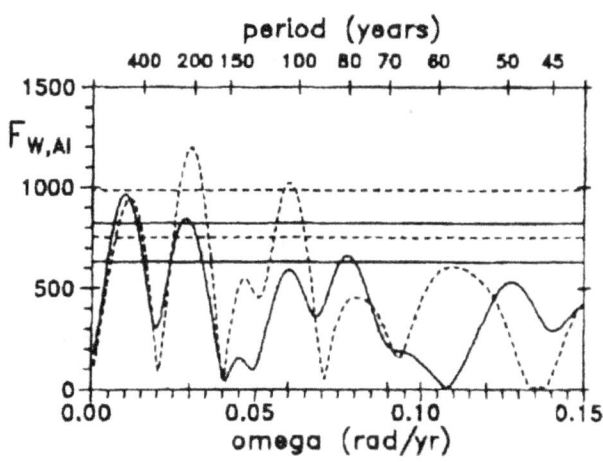

Fig. 4. The spectra of volcanic Al indices (solid) and annual sunspot numbers (dashed). Horizontal lines show the 95% and 99% confidence levels for the appropriate spectra.

We can see that the pairs of peaks are also harmonically related since
2 x 105 = 210, 6 x 105 = 630 and
$$79 \times (5/3)^2 = 219.5; (5/3)^2 \times 219.5 = 611.$$ We find familiar periods:

79 79 / 9 = **11.29**; 79 / 4 = 19.75 versus the JS conjunction period 19.85 ; 19 / 9 = 8.75 versus the lunar apsides cycle at 8.85.

105 105 / 9 = 11.67 versus Jupiter orbital period at 11.86; 105 / 5 = **21**; 105 / 6 = 17.5 versus the Hale peak at 17.9 years; 105 / 12 = 8.75, apsides cycle.

219 219 / 10 = **21.9**; 219 / 11 = 19.9 versus JS conjunction at 19.85; 21 / 12 = 18.2 versus hale 17.9; 219 / 17 = 12.9 versus JN conjunction at 12.8 years.

628 628 / 28 = 89 / 7 = **22.4**; 8 x 79 = 632; 105 x 6 = 630; 35 x 17.9 = 627; 29.46 x 21 = 619; 32 x 19.85 = 635.

We get 22.33, the Hale main peak and 19.85, the Hale cycle second peak and the JS conjunction period. The solar and planetary links are clear.

S4 References

1. Andrault D; Earth & Planetary Science Letters, 30th March 2016.
2. Report; 'How the Moon makes the magnetic Earth', Astronomy Now' May 2016.
3. Liritzis I et al; 'Probable evidence for periodicities in global seismic energy release'; Earth, Moon, and Planets, Vol. 60, 93-108 (1993)
    4. Bak P; 'How Nature Works: the science of self-organised criticality'; Chapter 9, Oxford University Press, 1997.
    5. Strestik J; 'Proc. ISCS 2003 Symposium, 'Solar Variability as an Input to the Earth's Environment', September 2003, Lomnica, Slovakia.
    6. Kauffman S; 'The Origin of Order'; Oxford University Press, 1993.

S5. Weather Variables, the Sun and the Planets

> Praise be to You, my Lord, through Brothers Wind & Air
> And fair and stormy, all weather's moods,
> by which You cherish all that You have made.

The Canticle of Brother Sun & Sister Moon: St. Francis of Assisi

5.1 Long Temperature Reconstructions: Tibet, Europe, Greenland, Antarctica

The growth of trees and the width of their annual growth rings depend on the 'distance' of growing conditions from the optimal state. For a given species, in a fixed location, this means deviations in rainfall and / or temperature across the seasons from year to year. Trees therefore preserve a record of weather variations over decades and centuries. In some instances where direct records are available the correlation of tree rings and for example, rainfall and temperature, has been shown to be good. Tree rings are useful weather proxies and have the advantage that they may stretch back centuries or millennia before instrumental records began. Here we examine the high resolution spectrum of a continuous tree ring record from the central-eastern plateau of Tibet reported by Liu Yu, et al [1].
That record is 2,485 years long and based on 7 sites. Such a long record provides a spectrum with well defined peaks for all cycles between 2 years and ~1,300 years. We will begin with the relationships of the very long cycles : 110, 199, 800, 1324 years. They are clearly and simply harmonically related.

$$1324 / (5/3) = 795 \text{ versus } 800 \text{ observed} ; 1324 / (8/5)^2 = 202 \text{ versus } 199$$

$$1324 / 12 = 110.3 \text{ versus } 110 \text{ observed}.$$

$$800 / 4 = 200 \text{ versus } 199; \; 800 / (5/3)^4 = 105 \text{ and } 800 / (13/8)^4 = 114.6 \text{ with a mean of } 109.6 \text{ versus } 110 \text{ observed}.$$

The appearance of consistent integer harmonics and Fibonacci ratio harmonics confirms the presence of a non-linear, quasi- periodic system. Smaller but still highly significant peaks occur at middling periods: **66.7, 38.9 and 29.5 years.** We see immediately and unequivocally that **66.7 / 3 = 22.33 years, the Hale solar activity series primary period.**

Also 38.9 / 2 = 19.5 close to the major 19.85 year peak in the Hale series which is also of course Cjs, the Jupiter-Saturn conjunction period. The 29.5 peak coincides exactly with Ps, the Saturn orbital period of 29.46 years. Note also that 1324 / 8 = 165.4 versus Pn = 165 years. The planetary signals could not be clearer.

Let us now return briefly to the very long tree ring cycles and look for solar signals. Note that the sunspot main cycle is 11.16 years and 11.16 x 10 = 22.33 x 5 = 111.6 versus the observed 110 tree ring peak. But also 1324 / 12 = 110.3 so the longest tree ring cycle is clearly linked to the Sun since 22.3 x 60 = 1338 years, just a 1% difference from the 1324 year tree ring peak. Given the finite width of that peak the pair is identical. That is not all: 22.3 x 9 = 200.7 compared with the observed 199 years. Also considering Cjs, 19.85 x 10 = 198.5 compared with the tree ring 199 year cycle. Considering Ps = 29.46 we see that 29.46 x 27 = 796 compared with 800 observed. The Uranus − Neptune conjunction period also shows up since 55.7 x 2 = 111.4 versus 110 observed.

Finally we note that 800 / (5 / 3)^3 = 174 and 800 / (13 / 8)^3 = 186 with a mean of 179.8 compared with Cjsn, the Jupiter-Saturn-Neptune conjunction period of 178.8 years. Recall the triple conjunctions are all in the range 171 to 181 years.

All the long tree ring cycles are solar and planetary cycle related. The shorter tree ring cycles are also very familiar: **23.8, 18.2, 6.74, 6.03, 4.92, 3.75, 3.0, 2.33, 2.1 years.** Note that **23.8** is 2 x 11.9 but Pj = 11.86 years and for Cjs 19.85 x 6 / 5 = 23.82.

18.2 is close to the large Hale series peak at 17.9 years but also 11.16 x 8 / 5 = 17.9 and for Ps, 29.46 / (5 / 3) = 17.7 years.

For period **6.74** we have Cjs again, 19.85 / 3 = 6.62; for the Wolf sunspot main cycle we have 11.16 / (5 / 3) = 6.7 and for Ps, 29.46 / (13 / 8)^3 = 6.83 ; 17.9 / (13 / 8)^2 = 6.78.

For the **6.03** cycle we have 17.9 / 3 = 5.97; using Pj we have 11.86 / 2 = 5.93 and for Cjs, 19.85 / 2 = 9.93, which appears in the Wolf sunspot spectrum, we get 9.93 / (5 / 3) = 5.97.

For **4.92** we note that 22.3 / (5 / 3)^2 = 4.84 ; 11.16 / (3 / 2)^2 = 4.92 ; 19.85 / 4 = 4.96 and for Ps, 29.46 / 6 = 4.91;

The 22.3 year third harmonic gives us 7.44 / (3 / 2) = 4.96 ; the harmonic mean estimate is 4.917 years.

For **3.75** note that 11.16 / 3 = 3.72 or 7.44 / 2 = 3.72 ; (11.86 / 2) / (8 / 5) = 3.71
$(19.85 / 2) / (13 / 8)^2 = 3.77$; for Ps, $29.46 / (5 / 3)^4 = 3.81$; the harmonic mean cycle estimate is 3.744 years.

For **3.0** considering Pj, $11.86 / 4 = 2.97$; $22.33 / (5 / 3)^4 = 2.9$;
$19.85 / (8 / 5)^4 = 3.03$
$7.44 / (8 / 5)^2 = 2.91$; $8.43 / (5 / 3)^2 = 3.04$.

For **2.33** years we note that the quasi biennial oscillation cycle is 2.375 years (see section 5);
But for Pj, $11.86 / 5 = 2.37$; $(19.85 / 2) / (13 / 8)^3 = 2.32$; $22.33 / 6 = 3.72$ as noted above but $3.72 / (21 / 13) = 2.304$.

For **2.1** years $8.44 / 4 = 2.11$; $(11.16 / 2) / (13 / 8)^2 = 2.11$;
$(11.86 / 2) / (5 / 3)^2 = 2.14$.

We see clear and consistent quantitative links between the tree ring widths and sunspot and Hale solar activity series across cycles from 2 to 1,324 years in duration. Planetary orbital signals are also very strong.
The value of the very long 2,485 years data series in unravelling these connections is immense and other long series will be sought for analysis later. The clarity and stability of the long term cycles in the tree ring series encouraged the authors to develop a temperature proxy prediction model. The model fit obtained was excellent over the period 464 BC to 834 AD, some 1,300 years. They then used the same model in a verification exercise for the period 835 AD to 1980 AD. The system is clearly stable over very long periods as the fit is again superb.

Interestingly there was nothing unusual about the 20th century weather in Tibet in comparison with natural variations over the last two millennia.

There is nothing currently to suggest unusual temperatures as far as the trees are concerned. The recent good growth conditions were matched exactly in growth peaks at 400, 870, 990 AD and other peaks were not far behind at various epochs.

So far as the author can establish there were very few 4 x 4s or CO_2 belching factories in early medieval Tibet or elsewhere on the planet.

It would be invaluable in the climate debate if similar very long term climate proxy data series could be reliably established for many locations. Then some of the 'warmist' hyperbole could be replaced with evidence and a sensible historical perspective. With Tibetan tree rings we at least have clear confirmation that the Sun can play a highly significant role in climate modulation and botanical productivity over many time scales.

Given the good model fit over centuries the authors could not resist making a forecast of tree growth conditions for several decades ahead. The results show a very significant decline in growing conditions from the ~2006 AD peak for the next fifty years followed by a modest recovery after 2068 …but not back to late 20th century levels.

As we noted the HadCRUT, RSS and UAH global data series showed a 'pause' in temperature increase after year 2000 which is still with us in 2016. The Tibet model predicts that halt after a rapid increase in ~1980-2000 based on 1,300 years of much earlier data. We saw that much of the temperature variation can be linked to natural solar variation. It is usually said the significant increase in manmade green house gas emissions began in the last two centuries since the industrial revolution.

But the model fit over this period, based just on natural cycles, is excellent, including the prediction of the ~2000 – 2015 halt. **Therefore Tibetan temperatures have not been affected by the 'massive' CO_2 increase of the last century.** Tibet is just one region. N.B. If Tibetan tree growth is mainly modulated by regional temperature, Tibet is in for a cooler time. Notice that the Little Ice Age of the 17th century (unique in the 2,500 year record) shows up as does the steady, three century, rise in temperatures until today.

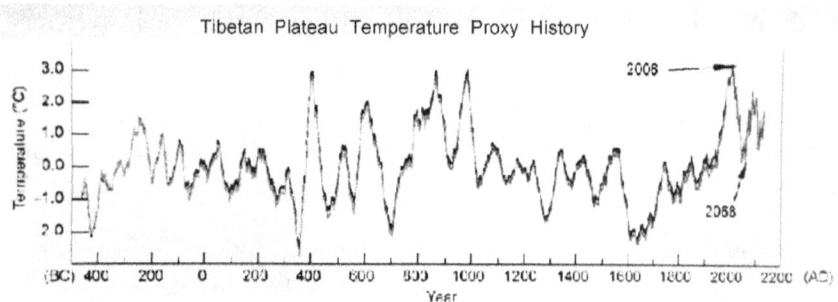

It would be interesting to know how Tibetan Plateau temperatures correlate with temperatures across Asia and beyond would it not? At the moment concern seems to centre on warming of the TP since it is the source of water for many major Asian rivers such as the Ganges. High TP temperatures and aridity are said to be bad news for India et al, with reduced water flows and maybe effects on the Asian monsoon (1A). However the model is predicting a *reversal* in rising temperatures and then a steep decline for sixty years.

What we need are similar long term temperature proxy time series and similar stable spectral models for other regions which can be fitted to pre-20th century data. Forecasts from such models from the last century to date can then be compared to actual global temperature series. Any difference should indicate the additional effect of manmade greenhouse gases... if any. This is the only *evidence based* way to determine AGH gases climate sensitivity. Such attempts are beginning to appear. Ludecke et al (22) have fitted spectral models to mid-European Instrumental records for the last 240 years and to 2,000 year long proxy data from Austrian stalagmites.

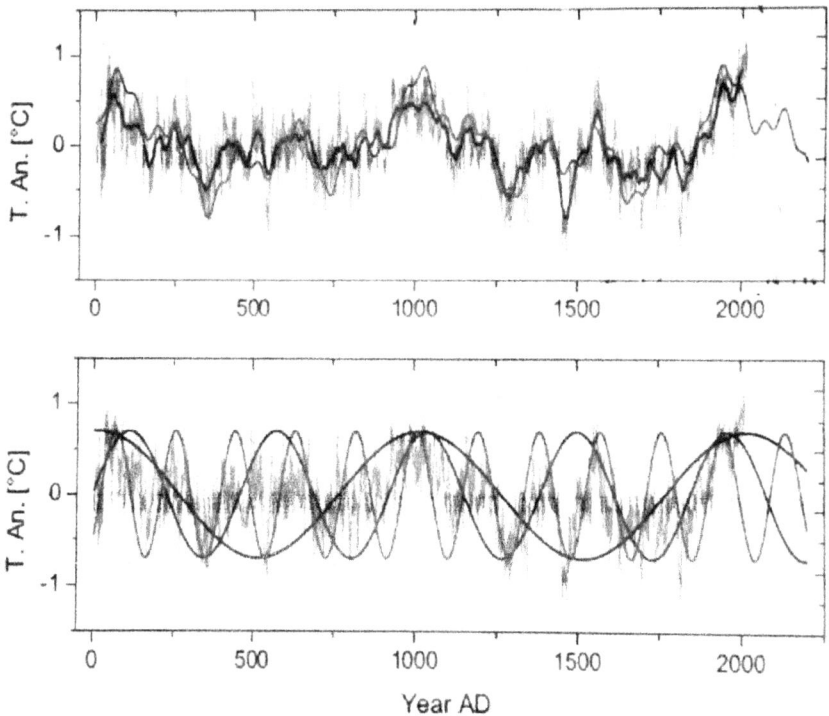

Similar models emerge including long cycles also seen in Tibetan, Be10 and CET temperature records. Most of the power is focussed in several sharp spectral peaks.

Several periods are again easily solar related (including 26 or 2 x 12.8, 32-33 or 3 x 11 and 4 x 8.1, 38-39 or 2 x 19.85, 66 or 3 x 22, 73 or 9 x 8.1, 101 or 9 x 11.2, 182 or 8 x 22.33). Using the six strongest periods the authors defined a Fourier model which is compared with the European temperature record and the fit is remarkably good ...including a temperature maximum in about 2005. *There is no sign of any failure of fit in the 20th century which could be attributed to CO2 etc.* Ludecke later studied 7 temperature proxy series and derived their G7 composite series. They found a strong cycle of 170-190 years and apparent harmonics of it. We appear to have the ~180 year solar cycle related to 3 and 4 planet conjunctions (J,S,U,N) centred on 179 years as noted earlier giving 179, 179 / 3 = 59.6, 179 / 9 = 19.87 years. Of course 19.85 is the JS conjunction period. 178.6 is also 22.33 x 8 and 17.9 x 10.

Using just this fundamental period and the two harmonics largely explains the temperature movements over two millennia (see above). Note that as in *Tibet the model predicts temperature falling back in stages towards the 17th century levels*. The 2000+ year spectral fit should not have been biased to any degree by the inclusion of mid 20th century data.

We note that the shortest period is 88 years which is just 22.33 x 4 = 89.32 years, often seen in weather time series and in the solar activity spectrum as the 'Gleissberg Cycle'. The other long periods are related to 88 but also 22 and 11 basic year cycles and the 19.85, Hale solar sideband and JS conjunction. Here are just the basic links:

88 = 4 x 22 = 8 x 11; 130 ~6 x 22= 132; 150 ~7 x 22 = 154, 5 x 29.46 = 147.3; 208 ~11 x 19.85 = 218, 7 x 29.46 = 206; 350 ~ 2 x 176, 12 x 29.46 = 353 ; 500 ~23 x 22 = 506; 700 ~8 x 88 = 704, 35 x 19.85 = 695; 1000 ~12 x 88 = 6 x 179 = 1056, 50 x 19.85 = 993; 1450 ~66 x 22 = 1452, 8 x 179 = 1432, 73 x 19.85 = 1430; 2200 ~ 25 x 88 = 2200, 12 x 179 = 2150.

The authors used these periods to create a solar activity potential, Phi, and a total solar irradiance (smoothed over a 22 year period) estimate using two methods: a Fast Fourier Transform with various periodicities and calibration windows of 4,000 and 6,000 years; a WTAR (autoregressive) method. They tested the fitted models by comparing the forecasts with known history after the various fitting periods. Figure A shows the forecasts against reality. The WTAR method predicts the correct general levels but in two of three cases underestimated the cyclical variations. The FFT method captures the levels over a 500 year window and the cyclical variations well. We might expect the FFT based predictions to be more reliable going forward from today.

These are shown in Figure B for both methods and a range of periodicities and calibration windows. Both methods predict a sharp fall in solar activity after 2000 AD which is happening. The minimum is reached in both cases at around 2060. WTAR predicts a minimum between the previous Dalton and Geissberg minima.

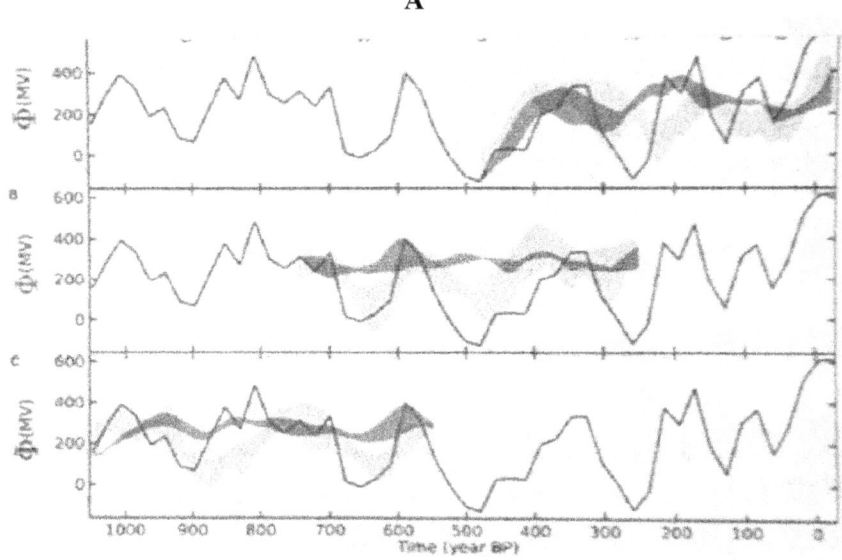

A

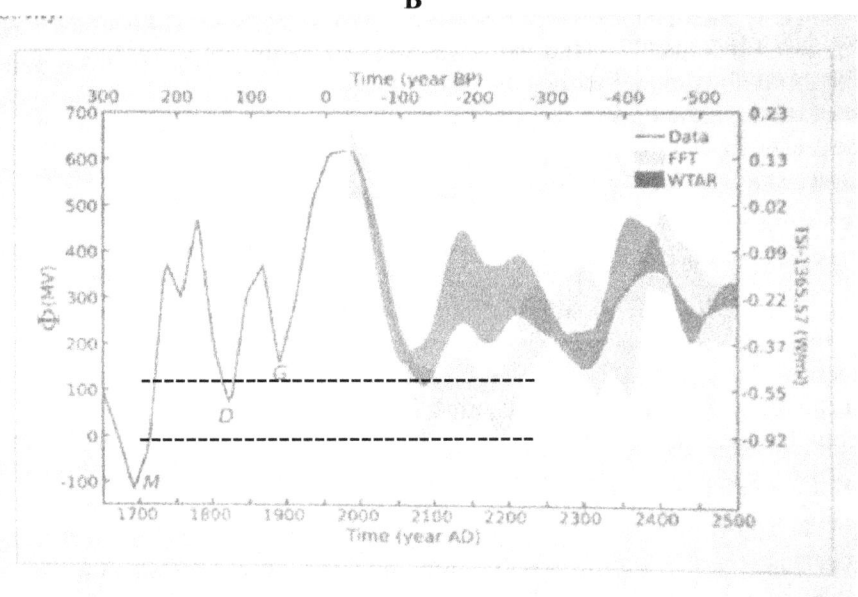

B

The FFT based model gives a minimum activity level half way between the Dalton minimum and the Maunder grand minimum in the Little Ice Age. For the FFT that minimum lasts until ~ 2200. Both models recover to 20th century solar activity levels of activity by ~2350. If the Sun has the influence expected from the previous several centuries we should predict a distinct cooling of the Earth at least back to early 19th century temperature levels.

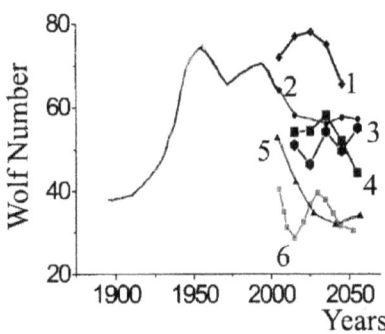

Figure 2. Actual Wolf number smoothed by 25 years (gray line with opened circles); forecast of sunspot number made in present work using 4 nearest neighbors, $d = 3, \tau = 1$ (line 4) and 5 nearest neighbors, $d = 4, \tau = 1$ (line 3); forecast of sunspot number made in present work using TISEAN program **nstep** with $d = 3, \tau = 1$ (line 2); forecasts of sunspot number made by Nagovitsyn and Ogurtsov (2003, line 6), Miletskij (2003, line 5), Vasiliev et al. (2002, line 1)

This is compatible with the Tibetan results and Ludecke's European results. The Russians too have used reconstructed solar times series from long C14 records (37). Ogurtsov reconstructed a solar signal using a physical 5 box model of the terrestrial carbon exchange system to avoid distortions in the C14 series. Ogurtsov used rescale analysis (R/S) and a de-trended fluctuation analysis to probe the properties of the long proxy series from 8005 BC to 1945 AD.

He found a high Hurst exponent value of 0.84 which confirms the presence of long term memory effects in the solar proxy. As he notes long term memory implies the possibility of long term forecasts of solar activity (and climate). He uses a non-linear forecasting technique which appears to recreate the internal dynamics from sets of time delayed data. This is a technique usually applied to dynamic systems suspected of hosting complex dynamic 'strange' attractors (see Appendix 1). The figure above shows the smoothed (25 years) mean Wolf sunspot number from 1900 to ~2000 and forecasts from various Russian workers. The key point is that none show any net increase in solar activity over the period 2000 to 2060 AD. Lines 2, 3, 4 are from Ogurtsov's analyses with various assumptions. These show a significant ~30% fall from the 1960s peak to 2060.

Lines 5 and 6 show an activity fall off of ~55%. Only line 1 shows a small increase and then a fall back to 1970s levels. We know from the low solar cycle 24 that line 1 can now be discounted.

Different disciplines with different histories inevitably learn to analyse time series with a variety of techniques. Digital Signal Processing for example has much in common with control engineering practice. Let us end with a DSP analysis of the HadCrut3 temperature series from 1900 to 2013 using various techniques to improve and test the results. HC3, of the ground based temperature series, avoids the 'adjusted' exaggerations of the others. Patterson used HC3 monthly series to create an unbiased annual series using box-car averaging to avoid smearing and aliasing effects (38). He created a power spectral density plot which highlighted several harmonically related spectral peaks. To avoid FFT artefacts he applied record periodisation which eliminates record end effects but preserves the peak positions. This procedure showed that the derived peaks are not artefacts of record length. Odd harmonics of a fundamental period applied until the 8th harmonic. He estimated the fundamental as ~171 years which is familiar. Recall that the Neptune –Uranus conjunction period is 171.3 years while the Jupiter-Saturn-Neptune triple conjunction is 179 years. Also 9 x Cjs = 14 x Cjn = 178.8; 5 x Csn = 13 x Ju = 179.4 years; 4 x Csu = 180.1 years and 8 x 22.33 = 178.64 years.

The method has picked up the multi-conjunction period despite the short 113 year record length. The periods picked up are 171, 57, 34.2, 24.4, 21.4 years. These are clearly approximations to 178.6, 59.55, 35.73, 25.5, 22.33. 21.4 is probably a reflection of $\sqrt{(22.33 \times 19.85)} = 21.2$, an unresolved Hale double peak Note that 59.55 is 3 x 19.85, the JS conjunction period and a major Hale solar period. 35.73 is 2 x 17.9, another Hale solar spectrum period. 25.5 is another planetary conjunction period and a smaller Hale period. The solar / planetary connections could not be clearer. Patterson not knowing what we know, does further checks to eliminate a statistical fluke result. He checks the residuals from the spectral model and finds nearly white noise.

He fits a fifth order autoregressive model to the data and calculates the expected PSD. Comparison with the spectral model shows a good match. Using simulation he carries out 500 trials of the AR model. 99.4% resulted in residual variance greater than that obtained from the actual temperature data. He concludes that the probability of the spectral model producing such a performance by chance is ~8%. He fits a further AR model to the FFT model residuals and uses this with the spectral model to project 100 future paths of global temperature for a 100 year period. The forecasts begin in 2013.

The dotted line is the mean temperature anomaly for 1950 -1965. The probable mean fall from present to 2040 is - 0.9 d C with a slow recovery to 2100 AD. This is the duration of 6 sunspot cycles or ~ 3 Hale cycles. That pattern we might call a grand solar minimum.

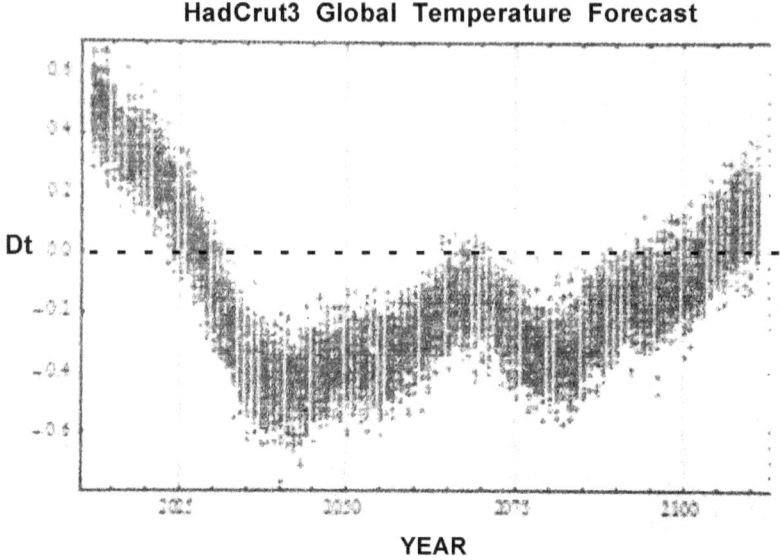

It is interesting that purely statistical models of both solar activity and temperature, based on very long data series, using a variety of established analysis techniques and expert assumptions, show similar significant *predicted declines* over this next century. The models know nothing of CO_2 climate sensitivity assumptions and the IPCC 'concensus' yet they explain much joint behaviour both ancient and modern. More work like this should clarify the CO_2 question once and for all.

5.2 USA Tree Ring Time Series

Tree ring time series, in some cases spanning centuries, are widely accepted as proxies for weather variables such as rainfall. Here we examine Colorado spruce tree ring spectra for 1673 – 1986; 1700 – 1986; 1800 – 1986; 1900 – 1986 Murphy (2). This data spans the period following the recovery of the solar cycle after the Maunder 'grand solar minimum'. The four spectra are quite consistent in terms of major peaks but resolution is lost as the data windows shorten. Five spectral peaks are significant at two standard deviations: 22.2, 14.2-15.3 for the various series lengths, 11.1, 5.3 and 3.0 years. Four peaks are significant at ~1.5 standard deviations: 30.3 –36 for the various series lengths, 8.4, 8.0, 6.0, 4.5, 3.7, 3.4, 2.5 years.

The largest peak is at 22.2 years, very close to the familiar Hale magnetic solar cycle at 22.33 years. The second harmonic at 11.1 years is also prominent in the longer data records. We also have $14.8 = 22.33 \times 2 / 3$ versus the 14.2 - 15.3 year average of 14.8; note that $Csu = 45.38$ years and $45.4 / 10 = 4.54$ and $22.33 / 5 = 4.47$ and $17.9 / 4 = 4.48$ versus the 4.5 years observed cycle. Also for the Hale third harmonic, $7.44 /(5 / 3) = 4.47$. The mean of 30.3 – 36 years is 33.3 and $22.33 \times 3 / 2 = 33.5$. Note also that $Csn = 30.3$ matching the prominent peak in the 1700 – 1986 data series. We again see Fibonacci harmonic ratios. The spectral match is very good.

As expected we also see other apparent planetary signals and sunspot harmonics $2 \times 11.86 = 35.5$ versus 36; $(8 / 7) \times 29.46 = 33.7$; for Cjs note $(5 / 3) \times 19.85 = 33.1$; $(2 / 5) \times 84 = 33.6$; $165 / 5 = 33$ versus the mean peak of 33.3 years. Also $22.2 \times phi = 35.9$; $17.9 \times 2 = 35.8$; $19.85 \times 3 / 2 = 29.8$ versus 30.3; we also have the conjunction period $Csn = 30.3$.

Also of course $(9 / 8) \times 19.852 = 22.33$ and $(3 / 4) \times 29.46 = 22.1$.

$(5 / 4) \times 11.86 = 14.8$; $29.46 / 2 = 14.73$; for Cjs $(3 / 4) \times 19.852 = 14.9$ versus a mean peak period of 14.8 years. For the 15.3 year peak we have from other planetary conjunction periods, $25.5 / (5 / 3) = 15.3$; $30.3 / 2 = 15.15$; $45.38 / 3 = 15.13$. 8.3 and 8.1 year peaks appear directly in the sunspot spectra. Also $11.86 \times 2 / 3 =$

2

7.92; $19.85 \times 2 / 5 = 7.94$; $11.86 \times 5 / 7 = 8.47$; $22.2 / (13 / 8) = 8.41$. $17.9 / 3 = 5.97$; $11.86 / 2 = 5.93$; $29.46 / 5 = 5.9$; $84 / 14 = 6.0$; $(19.85 / 2) / (5 / 3) = 5.96$ versus the peak at 6.0 years. $11.86 \times (2 / 3) \times (2 / 3) = 5.27$; $84 / 16 = 5.25$ versus the 5.3 year peak.

$(3/8) \times 11.86 = 4.4$; $(3/5) \times 29.46 / 4 = 4.42$, $19.85 / (13/8)^3 = 4.6$ versus the observed 4.5 year peak.

$11.1 / 3 = 22.2 / 6 = 3.7$; $(11.86 / 2) / (8/5) = 3.71$; $(19.85 / 2) / (5/3)^2 = 3.6$

$(17.9 / 3) / (13/8) = 3.68$; $8.4 / (3/2)^2 = 3.73$.

$22.2 / (8/5)^4 = 3.39$; $11.86 \times 2 / 7 = 3.39$; $(17.9 / 2) / (13/8)^2 = 3.39$ compared with the observed 3.4 year cycle.

$11.86 / 4 = 2.97$; $29.46 / 10 = 2.95$; $19.85 / (8/5)^3 = 3.02$; $17.9 / 6 = 2.98$; $29.46 / 10 = 2.95$; $11.86 / 4 = 2.97$ versus 3.0 years observed.

$22.2 / 9 = 2.47$; $19.85 / 8 = 2.48$. We will see 2.5 years in other weather variables later. This peak is also close to the QBO cycle of 2.38 years.

In many cases we have simple harmonic links to planetary periods often involving Fibonacci ratios. We cannot say whether there is some direct planetary 'influence' on the weather and tree ring cycles or if that 'influence' comes via the planetary modulation of solar activity. In the absence of a plausible physical theory we would assume a solar effect but we must also remember the probable direct planetary forcing of Earth-Moon dynamics. We look at this below.

$(5/3) \times 18.6 = 31$; $4 \times 8.85 = 35.4$; compared with the 33 year peak for the full 1673 – 1986 tree ring series. $18.6 \times (8/5) = 29.8$; $8.85 \times (3/2)^3 = 29.9$ compared with 30.3 for the 1700 – 1986 series.

$(6/5) \times 18.6 = 22.33$; $(5/2) \times 8.85 = 22.1$ versus the main, 22.2 year tree ring peak.

$(4/5) \times 18.6 = 14.9$; $(5/3) \times 8.85 = 14.75$; $1/8.85 - 1/22.2 = 1/14.72$ compared with the mean peak at 14.8 years. $18.6 \times 5/6 = 15.5$; $8.85 \times 8/5 = 14.16$;

$18.6 \times (2/\text{phi})^2 = 14.21$ compared with 14.2 for the period 1700 – 1986.

$1/8.85 - 1/18.6 = 1/16.88$ and $16.88/2 = 8.44$ compared with the observed 8.4.

$18.6 / (3/2)^2 = 8.27$. $18.6 / 3 = 6.2$; $(2/3) \times 8.85 = 5.9$;

$1/8.85 + 1/18.6 = 1/6.0$ versus the observed 6.0 year peak.

$(3/5) \times 8.85 = 5.31$; $(2/7) \times 18.6 = 5.32$; $1/8.85 + 1/11.86 = 1/5.1$ versus the peak at 5.3 years.

$8.85/2 = 4.43$ compared with 4.5 and $18.6/(8/5)^3 = 4.54$.

$18.6/5 = 3.72$; $8.85/(12/5) = 3.69$ compared with 3.7 years for tree rings.

$8.85/(13/8)^2 = 3.39$ compared with 3.4.

$18.6/6 = 3.1$; $8.85/3 = 2.95$ versus 3.0 years.

$18.6/(5/3)^4 = 2.43$; $8.85 \times 2/7 = 2.53$ versus the 2.5 year peak.

Simple rational fraction harmonics of the apsides and nodal cycles fit the spectrum well including several marking Fibonacci series ratios again. Some large peaks can also be seen to match interactions between the 8.85 year apsides cycle, the main 22.33 solar cycle and the Jupiter orbital period. The peak at 6.0 years matches an apsides / nodal cycle interaction as does the 8.4 year peak. As expected these related 'effects' are hard to unravel but everything observed seems to be traceable back to planetary forcing of some kind .

Summary

The tree ring spectrum is most simply interpreted in terms of rational fraction harmonics of the primary Hale cycle period of 22.33 years. However planetary period harmonics also fit well. We are confounded by the apparent relationship between solar activity and planetary effects since for example, $19.859 \times 9/8 = 22.34$ or $1/19.859 - 1/178.8 = 1/22.34$. The lunar cycle fit is also good and the large spectral peaks correlate with interactions between the apsides and Hale cycle or the apsides and Jupiter orbital period. We also noted that $18.6 \times 6/5 = 22.32$ and $8.85 \times 5/2 = 22.13$.

Despite this numerical confounding it is clear that the orbital dynamics of the gas giants are modulating solar activity, lunar orbital parameters and some terrestrial botanical and hence weather variables. This is worth exploring further.

5.3 The North Atlantic Oscillation

The NAO cycle in surface temperature is a major feature of Atlantic weather. Landscheidt (3) has analysed NAO data from 1825 to 2000 using 176 data points. He attempts to relate solar system centre of mass movements and the torque on the Sun and solar activity to the NAO series by inspection. He also carried out a MESA analysis on the data, noting a 13.3 year cycle. He notes that the rate of change of torque, dT / dt series generated from planetary movements over many centuries has a mean period of 13.3 years but a range of 9 to 14 years. However considering his NAO spectrum provides much more evidence for solar links. It turns out that planetary periods are also very clear in the NAO spectrum. Clear, strong peaks are seen at: 32, 13.3, 7.8, 5.1, 4 , 3.5, 2.94, 2.8 and 2.27 years.

The **32** year peak is broad. We note that for the Wolf sunspot series the main period is 11.16 and 11.16 x 3 = 33.4 years ; we also have the planetary conjunction period at 30.23^2 years and Ps = 29.46 ; Cjs gives, 19.85 x (21 / 13) = 32.1 ; 11.86 x (5 / 3) = 32.7 and 8.1 x 4 = 32.4.

Since **13.3** is Landscheidt's mean period for the dT / dt torque series it is worth a closer look. We note immediately that 22.33 / (5 / 3) = 13.39 and 8.1 x 5 / 3 = 13.4 and for Cjs, and the 9.9 year peak in Wolf sunspots, (19.85 / 2) x 4 / 3 = 13.24 ;

$8.4 \times (8/5)^2 = 13.4$; 29.46 / (3 / 2) = 13.1 ; 17.9 / (4 / 3) = 13.42.

We can see clear and precise links to the sunspot number series and the Hale magnetic series spectra. Since Landscheidt's torque proposal is about swinging the Sun about the solar system barycentre, the effect of planetary conjunction periods should be visible as the planets line up to swing the Sun around. Do the six pairwise and three triple conjunction periods show up in the NAO spectrum? They do. The most powerful pairwise conjunction cycle is Cjs at 19.85 years. We already noted that 19.85 / (3 / 2) = 13.23 matching the primary NAO spectral peak. The conjunction periods 12.8 and 13.8 straddle the observed 13.3 year peak of some width, and we note that their harmonic mean is √ (12.8 x 13.8) = 13.3. The conjunction period 30.23 years is also present as $30.23 / (3/2)^2 = 13.42$; similarly $45.38 / (3/2)^3 = 13.4$; $55.7 / (21/13)^3 = 13.22$.

All six pairwise planetary conjunctions are present in the spectrum as Fibonacci ratio harmonics.

The triple conjunctions have periods of ~171 to ~180 years. For Cjsn, the most powerful triple we have $(180/2)/(21/13)^4 = 13.22$ or $(180/3)/(5/3)^3 = 13.12$.

The shorter periods in the spectrum are 7.8, 5.1, 4, 3.5, 2.94, 2.8, 2.23. Some clearly relate back to the main **13.3** year peak : $5.1 \times (21/13)^2 = 13.31$; $4 \times (3/2)^3 = 13.5$; $2.94 \times (5/3)^3 = 13.4$; $2.8 \times (5/3)^3 = 12.96$; $2.22 \times 6 = 13.35$.

Also for the large **7.8** year peak we note that Pj, $11.86/(3/2) = 7.91$; $19.85/(8/5)^2 = 7.75$
$17.9/(3/2)^2 = 7.94$. The conjunction periods are prominent again
$12.8/(5/3) = 7.7$; $13.8/(4/3) = 7.77$; $30.23/4 = 7.56$; $55.7/7 = 7.95$
For the **5.1** year peak we see : $11.86/(3/2)^2 = 5.26$; $22.33/(13/8)^3 = 5.2$ or
$11.16/(3/2)^2 = 5$; $8.44/(5/3) = 5.08$. Considering the conjunctions :
$30.23/6 = 5.06$; $45.4/9 = 5.04$; $55.7/11 = 5.06$; $12.8/(5/2) = 5.12$; $19.85/4 = 4.96$.

For **4.0** we see : for Cjs, $19.85/5 = 3.97$; $11.16/(5/3)^2 = 4.02$; $11.86/3 = 3.95$;
$8.1/2 = 4.05$; $(12.8/2)/(8/5) = 4.0$; $13.8/(3/2)^3 = 4.08$; $55.7/14 = 3.98$
$30.23/(5/3)^4 = 3.95$; $45.4/(13/8)^5 = 4.01$.
For **3.5** years : $22.33/(8/5)^4 = 3.4$; $11.86/(3/2)^3 = 3.51$; $8.1/(3/2)^2 = 3.6$
$13.8/4 = 3.45$; $(19.85/2)/(5/3)^2 = 3.56$; $45.4/4 = 11.35$ and $11.35/$

(2 x phi) = 3.51.

For **2.94** years : 11.86 / 4 = 2.96 ; 29.46 / 10 = 2.946 ; 19.85 / (8 / 5)^4 = 3.02
12.8 / (13 / 8)^3 = 2.98 ; 13.8 / (5 / 3)^3 = 2.98 ; 30.23 / 10 = 3.02 .

For **2.8** years : 22.33 / 8 = 11.16 / 4 = 2.79 ; 8.43 / 3 = 2.81 ; 19.85 / 7 = 2.83
12.8 / (5 / 3)^3 = 2.77 ; 13.8 / 5 = 2.76 ; 30.23 / 11 = 2.76 ; 45.4 / 16 = 2.83 ; 55.7 / 20 = 2.79 ; 25.5 / 9 = 2.83.

For **2.23** years : 22.33 / 10 = 11.16 / 5 = 2.23 ; 19.85 / 9 = 2.21 ; 29.46 / 13 = 2.27 ;

17.9 / 8 = 2.24 ; 13.8 / 6 = 2.3 ; 45.4 / 20 = 2.27 ; 55.7 / 25 = 2.23.

This peak is also close to the QBO period of 2.37 years. Notice also that 13.3 / 6 = 2.22 years so the shortest period in NAO is a simple harmonic of the 13.3 year cycle which Landscheidt attributes to the dT / dt barycentre – Sun torque cycle.

It is interesting that all the NAO spectral peaks are harmonically related, often simply, to the planetary pairwise or triple conjunction periods. This is particularly so for the shorter periods 5.1, 2.8 and 2.23 years. It is tempting to conclude that Landscheidt is correct and that the solar dynamo or the process by which magnetic flux reaches the solar surface is modulated by a planetary torque effect. The matter might have been settled in the mid 1980s if Landscheidt had provided a high resolution spectrum for the dT / dt series, which he computed from planetary cycles for a window of 7,600 years, for comparison with NAO and other weather series. For longer solar activity cycles this was not done until 2012 as we discussed in section 3.6.

Even so it is remarkable that planetary conjunction periods show up in the NAO series having been 'processed' through the solar activity generator in the Sun and then through Earth's weather system, since both systems are clearly strongly non-linear. It is little wonder that simple cross-correlation or regression of solar activity and weather variables often fail to produce strong and time consistent links in models. Yet we see repeatedly that much of the spectral power in weather time series is related to solar cycles and their harmonics and thence to planetary periods.

Sceptics may only be convinced if physically plausible but hopefully simple, non-linear models can be formulated which can reproduce the harmonic structure in weather variables in detail (see Appendix 1). However with non-linear systems operating at the 'edge of chaos', or indeed if properly described by strange attractors, this may be very difficult to do. Even with an accurate non-linear model it may also be that prediction beyond a short term horizon is intrinsically impossible.

The NAO has a further significance to the climate change debate. The failure of the medium – long term climate models, favoured by the IPCC, to predict the current 18 year halt in 'global warming' has resulted in a belated appeal to 'natural' cycles leading to global cooling: cycles *not* included in the models even though such cyclical variations in temperature, for example, are larger than the 'warming trends' generated by the current climate models. The NAO is one such 'natural cycle' adduced by the CO_2 brigade and El Nino is another. It is ironic then that we have shown the dominance of solar activity cycles and planetary cycles in the NAO spectrum. We will do the same for the AMO and El Nino and the Pacific SOI, later. But of course the climate modeller's KNOW that the Sun is unimportant! We will explore all these issues in the last section of the book.

To be fair not all the climate modellers are evidence immune. At the UN's World Climate Conference in Geneva in September 2009 a few brave souls addressed the failure of the models (4). Mojib Latif proposed that the NAO and AMO were probably responsible for some of the strong warming of the previous decades. He also noted that the NAO was moving into a phase that will cool the planet. He attributed the greening of the Sahel after the droughts of the 1970s and 1980s to the NAO.

Vicky Pope of the UK Met Office also warned the conference that some of the dramatic Arctic sea ice loss in recent years was also down to the natural cycles, *not* man made warming. James Murphy also of the Met Office agreed and also fingered the AMO as responsible for Atlantic hurricane and Indian monsoon variability. The author suspects that there are many climatologists who are not fans of the IPCC political juggernaut but fear to speak up…it would be detrimental to funding applications! (Soon it may also be illegal to share such views in public! See App. 3).

5.4 The Atlantic Meridional (Multi-decadal) Oscillation

The AMO is associated with the intensity of storminess in the North Atlantic and the drought pattern in the Sahel region of Africa. As such it is of some interest to the current discussions about climate change. The figure below shows the AMO for the period 1856 to 2013 derived from de-trended values of North Atlantic sea surface temperature. Note that the temperature swings are around +/- 0.4 degrees C so the cycle amplitude is ~0.8 degrees, larger than the warming over the last century or so. Understanding the AMO is therefore of some importance. Recently McCarthy et al (5), by comparing the AMO index with sea level variation along the US east coast, have shown that the AMO is linked to ocean circulation. The authors tell us further that the Atlantic 'overturning circulation' is declining and as a result the AMO index will enter a negative phase with cooling surface waters. This is interesting since global surface temperature has not increased for ~18 years according to satellite data. Is a general cooling now on the way? Remember this was also predicted for the Tibet region from historic tree ring analysis.

Looking at the recent AMO cycles we can get a rough estimate of period by checking the crossover points for the three complete half cycles in the figure. We get an equivalent period length of 62 to 75 years with a mean value of ~66.5 years over 1896 to 1997. Note that 66.5 / 3 = 22.2 but the Hale solar activity cycle is 22.3 years. The prevailing explanation for the AMO is that it is an unforced, 'natural' oscillation intrinsic to the ocean. The possible role of the Sun has been proposed from time to time and rejected for reasons that are unclear to the author. Could it be that the 'Sun' is simply not politically correct and potentially prejudicial to an academic career? See what you think.

To begin recall that the Hale magnetic activity cycle contains major spectral peaks at 17.9, 19.85, 22.3, 25 years. We also noted the strong 3rd harmonic at 7.4 years and the cluster of sub-harmonics of ~60 – 70 years. We have 3 x 19.85 = 59.6; 3 x 22.3 = 66.9 ; 3 x 25 = 75, while the AMO spectra show peaks at a mean of ~66 years. This is suggestive. Several spectra are now available for areas surrounding the North and South Atlantic involving very long AMO proxy time series. These spectra between them cover the range from today back to 19,000 BP and so encompass the last ice age and the recovery into this interglacial. We might expect to see interesting changes to the AMO over this tumultuous series of events since changes in ocean circulation are claimed to be important in ice age modulation.

We will look for any pattern in the AMO spectra which might help us to choose between intrinsic and forced cycle behaviour.

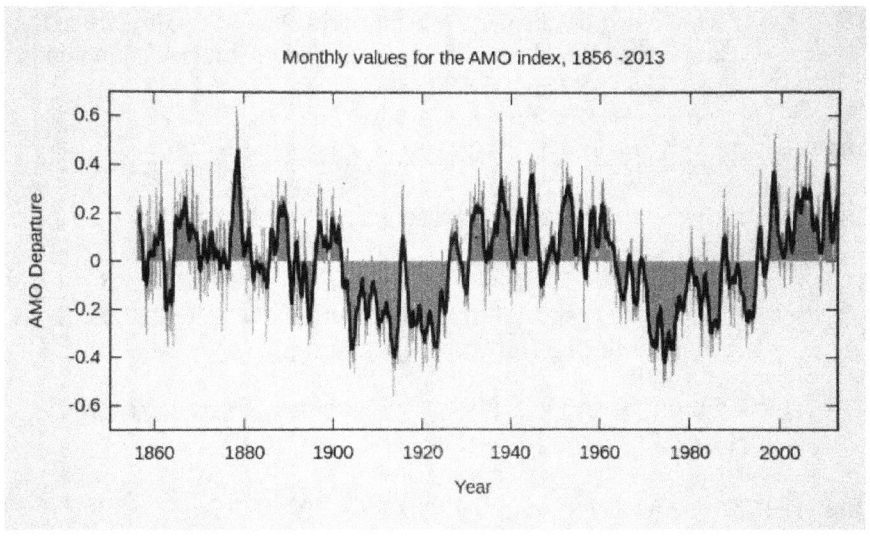

Chiessi et al (6) present Lomb-Scargle periodograms, which can accommodate unevenly spaced data, for two AMO proxy series taken at the DeoB6211-2 site in the south Atlantic off the River Plate region. One uses Delta oxygen 18 levels from the planktonic foraminifera Gloigerinoides ruber in sea bed cores. The second uses X ray fluorescence scans to determine titanium intensities in cores as a proxy for fluvial sedimentation input. These cores cover a period from 13,000-19,000 BP during the later phase of the last ice age until just before the sudden great thaw. The derived AMO spectra were similar in many respects but not identical, differing in enlightening ways which help confirm the solar influence.

G. ruber Delta Oxygen 18 from Site GeoB6211-2

The spectrum covers periods down to 60 years. The largest spectral peak (above 99% significance) was centred at 62.7 years but its width covered a range of a few years. A second large peak (95% significance) was centred at 159 years. Several smaller peaks were also present and above the equivalent red spectrum computed level. Remember we are looking for strong Hale periods such as 17.9, 19.85, 22.3, 25 and their harmonics.

Mean 62.7 +/- 2 years : 3 x 19.85 = 59.6; 3 ; 17.9 x 7 / 2 = 25 x 5 / 2 = 62.5; 3 x 22.3 = 66.9 and (59.6 + 66.9) / 2 = 63.2

67.3 yrs : 3 x 22.3 = 66.9
89.2 yrs : 4 x 22.3 = 89.2 ; 5 x 17.9 = 89.5.

> Over some periods the ~90 year period shows up directly in the Hale series. Recall also the various triple planetary conjunction periods at ~180 and 180 / 2 = 90.

101 yrs : 11.15 x 9 = 100.4 ; 19.85 x 5 = 99.3; 25 x 4 = 100.

159 yrs : 17.9 x 9 = 161, 22.3 x 7 = 156.3 ; 19.85 x 8 = 158.8.

182 yrs : 3 x 59.6 = 9 x 19.85 = 178.7 ; 17.9 x 10 = 179 ; 22.3 x 8 = 178.4
Also we have several triple conjunction multiples of J,S,U,N with periods of ~180 years.

215 yrs : 17.9 x 12 = 214.8 ; 11.16 x 19 = 212; 19.85 x 11 = 218.3 ; 159 x 4 / 3 = 212.

304 yrs : 59.6 x 5 = 19.85 x 15 = 17.9 x 17 = 304.3 ; 25 x 12 = 300 ; 182 x 5 / 3 = 303.3.

Each of the AMO proxy peaks corresponds very closely to 2 and usually 3 or 4 sub-harmonics of the main Hale solar activity cycles. The probability of this occurring by chance is very low and the solar forcing signal is strong, accounting for much of the spectral power.

Titanium Intensities at Site GeoB611-2

The Ti spectrum shares some of the above spectral peaks but also other peaks which are nevertheless simply solar related. In this case three peaks exceed 95% significance at 65 – 67, 266 and 354 years. Several other peaks are above the equivalent red spectrum background level.

60.3 yrs : 19.85 x 3 = 59.6.

Broad double peak 65 – 67 yrs : 3 x 22.3 = 66.9; 8.3 x 8 = 66.4

77.5 yrs : 11.15 x 7 = 78 ; 25 x 3 = 75.

110 yrs : 22.3 x 5 = 111.5 ; 9.93 x 11 = 109.6 ; 66.9 x 5 / 3 = 111.3

134 yrs : 66.9 x 2 = 22.3 x 6 = 133; 88.7 x 3 / 2 = 133.

182 yrs : 22.3 x 8 = 89.2 x 2 = 178.6 ; 19.85 x 9 = 178.7;

60.3 x 3 = 180.9.

266 yrs : 89.2 x 3 = 22.3 x 12 = 267.4 ; 17.9 x 15 = 268.5 ; 134 x 2 = 268.

354 yrs : 89.2 x 4 = 22.3 x 16 = 356.8; 25 x 14 = 350; 59.6 x 6 = 357.6; 266 x 4 / 3 = 354.6

489 yrs : 22.3 x 22 = 490.6 ; 17.9 x 27 = 483.3 ; 178 x (5 / 3)^2 = 492 ; 60.3 x 8 = 482.6.

Each of the AMO spectral peaks again corresponds very closely to harmonics of the Hale series and many Ti peaks are related via Fibonacci ratio harmonics. Comparing the G. rubber and Ti spectra we see that they are also harmonically related : 159 x 3^2 = 478; 304 x 8 / 5 = 486.4 ; 215 x (3 / 2) = 483.8; 215 = 8 / 5 x 134; 89.2 x 3 = 266; 89.2 x 4 = 350; both spectra have 182 year peaks. Fibonacci ratios are common again.

The two spectra differ but only in that spectral energy is moving between various harmonics of Hale solar activity cycle periods. The pattern could not be clearer. It is surprising that the authors, having done careful work to produce the spectra, do not explore harmonic relationships to try to unwrap possible underlying common cycles. But perhaps the authors accepted the 'consensus' view that the AMO is an unforced oscillation? It is interesting how often spectra which scream out 'Sun' are talked away and discounted.

We also have the long AMO proxy records for the North Atlantic prepared by Knudsen et al (8). The authors produced evolutive spectra for five sites surrounding the Atlantic Ocean: three in the arctic region and two in the tropics, based on records from today back to ~8,000 BP. These spectra are worth a close look and teach several lessons.

Agassiz Ice Cap Delta Oxygen 18 Record

The evolutive spectrum records all peaks at significance levels above 99% in the period range 50 to 100 years, the 'accepted' AMO frequency 'zone' which is a pity given that we know from the NAO that longer periods exist.

50 yrs : 25 x 2 = 50 ; 9.93 x 5 = 49.7; moderate SS no. peak at 52 years. Cycle duration in the spectrum : 1500 – 1800 BP.

58 – 60 yrs : 19.85 x 3 = 59.6 , 8.3 x 7 = 58.1.

Cycle prominent in : 1000-1700; 2700-3500; 4600-4800; 6600- 6900 BP.

66 – 68 yrs : 22.3 x 3 = 11.15 x 6 = 7.43 x 9 = 66.9 ; 8.3 x 8 = 66.4.
Cycle prominent : 1900-3200; 6000-8000 BP.

75 yrs : 25 x 3 = 75 ; 11.15 x 7 = 77.8 ; 7.43 x 10 = 74.3
Cycle prominent : 4600-5000 BP.

80 – 83 yrs : 19.85 x 4 = 79.4 ; 7.43 x 11 = 81.7 ; 8.3 x 10 = 83.
Cycle prominent : 5800-67000 BP.

~88 yrs : 22.3 x 4 = 7.43 x 12 = 89.2 ; 178 / 2 = 89 ; 17.9 x 5 = 89.5.
Cycle prominent : 4600-5500 BP.

~92 yrs : 8.3 x 11 = 91.3. Cycle prominent : 5700-6700 BP.

95 – 100 Yrs : 11.15 x 9 = 100.3 ; 19.85 x 5 = 99.3 ; 25 x 4 = 100 ;
8.3 x 12 = 99.6 ; 7.43 x 13 = 96.5.
Cycle prominent : 700-1100 ; 2800-3600 BP.

For ~5500 years out of 8000 (or 69%) years most of the spectral power was readily linked to harmonics of solar activity periods notably 59, 67, 89, 100 plus the latter's split sidebands at ~82 and ~92 years. So we have mainly power at 5, 6, 7, 8, 9 times the primary 'sun spot' cycle of 11.15 years. There are also hints of the harmonics of other strong periods like 7.43, 8.3 and ~25 years. Typically the AMO spectral periods remain strong for several hundred to two thousand years but sometimes with a slow drift in frequency and finally a larger shift to another mode. This looks like occasional shifts between phase locked states with some frequency modulation thrown in: the Atlantic Ocean is responding in a non-linear way to solar forcing over thousands of years.

NGRIP Ice Core Delta Oxygen 18

All spectral peaks of 95% significance or above are considered here.

50 yrs : 25 x 2 = 12.5 x 4 = 9.93 x 5 = 49.7; moderate SS No.
peak at 52 years.
Cycle prominent : 300-500 BP.

54-56 yrs : 11.15 x 5 = 55.7; 17.9 x 3 = 53.7
Cycle prominent : 2500-3700 BP.

60 yrs : 19.85 x 3 = 59.6; 7.43 x 8 = 59.4.

Cycle prominent : 6800-7800 BP.

64-66 yrs : 22.3 x 3 = 66.9 ; 8.3 x 8 = 66.4 ; 12.8 x 5 = 64.
Cycle prominent : 5000-8000 BP.

69-72 yrs : 9.93 x 7 = 69.5 ; 17.9 x 4 = 71.6.
Cycle prominent : 400-2000; 4500-5000 BP.

78 yrs : 11.15 x 7 = 78.1; 19.85 x 4 = 79.4; 12.8 x 6 = 76.8
Cycle prominent : 3200-4200, 5100-5600 BP

82 yrs : 7.43 x 11 = 81.7 ; 8.3 x 10 = 83.
Cycle prominent : 2800-3300 BP.

88 yrs : 22.3 x 4 = 11.15 x 8 = 89.2; 17.9 x 5 = 89.5
Cycle prominent : 6100-6600 BP.

For 5000 of the 8000 year record (or 62%) a strong cycle of 64-69 years persisted which matches the 3rd sub-harmonic of the Hale cycle of 22.3 years, that is, 66.9 years but with additional nearby peaks at multiples of 17.9, 9.93, 8.3 years. A period of ~78 years persisted for 1,600 years, ie the 7th sub-harmonic of 11.15 at 78.1 years and the 4th sub-harmonic of 17.9 years at 79.4 years. Cycles coexisted at ~55 and 82 years (~2500 – 3700 BP) which could be seen as sidebands since √ (55 x 82) = 67.1 compared with the strong 64 – 69 AMO cycle and the 66.9 year solar harmonic. This cycle was absent when the sidebands existed…the power shifted suddenly in frequency. The modulating period to create these sidebands would need to be ~ 330 years which compares with the 304 and 354 years AMO cycles found in the South Atlantic. Note that √ (304 x 354) = 328 years. There is a consistent pattern here of non-linear AMO solar forcing.

The cycles related to solar activity harmonics persisted overall, for 6600 years across the ~8000 year record (or 82%).

GISP2 Ice Core Delta Oxygen 18

50 yrs : 25 x 2 ; 9.93 x 5 = 49.7 ; 8.3 x 6 = 49.8 : moderate SS No. peak at 52 years.
Cycle prominent : 3900-4500 BP.

60 yrs : 19.85 x 3 = 59.6 ; 7.43 x 8 = 59.4 ; 180 / 3 = 60.
Cycle prominent : 2000-3000; 4300-5100 BP.

62-63 yrs : 12.5 x 5 = 62.5; 17.9 x 7 / 2 = 62.7 ; 8.1 x 8 = 64.8.

Cycle prominent : 5500-7500 BP.

66-67 yrs : 22.3 x 3 = 11.15 x 6 = 7.43 x 9 = 66.9 ; 8.3 x 8 = 66.4
Cycle prominent : 300-1100 BP.

70-71 yrs : 17.9 x 4 = 71.6.
Cycle prominent : 2000-3900; 7000-7500 BP.

75-77 yrs : 11.15 x 7 = 77.9 ; 25 x 3 = 12.5 x 6 = 75; 7.43 x 10 = 74.3 ; 8.3 x 9 = 74.7
Cycle prominent : 3000-3900 ; 5000-5800 BP.

87-90 yrs : 22.3 x 4 = 11.15 x 8 = 89.2 ; 9.93 x 9 = 89.4; 17.9 x 5 = 89.5 ; 7.43 x 12 = 89.4 ; 12.5 x 7 = 87.5 , 180 / 2 = 90.
Cycle prominent : 5700-7000 BP.

In this case the ~89 years 'Gleissberg Cycle' is clear and persists for 1,300 years. It is harmonically related to several strong periods in the sunspot number and Hale magnetic spectra. Again 6, 7, 8 multiples of the main Hale 22.3 year cycle are clear. The ~60 year cycle is long lived, nearly two thousand years, and is well known in the Hale spectrum as 3 x 19.85 years, the Jupiter – Saturn conjunction period. Overall solar related cycles are prominent for ~6300 years out of 8000 (or 79%). The long term persistence and dominance of particular subsets of solar periods suggests the potential for making decadal or longer forecasts of sub-regional AMO…providing we can model the forcing process.

Lake Chichancanab Delta Oxygen 18

All peaks of 90% significance and above are considered here. The spectrum is simpler than the others.

53 yrs : 17.9 x 3 = 53.7; moderate SS No. peak at 52 years.
Cycle prominent : 6300-6700 BP.

56-58 yrs : 11.15 x 5 = 55.8 ; 8.3 x 7 = 58.1
Cycle prominent : 2900-5700 BP.

60-61 yrs : 19.85 x 3 = 59.6 ; 7.43 x 8 = 59.4
Cycle prominent : 1500-3000 BP.

78-82 yrs : 11.15 x 7 = 78.1; 9.93 x 8 = 19.85 x 4 = 79.4 ; 8.3 x 10 = 83
Cycle prominent : 400-600; 6600-7200 BP.

The 56 – 61 year cycle persisted for ~4,000 years and appears to have shifted slowly between 55.8, the 5th sub-harmonic of the main sunspot cycle period of 11.15 years and 59.6, the 3rd sub-harmonic of 19.85 years from the Hale cycle, over this long interval. The strong ~79 year period appears twice for several hundred years near the beginning and end of the record but disappears for 6000 years! Probable solar related cycles occupied 5300 years of the 7200 year record in this case (or 74%).

Cariaco Basin Ti %

50 yrs :   25 x 2 = 12.5 x 4 = 50 ; 9,93 x 5 = 49.7; moderate SS No. peak at 52 years.
Cycle prominent : 6600-7500 BP.

53-54 yrs :   17.9 x 3 = 53.7 ; Cycle prominent : 3100-4100 BP.

56 yrs :   11.15 x 5 = 55.8 ; 8.3 x 7 = 58.1 .
Cycle prominent : 4200-5000 BP.

60 yrs :   19.85 x 3 = 59.6 ; 7.43 x 8 = 59.4.
Cycle prominent : 4500-6300 BP.

66-67 yrs :   11.15 x 6 = 22.3 x 3 = 66.9 ; 8.3 x 8 = 66.4
Cycle prominent : 4000-4300 ; 6800-7900 BP.

75 yrs :   25 x 3 = 75 ; 7.43 x 10 = 74.3 ; 8.3 x 9 = 74.7
Cycle prominent : 1000-200 BP.

79-80 yrs :   11.15 x 7 = 78.1 ; 19.85 x 4 = 79.4
Cycle prominent : 300-600 ; 1400-2000; 7500-7900 BP.

Broad band of spectral energy between **80-100** years over the period 300 to 600 BP. Also a band at **~88 years** close to the Geissberg Cycle of 89.2 (or 22.3 x 4) years, over the period 5700- 6800 BP.

90-96 yrs :   a strong wide band lasting from 7500-8000 BP.
Also a 96 year cycle from 300-900 BP. We have 8.3 x 11 = 91.3 and 7.43 x 13 = 96.6 years but no major solar periods involved.

99-101 yrs :   11.15 x 9 = 100.3 ; 19.85 x 5 = 99.3 ; 8.3 x 12 = 99.6
Cycle prominent : 300-600; 1800-2300 BP.

Again we have spectral power at 5, 6, 7, 8, 9 times the basic solar 11.15 year cycle. The 60 years, 3 x 19.85, cycle persists for 1,800 years. Lesser solar peaks explain the other AMO periods such as 50, 54, 75 and 96 years. Overall probable solar forced periods occupy ~6300 years of the 7700 year record (or 82%).

Summary

Across the five, arctic to tropical AMO proxy series, 'solar' direct or harmonic periods are present in the spectra for 63 – 82 % of the record length over the last 7200 – 8000 years. The average is 76%. However the authors restricted the spectra to 50-100 year periods because this is the 'accepted' AMO range. This is a pity. Given the observed shifts in spectral energy across the 50 – 100 year window one wonders if, for the 24% of the time when solar signals were 'absent', power had simply shifted beyond the period window studied. We noted for example that the South Atlantic AMO proxy spectra had much activity at periods greater than 100 years, easily related to solar harmonics: 159 or 17.9 x 9 and 19.85 x 8; 182 or 59.6 x 3 and 89.2 x 2; 215 or 17.9 x 12 ; 304 or 59.6 x 5 and 17.9 x 17 ; 111 or 22.3 x 5 ; 134 or 66.9 x 3 and 22.3 x 6; 266 or 89.2 x 3 and 66.9 x 4 ; 354 or 89.2 x 4 and 17.9 x 20 and 59.6 x 6 ; 488 or 59.6 x 8 and 22.3 x 22.

Across seven proxy AMO series covering several areas around the Atlantic Ocean from the present back to 19,000 BP, we have found strong and consistent evidence of harmonics of the largest solar activity periods in sunspot number and the Hale spectra. Periods like ~60, 67, 89 and occasionally 75-78 years and their harmonics dominate the AMO spectra with cycle periods up to ~490 years in the South Atlantic. The 19.85 year Jupiter – Saturn conjunction period is also prominent in sub-harmonics like ~60 years. The five North Atlantic evolutive spectra demonstrate again the climate system's ability to respond non-linearly to solar forcing but not stably. A strong response at a particular or a few periods, may persist for several centuries and exceptionally, 3000 years, perhaps with a slow drift in period length but then, suddenly, shift to another period (also solar related). These changes in response are not apparently consistent from location to location so each area is responding specifically to the forcing. Nevertheless it would be dangerously naïve to assume the effect of the Sun 'averages' out in some way. The good news is that 'local' long lasting, stable solar signal responses suggest the potential for good 'local' forecasting once the response model is understood.

This may be very useful. In the current climate models lack of high spatial frequency, that is the use of big computation cells, is a problem.

You cannot successfully model low clouds or ocean interchanges or the effects of local topography, in a cube with 100 Km sides or even 10 Km sides. Empirical models are preferable given current limitations.

Recently direct evidence for an AMO - solar variation connection emerged from a simple regression study for the period 1780 – 1980 published in Nature (Knudsen et al, 24). The analysts reconstructed a long history of AMO from 12 tree ring series and calibrated it with instrumental Atlantic sea surface temperatures. Various solar proxies were developed from long Be10 and C14 records. Altogether nine reconstructions of solar and volcanic variations were created. All were very similar. Using solar and volcanic forcing together yielded one AMO model with R = 0.82 and p = 0.0007. The fit is visibly good over 200 years. It is also interesting, given what we have already established the clear correlation of earthquake energy release and solar cycles, that the solar and volcanic forcing series are very similar in form. The model fit post 1900, the period of supposed greenhouse gas warming, is as good as for the late 18th century – mid 19th century ...before significant CO_2 emissions growth. The fit is dominated by a 60-90 year cycle (with a small up trend of ~1/2 degree C per century) both solar – volcanic related. This fit rather suggests the role of greenhouse gases in 20th century ocean and global temperature increases has been overestimated.

New evidence for solar connections is emerging continuously. Krahenbuhl (23) has looked at the North American region in terms of the AMO, ENSO and the PDO, solar variability indicators, galactic cosmic rays (from neutron monitor time series), and cloud cover at various levels. Correlation analysis confirms that the AMO exerts the greatest control over cloud cover (and hence albedo) in the NA region. ENSO and the PDO also show significant effects. GCR are positively correlated with low maritime cloud cover which is consistent with the Svensmark cloud heresy.
In other (non-maritime) areas solar activity measures correlate most strongly with cloud cover. The results are complex and need careful study but a clear message is that the influence of the Sun is significant. Since we have shown here that the AMO itself is also strongly solar related (along with ENSO and the PDO) this suggests NA regional cloud cover *is* strongly controlled by the Sun. The author notes suggestions that long term solar activity acts as a 'chaotic attractor' for some climate modes in the ocean and atmosphere (see Appendix 1 and section 3.7).
The appearance of Fibonacci ratios and other irrational harmonics linking AMO (and ENSO and the PDO, etc, etc) and solar activity periods is compatible with such a near chaotic, non-linear system.

5.5 Wavelet Analysis of Central England Temperature

Reference 6, section 3, also provides a wavelet analysis for the classic CET time series over the period 1659 to 2007, from the last few decades of the Maunder Minimum in solar activity to the present. Wavelet analysis is another modern spectral technique yielding high resolution spectra and a natural way of tracking the time evolution of the relative importance of different cycle periods. We begin with the short periods. The mean enhanced power wavelet spectrum of CET shows prominent peaks at the odd harmonics of the Hale cycle of ~22.3 years. The highest power occurs at the highest harmonics. This is an important observation about non-linear physical systems: the strongest system response may be a long way from the basic forcing signal frequencies. If you only look for a response at the apparent main forcing frequency a great deal of behaviour will be missed. That seems to happen frequently these days. In terms of the planetary periods, similarly

	J 11.86	S 29.46	JSconj. 19.85
9th H 2.48	$1/5 = 2.4$	$1/12 = 2.46$	$1/8 = 2.48$
7th H 3.2	$(2/3)(2/5)$ $= 3.16$	$(1/2)(5/3)^3$ $= 3.19$	$(2/5)^2 = 3.18$
5th H 4.46	$3/8 = 4.45$	$(3/5)/4 = 4.43$	$(2/3)/3 = 4.41$
3rd H 7.44	$5/8 = 7.41$	$1/4 = 7.37$	$3/8 = 7.443$

The simple harmonic relationships of Jupiter and Saturn orbital periods to the strongest CET period peak is striking.

We also note a clear CET peak at ~25.2 years compared with a Hale MESA peak of
25.5 years which we parsed as $(22.1 \times 29.46)^{½} = 25.5$. The main CET peaks are simply related to the orbital periods of J and S and often in Fibonacci series ratios.

There is some sign as expected of lunar cycles including $8.85 / (5/3)^2 = 3.2$;

$18.6 \times (2/3) / 5 = 2.48$; $8.85 \times 2 / 7 = 2.52$; $8.85 / 2 = 4.43$; $18.6 \times 2 / 5 = 7.44$. We have the usual confounding of effects.

The main power peak is at 2.48 years compared with the prominent peak we note in the Southern Oscillation Index (El Nino) MESA spectrum at 2.5. We found evidence there of strong harmonic links to Pj, Ps, Cjs, Cjn, Csn and Csu (see below). We also have moderate CET peaks at 3.2 and 4.5 compared with 3.6 and 4.9 years in the SOI spectrum. Both SOI and CET are clearly linked to solar activity variation but obviously in some complex non-linear way. Reference 5 spectra also point to moderate peaks at 14.4 ; 36 ; 65 ; 102 ; ~200 years. (Recall that the Colorado tree ring spectrum contained prominent peaks at 14.2 and 36 years). In relation to the solar activity MESA spectral peaks note that $25.5 \times 4 = 101.2$; $22.33 \times 3 = 66.9$; $17.9 \times 2 = 35.8$; $22.33 \times 2 / 3 = 14.9$ years; $19.85 \times 5 = 99.3$ years. Also for the planets

Spectral peak	J 11.86	S 29.46	JSconj. 19.85
200 +/-15	$(8/5)^6 = 199$	$(13/8)^4 = 200.4$	$10 = 199$
102 +/-10	$2(13/8)^3 = 101.8$	$(3/2)^3 = 99.4$	$(3/2)^4 = 100.5$
65 +/-8	$2(13/8)^2 = 62.6$	$(1/2)(13/8)^3 = 63.2$	$2(13/8) = 64.5$
36 +/- 6	$3 = 35.6$	$(5/2)/2 = 36.8$	$2 = 39.7$
14.4 +/-1	$(3/5)2 = 14.2$	$1/2 = 14.7$	$(3/2)^2 / 2 = 14.8$
7.5 +/-1	$5/8 = 7.43$	$1/4 = 7.4$	$(8/13) = 7.5$

Given the uncertainty bounds the fit is again good and combined with the main Hale cycle harmonics analysed above it again points to planetary forcing of Earth's weather. Note for interest that $8.85 \times (5/3) / 2 = 7.4$; $8.85 \times 13 / 8 = 14.4$; $8.85 \times 4 = 35.4$; $8.85 \times (13/8)^4 = 62$; $8.85 \times (13/8)^5 = 100.2$; $8.85 \times 2 \times (13/8)^5 = 200.4$. The apsides cycle is also present in the spectrum linked by Fibonacci factors again.

Also for the nodal cycle, $(18.6/2) / (21/13)^5 = 102.3$; $18.6 \times (21/13)^5 = 204.6$, etc, etc.

Analytical techniques for non-stationary and non-linear times series are continuously evolving and fortunately the analysts are attracted to the world's longest instrumental temperature record, CET.

Yang & Zhou have recently looked at CET again using Slow Feature Analysis (SFA) and wavelet analysis to derive a driving force model from the monthly data which is non-stationary over the period 1659 to 2013.

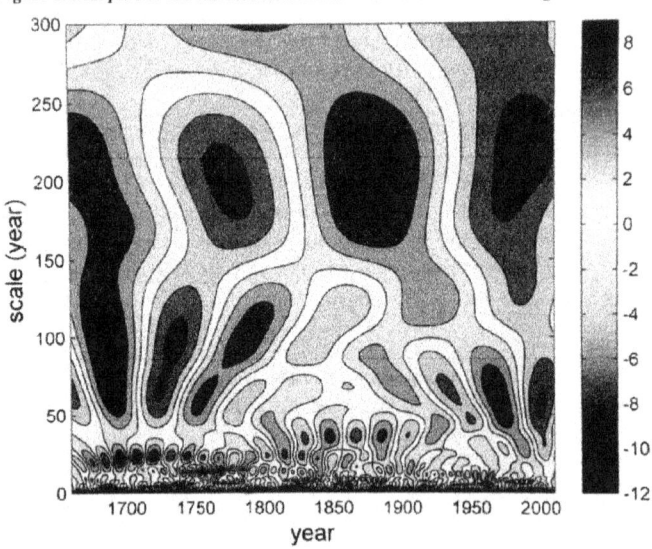

Figure 3: The time-averaged power spectrum of the driving force.

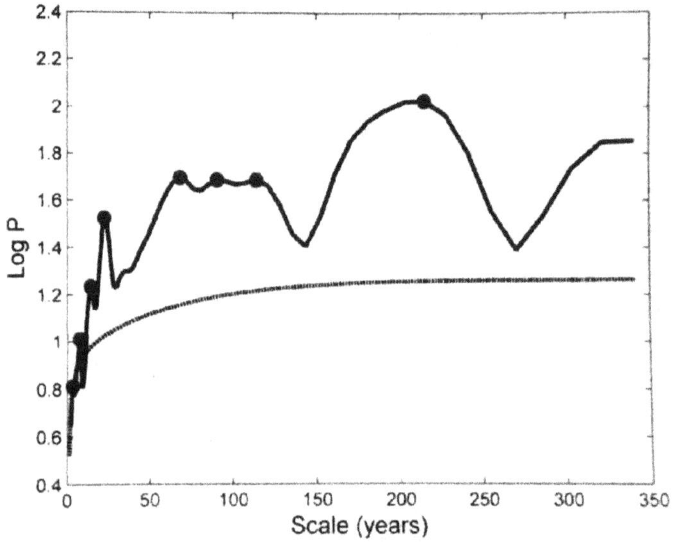

Figure 2: Real part of the wavelet transform coefficient for the driving force.

Figures 3 and 2 show the time-averaged power spectrum and the wavelet transform of the SFA reconstructed driving force above.

The time averaged spectrum is easiest to interpret. The small circles pick out key periods: 3.36, 7.5, 14.5, 22.6, 67.7, 90.4, 113.9, ~215. All are above the 95% statistical significance line marked in Figure 3. Yang picks out 22.6 as the Hales solar cycle and decides that 3.36 is related to the El Nino SO cycle in the Pacific. Some other peaks he sees as harmonics or interactions of these two periods. Let us look a bit closer as usual for underlying cycles in other phenomena.

3.36 $19.85 / 6 = 9.925 / 3 = 3.31$; $11.15 / (3/2)^3 = 3.3$; $8.85 / (13/8)^2 =$ 3.35. We can say that 3.36 may be associated with El Nino but it is also related to the 19.85 year Jupiter-Saturn conjunction which is also a Hale solar activity spectral peak.

7.5 $22.33 / 3 = 7.44$; $29.46 / 4 = 7.37$; $11.86 \times (5/8) = 7.41$; $18.6 \times (2/5) = 7.44$.

14.5 $22.33 \times (2/3) = 14.9$; $29.46 / 2 = 14.7$.

22.6 Hale cycle 22.33.

67.7 $3 \times 22.33 = 67$.

90.4 $4 \times 22.33 = 89.3$; $3 \times 29.46 = 88.4$; $11.86 \times (5/3)^4 = 91.2$; $8.85 \times 10 = 88.5$.

113.9 $5 \times 22.33 = 111.5$; $6 \times 18.6 = 111.6$; $9 \times 12.8 = 115.2$; $13 \times 8.85 = 115$. this period is strong in the first half of the record but then fades as Figure 2 shows us.

215 $19 \times 11.15 = 211.9$; $18 \times 11.86 = 213.5$; $24 \times 8.85 = 212.4$. This peak is wide reflecting a slight drift in frequency over three centuries.

The relationship to solar activity, planetary and lunar orbital dynamics is very clear in the reconstructed driving force. The detailed wavelet spectrum adds fascinating detail to the above analysis. We can see that the power centred at the mean period ~215 years is not constant but modulated in a regular modulation pattern as are some other periods. The long modulation cycles are at 140, 250, 420, 700 and ~1000 years. The first three are simply related harmonically : $420 / 140 = 3$; $700 / 140 = 5$; $700 / 420 = 5/3$; $420 / 250 = 5/3$; $1000 / 250 = 4$. We note that $140 = 7 \times 20$; $21 \times 20 = 420$, $35 \times 20 = 700$, reflecting the 19.85 year JS conjunction.

Note that 11 x 22.3 = 245.3; 14 x 17.9 = 250.6; 19 x 22.33 = 424; 21 x 19.85 = 417; 5 x 84 = 420, where 84 is the Uranus orbital period; 4 x 178 = 712.

We still seen to have links to the sun and planetary dynamics in the long modulation or beat cycles lasting for centuries. Phases between the cycles can also vary over time in occasional discontinuities such as around 1900. The origin of the longest period of ~1,000 years is not clear but Yang assigns it to a manmade greenhouse gas trend mimicking a long cycle. We could just as easily note that for the 90 year period observed, 11 x 90 = 990 years, or more precisely,11 x 89.32 = 982.5 and 4 x 22.33 = 89.32 ; 34 x 29.46 = 1002; 12 x 84 = 1008, from Uranus' 84 year period or 6 x 165 = 990, from Neptune's period. We have a near conjunction of gas giant planets with Jupiter in anti-phase to the others. In fact the Uranus-Neptune conjunction period is 171 years giving us 6 x 171 = 1026 years and this is 86.5 Jupiter periods exactly. Starting in phase, Jupiter would be in anti-phase 1026 years later. 46 x 22.33 is 1027 years.

5.6 Total Ozone Variation

Kane et al (8) report MESA analyses of total ozone for ten locations finding familiar peak periods in narrow ranges. The data records were for 1957 to 1982 for nine locations and 1932 to 1971 for Arosa in Switzerland. At Arosa a 16 year period was found in addition to those below.
The periods found were :

6 months, semi-annual ; annual.

Quasi Biennial Oscillation, QBO at **2.37** years : for PJ, $11.86 / 5 = 2.37$;
$11.1 / (5/3)^3 = 2.4$; $(19.85 / 2) / (21 / 13)^3 = 2.37$; for Ps,
$(29.46 / 3) / (21 / 13)^3 = 2.35$
$8.1 / (3/2)^4 = 2.37$; $17.9 / (5/3)^4 = 2.33$; $12.8 / (3/2)^4 = 2.5$

For **3.75** years (3.5 – 4) : $22.33 / 6 = 3.72$; $17.9 / 5 = 3.58$; $19.85 / (3/2)^4 = 3.91$
$11.86 / (3/2)^3 = 3.53$; $29.46 / (5/3)^4 = 3.82$

For **6.5** years (6 – 7) : $19.85 / 3 = 6.62$. $22.33 / (3/2)^3 = 6.62$;
$17.9 / (5/3)^2 = 6.45$

$12.8 / 2 = 6.4$.

For **10.5** years (10 – 11) : $19.85 / 2 = 9.93$; $22.33 / 2 = 11.16$;
$17.9 / (5/3)^2 = 10.8$; Wolf sunspot major peak 10.5 ;
$29.46 / (5/3) = 10.6$

For **~16** years : $19.85 / (5/4) = 15.9$; $11.86 \times (4/3) = 15.8$; conjunction, $25.5 / (8/5) = 15.94$; conjunction , $12.8 \times (5/4) = 16$; for sunspot peaks, $10.5 \times (3/2) = 15.8$; $9.4 \times (5/3) = 15.7$; $8.1 \times 2 = 16.2$

We have a mixture of solar activity and planetary harmonics and more evidence of non-linearity in the Fibonacci ratio harmonics. The QBO is also apparent.

5.7 Northern Hemisphere Mean Annual Surface Air Temperature

Schonwiese (9) organised an air temperature data series for the long period 1781 to 1980 and created a MESA spectrum with well defined, statistically significant peaks with periods from ~2.1 to ~50 years. Solar and planetary periods show up clearly across this wide frequency range.

For ~ mean **45** years (40 to 60) : for the Hale cycle, 22.33 x 2 = 44.7 ; for the Cjs, 19.85 x 2 = 39.7 ; for Pj, 11.86 x 4 = 47.5 ; for Ps, 29.46 x 2 = 58.8 ; for the Hale 17.9 year peak, 17.9 x (5 / 2) = 44.75. The peak is broad so we cannot be precise in this case.

For mean **24.7** years (21 to 30) : Ph = 22.33 ; Ps = 29.46 ; 8.1 x 3 = 24.3 planetary conjunction period C , 25.5 ; 2 x Pj = 2 x 11.86 = 23.72

For **17.3** years (16 to 18) : 17.9 Hale series solar period ; for Pj , 11.86 x (3 / 2) = 17.7 ; Wolf cycle 8.44 x 2 = 16.9 ; for Ps, 29.46 / (5 / 3) = 17.67.

For mean **12** years : Pj = 11.86 ; for Cjs, 19.85 / (5 / 3) = 11.91 ; for Ps , 29.46 / (5 / 2) = 11.8

For **8** years : Wolf sunspot cycle 8.1 ; for Cjs, 19.85 / (5 / 2) = 7.94 ; conjunction period, 12.8 / (8 / 5) = 8 ; for Pj, 11.86 / (3 / 2) = 7.91

For **7.63** years : for Ph, 22.33 / 3 = 7.44 ; for Pj, 11.86 / (8 / 5) = 7.41 ; for Cjs,
$$19.85 / (21 / 13)^2 = 7.61; \text{ for Ps }, 29.46 / 4 = 7.37$$

For **5.9** years : for Pj, 11.86 / 2 = 5.93 ; 17.9 / 3 = 5.96 ;
for Cjs , $19.85 / (3 / 2)^3 = 5.88$
$29.46 / (3 / 2)^4 = 5.83$

For **5.38** years : $22.33 / (8 / 5)^3 = 5.41$; 11.1 / 2 = 5.55 ; 8.1 / (3 / 2) = 5.4
$11.86 / (3 / 2)^2 = 5.27$; $(29.46 / 2) / (5 / 3)^2 = 5.31$

For **3.9** years : 19.85 / 5 = 3.96 ; 11.86 / 3 = 3.95 ; 11.165 / 3 = 3.74 ;
$29.46 / (5 / 3)^4 = 3.84$

For **2.57** years : $11.165 / (13 / 8)^3 = 2.6$; $17.9 / 7 = 2.56$; $19.85 / (5 / 3)^4 = 2.58$

For **2.38** years : the QBO atmospheric cycle is 2.375 years (see below) ; $11.86 / 5 = 2.37$; $(19.85 / 2) / 21 / 13) = 2.37$; $11.165 / (5 / 3)^3 = 2.393$; $17.9 / (5 / 3)^4 = 2.33$

Also the Chandler Wobble is 1.186 years and $2 \times 1.186 = 2.372$.

For **2.1 – 2.2** years $11.165 / 5 = 2.22$; $19.85 / 9 = 2.2$; $(11.86 / 2) / (5 / 3)^2 = 2.14$; Sunspot peak, $8.44 / 4 = 2.11$; Hale sunspot peak , $17.9 / 8 = 2.23$; $(29.46 / 2) / (8 / 5)^4 = 2.23$.

Although the QBO peak is small it is harmonically closely related to the larger peaks :

$2.37 \times 5 / 3 = 3.94$ versus 3.9 years in the temperature spectrum. Also $2.37 \times (3 / 2)^2 = 5.33$ versus 5.38 ; $2.37 \times 2 \times (21 / 13) = 7.66$ versus 7.63 ; $2.37 \times (3 / 2)^3 = 8.0$; $2.37 \times (3 / 2)^4 = 12.0$; $2.37 \times (5 / 3)^4 = 18.2$ and $2.37 \times (13 / 8)^4 = 16.6$ giving a harmonic mean of 17.36 versus observed 17.3 years.

How should we interpret this? Most of the temperature spectral power is in simple solar and planetary period harmonics suggesting a direction of 'influence' from solar activity to air temperature. The 'QBO' 2.37 year minor peak is an echo of the stronger solar periods. In the tropics we will see that the QBO itself is expressed more strongly in weather records but that it too is also solar related.

For interest and to confirm the small QBO period is real in NH air temperature we can also see that the QBO sideband period of 1.73 years is hidden in that spectrum since :

$2.15 / (5 / 4) = 1.72$

$2.57 / (3/2)^2 = 1.713$; $3.9 / (3/2) = 1.733$; $5.38 / 3 = 1.79$; $5.9 / (3/2)^3 = 1.747$;

$8 / (5/3)^3 = 1.73$; $12 / (13/8)^4 = 1.72$; $17.3 / 10 = 1.73$.

We again have a clear, consistent pattern of integer harmonics and Fibonacci ratio harmonics of the QBO sideband. This illustrates again the problem of indirect frequency responses in strongly non-linear systems. Often an important physical driving signal will be hidden in the systems response to its harmonics with the primary period not visible at all or much reduced.

5.8 Tropical Weather and the QBO

A prominent feature of the tropical weather system is the Quasi-Biennial Oscillation or QBO. The QBO refers to equatorial zone stratospheric winds which blow either easterly or westerly in a regular cycle of 28 to 29 months with a mean period of 29.5 months or 2.375 years. The QBO is said to originate in atmospheric waves generated in the tropical troposphere (lower atmosphere) which travel upwards into the stratosphere(10). The QBO period shows up in tropical weather records including rainfall and equatorial temperature anomalies. It is also interesting to note that 2.375 / 2 = 1.1875 years compared with the Chandler Wobble period of 1.185 years. Baldwin in his major QBO review paper (10) also notes that an ~11 year cycle is associated with the QBO in some instances. We should not be surprised by now. Can we link the QBO more formally to the solar activity cycles or indeed to the planets? We can on closer inspection. Consider first sunspot and Hale periods.

$Ph / 2 = 22.3 / 2 = 11.15$ and $11.15 / (5 / 3)^3 = 2.4$ compared with the QBO's ~2.38.

The Hale sunspot series also has a strong cycle of 17.9 years and $17.9 / (5 / 3)^4 = 2.33$

The Wolf sunspot series has a cycle of 8.1 years and $8.1 / (3 / 2)^3 = 2.4$.

The Hale series has a cycle of 19.85 years and the Wolf sunspot series one of 19.85 / 2. So we have $(19.85 / 2) / (21 / 13)^3 = 2.36$.

Note the clear presence of Fibonacci harmonics. Of course 19.85 is also Cjs, the Jupiter – Saturn conjunction period. We also see immediately that Pj, Jupiter's orbital period gives us $11.86 / 5 = 2.372$. The QBO marks the 5th harmonic of Jupiter's year in Earth's weather system. For Saturn Ps = 29.46 years and the 3rd harmonic is $29.46 / 3 = 9.82$ giving us
$9.82 / (21 / 13)^3 = 2.33$

The observed variation in the QBO is 2.33 to 2.42 years so the match is very good. Considering the Earth-Moon system as a whole we see, considering the 18.6 year nodal cycle, $(18.6/3)^2/(21/13) = 2.377$ versus 2.375.

Baldwin reports that the QBO spectrum also shows interactions between it and the Earth's orbital year. That is $1 +/- 1/2.375$ which yields periods of 1.73 and 0.74 years

It is no surprise that $11.2/(8/5)^4 = 1.71$; $19.85/(3/2)^6 = 1.74$; $11.86/(21/13)^4 = 1.74$ and $29.46/(8/5)^6 = 1.75$. The solar and planetary links are again remarkably clear.

The QBO and its sidebands are a tropical phenomenon and we might expect it to show up in low latitude weather systems as Baldwin suggested. Next we look at such systems in the Atlantic and Pacific.

5.9 QBO & QTO Signatures in Brazilian Rainfall

Kane (11) presents a comprehensive review of rainfall and zonal wind speed data for NE Brazil at several locations based on high resolution MESA analyses. They show familiar features including the QBO and other periods. We will consider the rainfall record for the coastal Fortaleza region. Twelve month running means of rainfall sampled at three month intervals were used giving four smoothed records per annum over the period 1951 – 1990. The following cycles were observed at or above the 95% significance level : 28, 11.6, 5.2, 3.5, 2.49, 2.33, 2.07 years. The first point of interest is that these periods are harmonically related.

$(28 / 2) / (5 / 3)^2 = 5.06$; $11.6 / (3 / 2)^2 = 5.16$; $5.2 / (3 / 2) = 3.47$; $5.2 / (5 / 2) = 2.08$

$3.5 / (3 / 2) = 2.333$; $11.6 / (5 / 3)^3 = 2.51$; and for a smaller peak at 4.4 years, $11.6 / (13 / 8)^2 = 4.4$.

The consistent presence of Fibonacci ratio harmonics points again to a strongly non-linear system. The period of 2.333 is close to the main QBO period of 2.37 and so the QBO is harmonically related to the other spectral peaks in rainfall. This pattern extends to the QBO- annual sideband at 1.73 years since :

$1.73 \times 2 = 3.46$ versus observed 3.5 years ; $1.73 \times 3 = 5.19$ versus 5.2.
$1.73 \times (21 / 13)^4 = 11.75$ versus 11.6 ; $1.73 \times 10 \times phi = 28$ versus 28.

Can we relate NE Brazilian rainfall to solar and planetary cycles? We can. For the 28 year cycle recall that $Ps = 29.46$ years. The broad 11.6 year peak is very close to $Pj = 11.86$ years but also covers the 11.16 year primary Wolf sunspot series cycle.

We also note that $22.3 / phi^3 = 5.26$ versus 5.2 years observed ; $11.86 / (3 / 2)^2 = 5.26$.

$8.44 / phi = 5.2$; $17.9 / (3 / 2)^3 = 5.3$.

$(11.165 / 2) / (8 / 5) = 3.49$ versus 3.5 ; $8.1 / (3 / 2)^2 = 3.6$; $17.9 / 5 = 3.57$

$19.85 / 8 = 2.48$; $22.3 / 9 = 7.44 / 3 = 2.498$ versus 2.49 observed ;

$17.9 / 7 = 2.55$

$8.43 / (3/2) = 2.5$; $(8.1/2) / (13/8) = 2.492$.

QBO = 2.37 versus 2.333 observed ; $11.86 / 5 = 2.37$; $11 / (5/3)^3 = 2.4$

$(7.44 / 2) / (8/5)^4 = 2.33$; $17.9 / (5/3)^3 = 2.32$; $8.1 / (3/2)^3 = 2.4$

$8.1 / 4 = 2.03$ versus 2.07 observed ; $8.44 / 4 = 2.11$; $11.16 / (3/2)^4 = 2.2$

For the QBO 1.73 year sideband we note :

$P_j = 11.86$ and $11.86 / (21/13)^4 = 1.74$ years ; $11.2 / (8/5)^4 = 1.71$

$8.1 / (5/3)^3 = 1.75$; the 3rd Hale harmonic is 7.44 and $7.44 / (13/8)^3 = 1.733$.

Solar and planetary harmonics and Fibonacci ratio harmonics are frequent and consistent throughout the rainfall spectrum and we can see the clear links back to the QBO and its sidebands. The QBO and NE Brazilian coastal rainfall are solar related and since the QBO is given by $11.86 / 5 = 2.372$ years, and considering the other links identified, QBO is clearly influenced by the planetary orbital periods.

5.10 El Nino & the Southern Oscillation

The El Nino warming phenomenon of the tropical Pacific has been implicated in wide ranging weather variations elsewhere so its origin is of some interest. It is also 'blamed' for the failure of the IPCC promoted climate models to forecast the current 18 year hiatus in 'global warming'. It is not a simple cycle but varies in duration from 2 to 7 years. Closely related to El Nino is the Southern Oscillation Index which measures the surface pressure difference between Darwin, Northern Australia and Tahiti in the central Pacific. A high resolution MESA spectrum of the SOI is now available and we will analyse this here (12).

The spectrum consists of a large peak at 2.5 years and five medium amplitude peaks at 13.1, 4.9, 3.6, 2.1 years. There are smaller but clear peaks at 6.7, 2.8, 1.7 and 1.5 years which are harmonically related to the larger peaks.

$13.1 / 2 = 2.5 \times 8 / 3 = 6.7 \quad 2.1 \times 2 \times (2/3) = 2.8 \quad 2.5 \times 2 / 3 = 1.67$

$2.5 \times 3 / 5 = 1.5$. Fibonacci ratios are already apparent.

The QBO period is not apparent but note that $2.375 \times 2 \times (5/3)^2 = 13.18$ and $2.375 \times 2 = 4.75$ while $2.375 \times 3 / 2 = 3.56$ compared with the large SOI 13.1, 4.9 and 3.6 year peaks.

We begin with possible solar activity links.

For 13.1 years $19.852 \times 2 / 3 = 13.2$ or $9.93 \times 4 / 3 = 13.2$ from the SS intensity spectrum and $10.51 \times 5 / 4 = 13.1$ and $9.4 \times 7 / 5 = 13.16$ and $8.13 \times (21/13) = 13.1$.

For the observed 6.7 years $19.852 / 3 = 6.62$; $9.93 \times 2 / 3 = 6.62$; $9.4 \times 5 / 7 = 6.71$
$17.88 / (13/8)^2 = 6.77$.

Also $1/2.5 - 1/2.1 = 1/13.12$ and $1/3.57 - 1/4.9 = 1/13.1$

For 4.9 years $19.852 / 4 = 9.93 / 2 = 4.96 \quad 11.16 \times (2/3)^2 = 4.96$

$8.13 / (5/3) = 4.88 \quad 7.33 \times 2 / 3 = 4.88$

For 3.6 years $22.33 / 6 = 11.16 / 3 = 7.43 / 2 = 3.71 \quad 17.9 / 5 = 3.58$

For 2.8 years   11.16 / 4 = 22.33 / 8 = 2.79   19.852 / 7 = 2.83

2.5 years   22.33 / 9 = 7.43 / 3 = 2.48   19.852 / 8 = 9.93 / 4 = 2.48

10.5 / (13 / 8) = 2.45   30.3 / 12 = 2.52   8.4 / (3 / 2)^3 = 2.49

17.9 / 7 = 2.55   25.5 / 10 = 2.55   (8.13 / 2) / (13 / 8) = 2.5.

Considering the main peaks from the solar Hale and sunspot intensity spectra.

For 2.1 years   8.43 / 4 = 2.11   10.5 / 5 = 2.1   9.4 x 2 / 9 = 2.09

For 1.7 years   8.43 / 5 = 25.5 / 15 = 1.69   22.2 / 13 = 1.71

For 1.5 years 1 / 2.5 + 1 / 3.6 = 1 / 1.48   7.43 / 5 = 1.49

All peaks can be related to the Hale magnetic cycle or the sunspot activity cycle via simple rational fractions and harmonics. The links via some non-linear forcing mechanism are very clear. Let us now explore the 'direct' planetary periods.

For 13.1 years   29.46 x (2 / 3)^2 = 13.09   11.86 x 11 / 10 = 13.05

19.859 x 2 / 3 = 13.2   and   Cjn = 12.8   Cun gives 55.7 / phi^3 = 13.14

For 6.7 years   29.46 / (13 / 8)^3 = 6.8   19.859 / 3 = 6.63

For 4.9 years   29.46 / 6 = 4.91   11.86 x 5 / 12 = 4.94   19.859 / 4 = 4.96

For 3.6 years   29.46 / 8 = 3.68   11.86 x (3 / 5) / 2 = 3.56

For 2.5 years   29.46 / 12 = 2.46   Cjs gives 19.859 / 8 = 9.93 / 4 = 2.48

11.86 / (5 / 3)^3 = 2.55   Cjn gives 12.8 / 5 = 2.56

 Csn gives 30.23 / 12 = 2.52   **Csu gives** 45.38 / 18 = 2.52

For 2.1 years   29.46 / 14 = 2.1   13.8 / (8 / 5)^4 = 2.1

For 1.7 years   11.86 / 7 = 1.694

All the peaks are simply related to the orbital periods of Jupiter and Saturn, their conjunction period and other pairwise conjunction periods. But of course the JS conjunction period also shows up strongly in the sunspot and Hale spectra.

Lastly we examine the lunar apsides and nodal cycles of the Moon.

For 13.1 years	8.85 x 3 / 2 = 13.2	18.6 x 5 / 7 = 13.3
For 6.7 years	8.85 x 3 / 4 = 6.64	18.6 / ((3 / 5) x (3 / 5)) = 6.705
For 4.9 years	8.85 / (5 / 3) x 3 = 4.92	1 / 8.85 + 1 / 11.16 = 1 / 4.93
For 3.6 years	8.85 x 2 / 5 = 3.55	18.6 / 5 = 3.57
	3	4
For 2.5 years	**8.85 x (2 / 3) = 2.62**	**18.6 x (3 / 5) = 2.41**
For 2.1 years	8.85 / 4 = 2.2	18.6 / 9 = 2.07
		4
For 1.7 years	18.6 / 11= 1.69	8.85 / (3 / 2) = 1.74
For 1.5 years	8.85 / 6 = 1.48	

There seem to be simple Fibonacci ratios linking peaks at 13.1, 6.7, 3.6 years to the 8.85 year lunar apsides cycle but also as 1 / 4 and 1 / 6 harmonics. The 4.9 peak seems to be an interaction between the apsides cycle and the 11.16 sunspot main peak but as we noted 8.85 x 5 / 4 = 11.1 and 18.6 x 3 / 5 = 11.16 years. The nodal cycle, 18.6 years, appears in simple form only as odd 1 / 5, 1 / 9 and 1 / 11 harmonics. The effects are again confounded but the solar activity and planetary links to the Southern Oscillation Index and El Nino are clear whatever the causal pathway.

As we noted earlier for the NAO, El Nino and the SOI have been 'blamed' for the failure of the IPCC climate models to predict the halt in 'global warming' despite the continued growth in man made CO_2 output into the atmosphere. These 'natural' cycles it seems have produced cooling not included in the models. But surely if their cooling effects can actually halt or reverse the dreaded CO_2 warming effect they should have been taken into account from the beginning? This perhaps tells us something in general about the IPCC approved climate models.

Our analyses here have also showed us the clear links of the SOI and the NAO 'natural cycles' to solar activity and even to planetary forcing.

Perhaps it is time for the approved climate modellers to look again at the Sun with fewer pre-conceptions? Let us help them further. We met Scarfetta in section 3.6 in relation to the analyses linking solar activity to planetary dynamics and global temperature. This group later went on to look more closely at the influence of El Nino on temperature trends, the effect of removing El Nino events from several ground and satellite global temperature records and to compare the IPPC CMIP 5 models with their own semi-empirical models based on planetary dynamics driving of temperatures. This work provides several insights. Their first action was to consider the NINO 3.4 index (east central tropical Pacific) over the period 1950 to 2017 shown below. This series was correlated with six global temperature records: NCDC, HadCrut4, GISS.250, GISS.1200, RSS, MSU(UAH). The last two are satellite record based, the others from multiple ground stations and therefore subject to sampling biases, heat island trends, and various questionable 'adjustments' again accused of biasing the record.

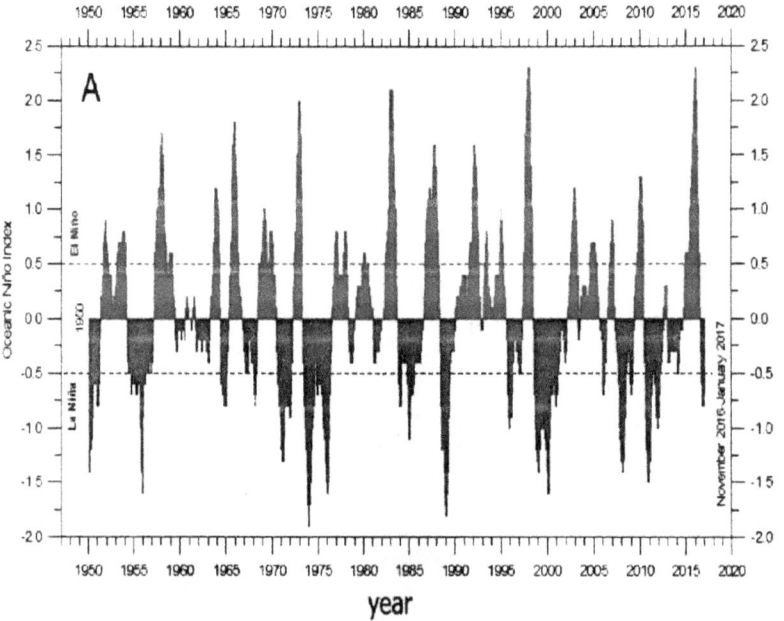

The highest correlations were found with the RSS and MSU series reaching R=0.71 for a time lag of 3 months. A range of non-linear models was then constructed of the form

$$T(t) = C + Bt + (a1.ENSO + a2.ENSO^2 + a3.ENSO^3 + a4.ENSO^4 + ...)$$

The order of the polynomial was changed until the trend coefficient b stabilised. This occurred for all temperature series for polynomial order 3. The original trends in all the temperature series were significantly reduced by adding in the El Nino series as a possible causative factor as well as removing the large El Nino spikes at particular points such as 2010 and 2016. The results for temperature series RSS and MSU are shown below. The record did not contain the second El Nino spike of 2017-18 nor the fall back in temperatures from 2019 on.

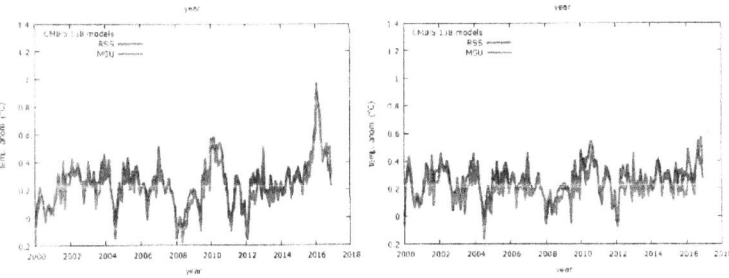

Figure 5: (left) Original temperature records (blue) against the CMIP5 mean simulations from 138 GCMs. (right) The original temperature records are filtered off of their estimated NINO3.4 signature depicted in Figure 4. respectively

For the satellite series temperature trends are greatly reduced.

RSS   0.1 d C / decade to   0.033 d C / decade

MSU 0.11 "   "   " 0.033 "   "

HadCru4   0.162 to 0.112   GISS.250   0.144 to 0.103

The ground station records start at higher growth rate and do not fall as strongly after El Nino filtering. Even so the impact is marked. The El Nino filtered satellite trends are now 4X lower than the ground based temperature record trends. The authors then go on to consider their earlier findings of planetary signals in temperature series like HadCrut3 at 61, 30-34, 21, 14.5, 10.4, 9.1, 8.25, 7.5, 6.54, 5.9, 5.2. We noted that 61 is 3 x 19.85; 30-34 includes 11.15 x 3 = 33.5, the sunspot primary period; 21 is just the Hale cycle at 22.3; 14.5 is 22.3 x 3 / 5; 10.4 is the mean of √(11.15 x 9.925) = 10.5; 9.1 is the apsides lunar cycle of 8.85 years; 8.25 appears in the SS and Hale solar spectrum; 7.5 is 22.3 / 3 = 7.44; 6.54 is 19.85 / 3 = 6.6, from the JS conjunction period; 5.9 comes from 22.3 / 4 = 5.58; 5.2 is 10.5 / 2 from the mean of the 11.15 & 9.92 SS cycles.

The authors leave out significant shorter periods. They do include volcanic activity and manmade forcing. Their model is marked as AM in the above figure. It matches the various temperature series discussed rather well

Scarfetta looks at coherency maps for global, land, sea, northern, southern hemisphere temperature series spectra and finds that 10 of the 12 shared periods in these records and the 'astronomical' spectral periods are coherent. On this basis the new paper uses a few key periods of ~11, ~20 (19.85), ~60 (3 x 19.85), 115 (5 x 22.3; 11.86 x 10; 4 x 29.46 = 118; 6 x 18.6; 13 x 8.85 = 115) and 980 (6 x 165 = 990, Neptune period; 33 x 29.46 = 973) years.

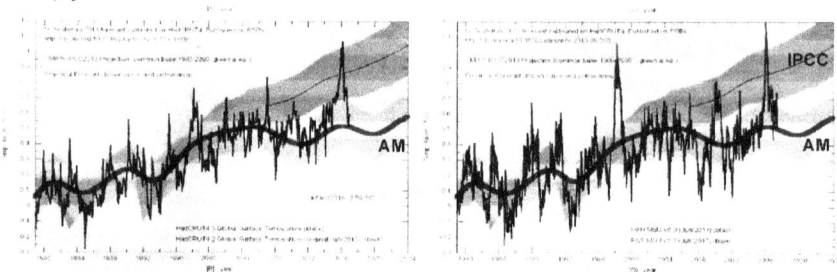

Figure 6. Performance of the semi-empirical models (yellow area) based on astronomical oscillations proposed in 2011 ([A] e [C] diagrams) and 2013 ([B] e [D] diagrams) [15,19] versus that of the CMIP5 models (green area). [A] and [B] use the HadCRUT temperature records available in 2011, 2013 and up to Dec/2016. [C] and [D] use the UAH and RSS records available up to Jun/2017. Data well agree with the Scafetta's forecast

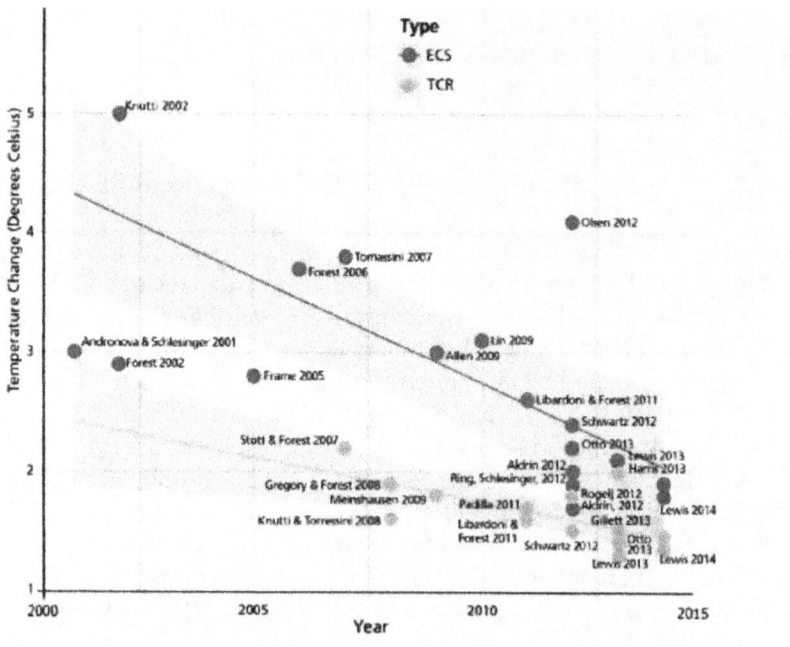

. The mean of the CMIP5 model set is marked IPCC. These models have been running hot since the last large El Nino event in 1999.

Using the shorter confirmed astronomical forcing periods in the model would take out the El Nino spikes around 1999 and 2016 and drop the trend further.

The CMIP5 model climate sensitivity ranged from 1.5 to 4.5 d C in 2013 (much reduced from the earlier 1 to 6 d C model set). Scarfetta suggests from their semi-empirical model, a climate sensitivity as low as 0.75 d C. They point out quite accurately that the tendency of the last two decades is for independent measures of CS to fall steadily. By 2017 some energy balance models were giving ECS as < 1 d C. Spectroscopic methods and radiative transfer equations were all giving ECS as < 1 d C.

The estimates of equilibrium and transient climate sensitivity in the figure above have fallen by a factor of 2.3 X and 5 X respectively. It appears that the poor fit of the IPCC models and their retreat from any certainty about the values has allowed objective measurements to slowly percolate into the public domain. The lowest estimates take us within reach of a feedback neutral, basic physics, 1 d C per CO_2 doubling effect. The assumption of large positive feedback processes existing are evaporating or at least that such processes are compensated by natural negative feedback processes... probably involving clouds and possibly biological feedbacks.

5.11 The Long Game : Oxygen Isotope Records

Delta Oxygen 18 levels (i.e. $O^{18} - O^{16}$ ratios) recovered from marine sediments are an accepted proxy for sea surface temperature. Sharma (13) has compared long term Oxygen 18 records with the equivalent Beryllium 10 record for the same 200 Kilo year epoch. He concluded tentatively that 'temperature' variations on cycle scales of ~100 Kyr is significantly correlated with the 'solar modulation factor' derived from Beryllium 10 variation. In other words he says the Sun drives the ice ages rather than the very small changes in solar radiation received by the Earth caused by changes in the orbit and axis inclination. These Croll – Milankovitch cycles also contain periods of ~100 Kyr. The issue of ice age causation is discussed below. First we will look in more detail at Oxygen 18 records for site 607 (in the mid-latitude north Atlantic) for the periods 0 to 600 Kyr BP and 0.8 million to 2.75 million years BP and site 846 for the period 1.0 to 1.94 million years BP to see what we can learn of the recent ice ages (14,15,16).

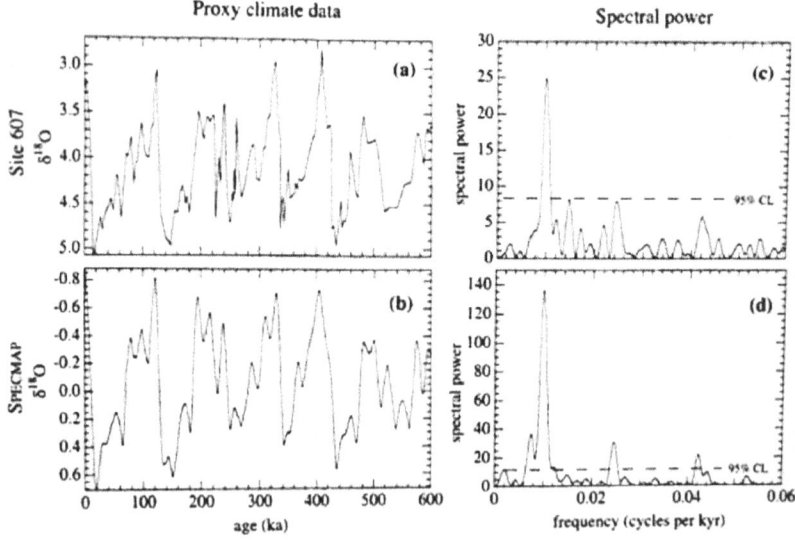

The recent (0 – 600 Kyr) site 607 series yields a spectrum with significant peaks at 100, 69.4, 41 and 23.6 Kyr.

The 100, 41 and ~23 Kyr cycles are just those predicted from the astronomical forcing theories. What about the 69.4 peak? Note that 3 x 23.6 = 70.8 Kyr. But also $1/41 - 1/100 = 1/69.5$. This peak is simply related to all three astronomical cycles. In fact we will see that over longer time scales spectral energy is redistributed between frequencies in a major way which complicates interpretation as usual. Let's look at the longer time span spectra for sites 607 and 846.

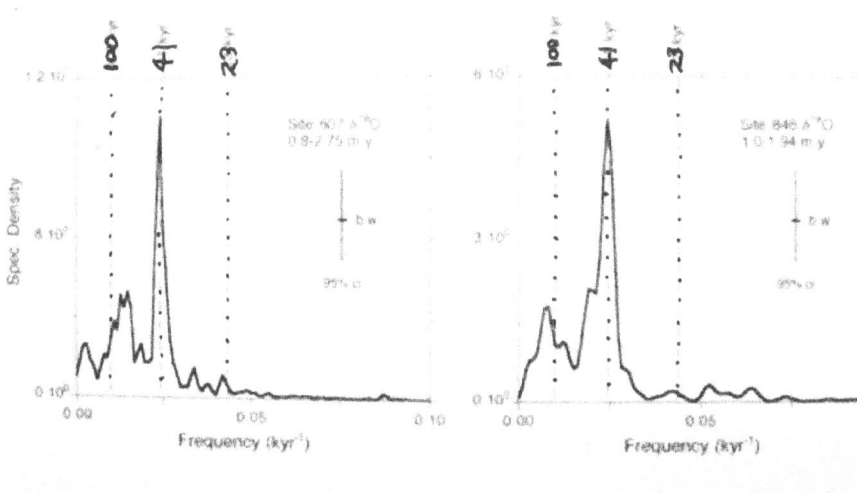

We have :

Site 607 (0.8 – 2.75 Myr) : 360, 71.3 av. (66.6, 75.9 twin peaks), **41**, 29.3, **23.7** Kyr.

Site 846 (1.0 – 1.94 Myr) : 132, 50.6, **41**, **23.7**, 19.4, 15.6 Kyr.

The dominant period in both spectra is the period of 41 Kyr corresponding to the Croll cycle in the tilt of the Earth's spin axis. Both spectra also show a small peak at a period of 23.7 Kyr, close to the axis precession cycle of 26 Kyr which is modified by alterations in the orientation of the earth's orbit (due to the other planets) yielding a final cycle of ~23 Kyr. The ~100 Kyr cycle in changes in the orbital shape (eccentricity) which was prominent in the last three glacial cycles, is not apparent here. That cycle alters the season at which the earth is nearest and furthest from the Sun. Lets look at the spectra in more detail in the usual way.

Site 607   The largest peak is at 41 Kyr.

All the 'temperature' proxy periods are harmonically related in simple and familiar ways :

3 x 23.7 = 71.1 versus observed 71.3 (mean) Kyr.

5 x 71.3 = 356.5 versus ~360.

And 15 x 23.7 = 355.5 versus ~360.

   Several cycles are integer sub-harmonics of others. In the same vein note that 9 x 41 = 368 and 12 x 29.3 = 352. There is further evidence of non-linear forcing in the presence of Fibonacci ratio harmonics.

66.6 / 41 = 1.624 = 13 / 8

66.6 / 29.3 = 2.27 = 1.505^2 = $(3 / 2)^2$

75.9 / 29.3 = 2.59 = 1.61^2 = $(21 / 13)^2$

75.9 / 23.7 = 3.203 = 2 x (8 / 5)

41 / 29.3 = 1.4 = 7 / 5

   The main driving periods at 41 and 23.7 Kyr have Fibonacci 'echoes', the sign of a non-linear system near the transition to chaos. The oxygen time series records look like semi-chaotic limit cycles. This is good news if correct : although chaotic such systems are constrained to move between tight limits. Does this imply that temperatures (in the current ice epoch) cannot suffer 'runaways' either upwards or downwards? This is something worth exploring given the current hysteria about man made climate change.

Where was the 100 Kyr cycle during this period? It manifested only as interaction 'sidebands' of the main 41 and 23.7 Kyr cycles.

1 / 41 − 1 / 100 = 1 / 69.5 versus the observed 71.3 Kyr (mean peak)

1 / 41 + 1 / 100 = 1 / 29.1 versus 29.3 Kyr

1 / 23.7 − 1 / 100 = 1 / 31 versus 29.3 Kyr

100 x 2 / 3 = 66.6 versus 66.6   and   100 x 2 / 75.9 = 2.635 = 1.623^2

$= (13/8)^2$. Also note that $1/100 - 1/360 = 1/78$ versus the 75.9 Kyr (split peak)

Site 846

This site covers the most recent ~2 million years and presents an apparently different set of medium peaks compared with site 607 covering ~3 million years. However both 41 and 23.7 Kyr cycles are present and all the periods are harmonically simply related.

$132 / 41 = 2.61 = 1.615^2 = (21/13)^2$

$132 / 50.6 = 3.22 = 2 \times 1.61 = 2 \times (21/13)^2$

$41 / 50.6 = 0.81 = 1.62 / 2 = phi / 2$

$50.6 / 19.4 = 2.608 = 1.615^2 = (21/13)^2$

$132 / 19.4 = 6.8 = 1.6151^4 = (21/13)^4$

$23.7 / 15.6 = 1.52 = 3/2$

Or $15.6 / 23.7 = 0.658 = 0.811^2 = (phi/2)^2$

$50.6 / 15.6 = 3.244 = 2 \times 1.622 = 2 \times (13/8)$

   The Fibonacci ratio harmonics are again strong. It is quite remarkable that our spectral analyses have found such 'golden section' relationships in solar and terrestrial phenomena on time scales from years to tens of thousands of years. Nonlinearities are manifesting in very similar forms again and again. Systematically applying the best tools for non-linear analysis now available would pay great dividends across the whole climate – weather arena.

Where is the 100 Kyr astronomical cycle in this case? It appears as interaction frequencies with the other driving frequencies.

$100 / 2 = 50$ versus observed 50.6 Kyr

$1/100 - 1/360 = 1/138$ versus 132

$1/23.7 + 1/100 = 1/19.2$ versus 19.4

$132 / 100 = 1.32 = 4 / 3$

$100 / 19.4 = 5.15 = 1.505^4 = (3/2)^4$

Note also that $1 / 23.7 - 1 / 41 = 1 / 52.3$ versus 50.6
and $1 / 23.7 + 1 / 41 = 1 / 15.2$ versus 15.6

The spectral energy in the 100, 41 and 23 Kyr cycles shifts between sets of related frequencies but is conserved over time. This is very like what we observed in the evolutive spectra of the (decadal scale) Hale solar activity time series and elsewhere. Comparing the two sites emphasises this point. Despite spectral differences (presumably reflecting changes during the million year data window difference) the frequencies are all simply and consistently related.

$360 / 132 = 2.73 = 1.654^2 = (5/3)^2$

$360 / 50.6 = 7.11 = 1.63^4 = (13/8)^4$

$132 / 66.6 = 1.983 = (2/1)$

$75.9 / 50.6 = 1.5 = (3/2)$

$66.6 / 50.6 = 1.314 = 1.622^2 / 2 = (13/8)^2 / 2$

$66.6 / 19.4 = 3.43 = 1.506^3 = (3/2)^3$

$66.6 / 15.4 = 4.27 = 1.622^3 = (13/8)^3$

Across both sites from 0 to ~2.75 million years BP the 'climate system' is responding in a consistent way, involving Fibonacci harmonics, to forcing signals containing Croll – Milankovitch theory periods of ~23, 41 and 100 Kyr. This is remarkable, compatible with a physical non-linear response mechanism and beyond any question of coincidence.

But what is the causal pathway? The changes in insolation caused by small changes in the shape of the Earth's orbit and by small changes in axis tilt have been calculated and also seem 'too small' to precipitate a shift between glacial and inter-glacial conditions. While these astronomical cycles churn on slowly over several tens of thousands of years the transition from glacial to inter-glacial conditions is typically very fast: no more than a few centuries.

It should however be remembered that a cycle containing several significant harmonics of that cycle and the right phase relationships, can sum to a wave form like a saw tooth with a very steep rise and slower fall: very like the recent ice age cycles (see Appendix 1, Figure 5). Also there are questions about the relative phasing of the cycles and temperature: changes do not always line up neatly. On the other hand we are dealing with a highly non-linear system perhaps on the 'edge of chaos'. Is frequency modulation in play, which would change the phasing? We noted in Appendix 1 the work of Rial (17) who for certain time windows and locations found that the spectra of proxy temperature series showed peaks which clearly derived from phase modulation of the Croll-Milankovitch cycle periods. We also noted earlier work which used a non-linear phase oscillator model of ice volume forced by the Milankovitch cycles to reproduce the pattern of ice ages for the last two million years including the switch to the dominant ~100,000 year cycle of the last five ice ages (18). In this case the author concluded we have a forced 'strange non-chaotic attractor' hidden in the climate system. Perhaps alternatively or in addition, we have something in the climate system acting like a non-linear relay switch?

As the author was concluding this book he discovered a paper by Ditlevesen (19) which probably confirms the form of the underlying non-linearity in the Croll –Milankovitch forcing and explains how the 'weak' 100 Ky eccentricity signal is dominant for the last several ice ages…it is not. The 100 Ky signal in the recent ice ages is generated from the other 23 and 41 Ky periods through a change in the climate bifurcation structure around one million years ago. This is possible since the three orbital / axis periods are clearly related numerically in familiar 'resonant' patterns.

$41 / 23.7 = 1.73 = \sqrt[3]{3}$; $23.7 \times 3 \times \sqrt[3]{2} = 100.55$; $41 \times \sqrt{3} \times \sqrt{2} = 100.43$
Also $100.43 / 23.7 = 1.6182$ = phi, an old friend.

Ditlevesen created a simple model with a deterministic component based on the C-M cycles and a stochastic component which accounted for the noise and intrinsic dynamics in the climate system. The proposed bifurcation structures, moving from two states to three states is shown below. Using this model and the 23.7, 41, 100 Ky orbital / axis forcing periods he was able to reproduce the main features of the long term isotope (ice volume / temperature proxy) spectrum before and after the transition between 41 Ky and apparent 100 Ky dominance in the climate records. This is shown below. The interesting point here is the existence of two natural fixed points in the system between which the climate has moved under orbital / axis forcing for over two million years even when the underlying bifurcation model changes. As Ditlevesen says the climatologists should be looking for the underlying physical mechanism but he doubts that the current generation of climate models are capable of exploring such non-linear behaviour.

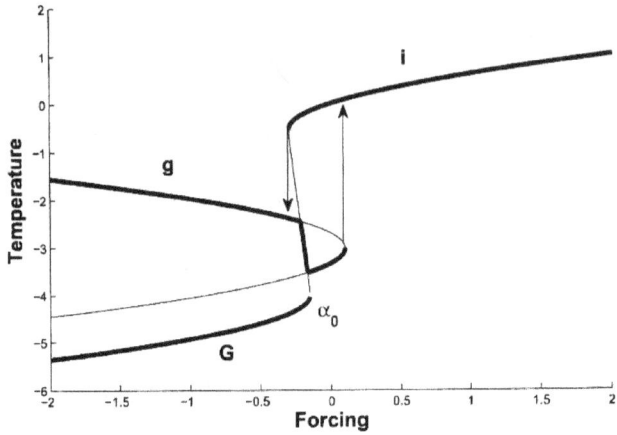

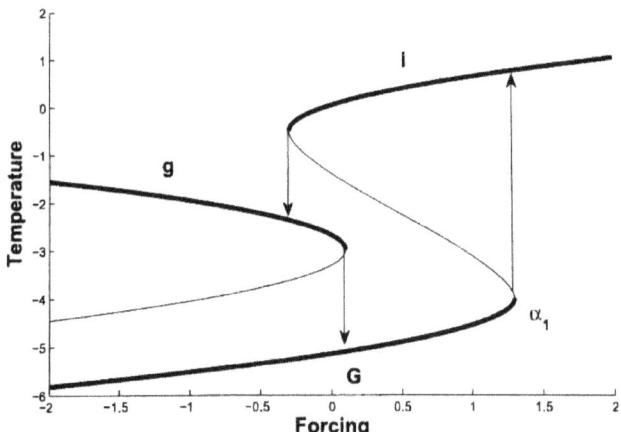

Remember the incredibly rapid rise in temperatures and CO_2 at the end of each recent ice age, for over a million years, has ended abruptly at close to the current interglacial temperature level. If these stable fixed points for global temperature limits exist, which now seems highly likely, the key question for the 'CO2 means warming' consensus is: what has changed in the last century to change these long lasting, bi-stable dynamics? It cannot be a small increase in CO_2 Since over half the CO_2 we have put out in the last century has not ended up in the atmosphere, the stabilising feedbacks must surely still be operating.

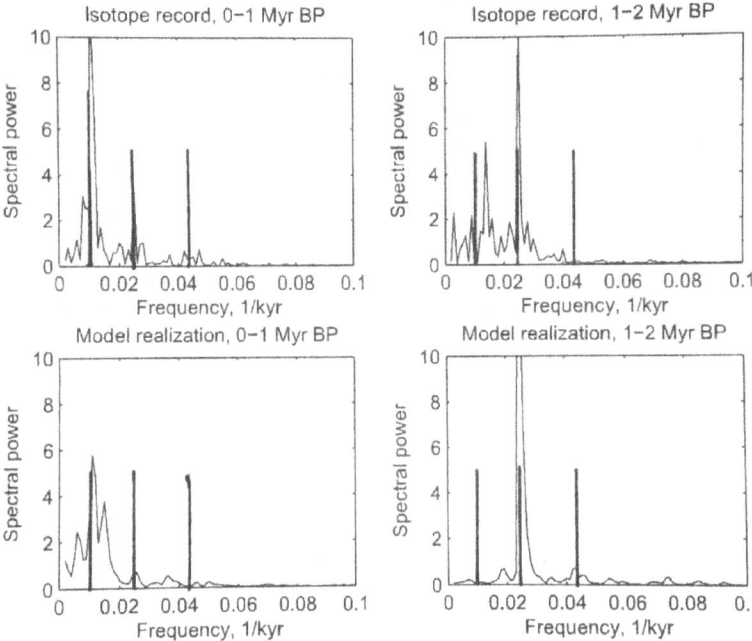

But we also saw earlier, on shorter times scales that the planets not only modulate Earth - Moon orbital properties, like the nodal cycle, and earthquake energy release but apparently modulate solar activity directly, which in turn modulates Earth's weather.

Are the oxygen isotope records similarly reflecting very long period changes in the output of the Sun driven ultimately by planetary dynamics? We know there are 'grand minima' in solar activity on the scale of several decades. Could there be occasional, much longer minima? We also know now that 20% of Sun like stars sampled so far **do not** exhibit detectable magnetic and presumably sunspot, variation. Perhaps Sharma is correct and the Croll – Milankovitch cycles are simply reflecting planetary dynamics also driving similar, long term cycles in the Sun, rather than the Milankovitch cycles having a direct effect on climate. Or perhaps both mechanisms have an effect? Whatever the pathway, the new evidence tells us there is a real and very significant effect here on our long term climate ...one which has been operating for millions of years and has important implications for the current climate debate.

5.12 The Pacific Decadal Oscillation

The PDO is a cycle of sea surface temperatures for the equatorial and northern hemisphere zones of the Pacific Ocean on a timescale of decades. During equatorial warm phases easterly winds dominate and the North Pacific cools strongly. During equatorial cool phases westerly winds dominate and the North Pacific warms. The PDO has been suspected of strongly influencing global climate. Aban et al (20) provide high resolution MEM spectra for the PDO and Normal Difference Vegetation Indices (NDVIs), for Africa, Australia and South America and demonstrate very similar spectral structures in all cases.

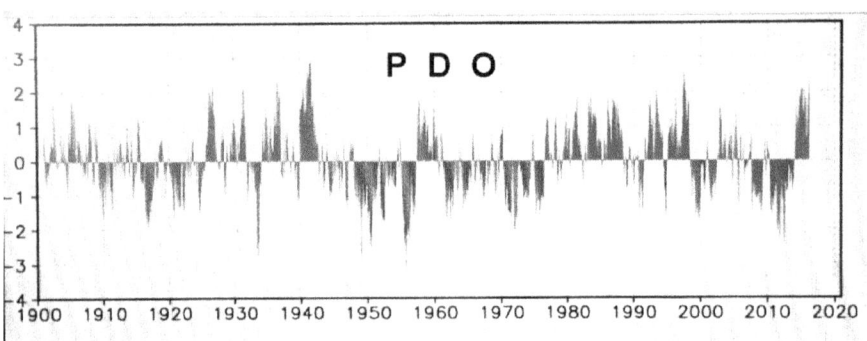

The largest common spectral peaks include 28.67, 14.33, 9.56, 7.17, 2.58, 2.38, 1.43, 1.37, 1.25 years. 7.17 and 2.58 dominate. 9.56 and 1.43 are smaller. We see immediately that all these cycles are simply harmonically related via integers, phi or other common non-linear factors (see sec.3.7).

$28.67 / 14.33 = 2.0$; $14.33 / 7.17 = 2.0$; $14.33 / 9.56 = 3 / 2$; $9.56 / 2.38 = 3.01$; $7.17 / 1.43 = 5.01$; $9.56 / 1.37 = 6.982$; $2.38 / 1.43 = 5 / 3$; $7.17 / 2.58 = 2.78 = 1.667 = (5 / 3)^2$; $9.56 / 1.43 = 6.685 = (8 / 5)^2$; $9.56 / 7.17 = 1.33$ but 1.325 is the plastic ratio; $2.58 / 1.43 = 1.81$ but 1.839 is the Tribonacci constant; $2.38 / 1.37 = 1.736$ but $\sqrt{3} = 1.732$; $7.17 / 1.37 = 5.23 = (\sqrt[3]{3})^4$.

For the two main peaks we see that $4 \times 7.17 = 11 \times 2.58 = 28.7$ versus 28.67 and $9 \times 7.17 = 25 \times 2.58 = 64.5$. We don't see this peak possibly because the data series is too short. It is close to the dominant AMO period of ~60-70 years.

Can we relate the PDO to solar activity cycles? We can in several ways. Consider first the main 11.16 years sunspot cycle.

$28.67 / 11.16 = 2.57 = 1.602^2 = (8/5)^2$; $14.33 / 11.16 = 1.284 = 1.603^2 / 2 = (8/5)^2 / 2$; $9.56 / 11.16 = 0.857 = 1.604^2 / 3 = (8/5)^2 / 3$; $7.17 / 11.16 = 0.642 = 1.603^2 / 4 = (8/5)^2 / 4$; $11.16 / 2.58 = 4.32 = 1.628^3 = (13/8)^3$
$11.16 / 2.38 = 4.684 = 1.67^3 = (5/3)^3$; $11.16 / 1.43 = 7.804 = (5/3)^4$.

The Fibonacci ratio harmonics are clear, accurate and consistent. The Hale 22.33 year cycle is not visible but its sidebands are. Consider the strong 17.9 year peak.

$28.67 / 17.9 = 1.602 = 8/5$; $14.33 / 17.9 = 0.8 = 4/5$; $7.17 / 17.9 = 0.4 = 2/5$; $2.58 / 17.9 = 0.144 = (8/13)^4$; $17.9 / 9.56 = 1.87$ cf Tribonacci 1.84; $17.9 / 1.43 = 1.324^9$ cf 1.325, the plastic ratio.

The other main Hale sideband is at 19.85 years and
$28.67 / 19.85 = 1.443$ cf $3^{1/3} = 1.442$; $14.33 / 19.85 = 0.721 = (3)^{1/3} / 2$; $7.17 / 19.85 = 0.36 = (3/5)^2$; $19.85 / 2.58 = 7.694 = 1.665^4 = (5/3)^4$; $19.85 / 9.56 = 2.076 = 1.441^2 = 3^{2/3}$; $(19.85 / 2) / 1.43 = (13/8)^4$.

The pattern is again clear. Recall that 19.85 is also the Jupiter – Saturn conjunction period. Perhaps other planetary conjunctions are present.
$28.67 / 12.8 = 2.24 = 1.497^2 = (3/2)^2$; $12.8 / 14.33 = 0.893 = 2(2/3)^2$
$12.8 / 9.56 = 1.33$ cf plastic ratio 1.325 ; $12.8 / 7.17 = 1.785 = 1.336^2$ cf plastic ratio of 1.325.
$12.8 / 2.58 = 4.97 = 1.492^4 = (3/2)^4$

$14.33 / 13.8 = 1.0385 = (3/2)^{2/3} / 3$; $13.8 / 9.56 = 1.443 = 3^{1/3}$; $13.8 / 7.17 = 1.925 = 1.33 \times 1.446 = 1.325 \times 3^{1/3}$

In all cases there are clear Fibonacci ratio and other common non-linear system irrational factors. What about the other planetary conjunction periods? $25 / 9.56 = 2.615 = $ phi ; $25 / 7.17 = 3.48 = (3/2)^2$; $25 / 2.58 = 9.65 = 1.382$ cf 1.381; $30.3 / 7.17 = 4.226 = 1.6168^3 = $ phi ; $30.3 / 2.58 = 11.4 = (3/2)^6$; $30.3 / 9.56 = 3.16 = 1.467^3$ cf 1.466; $45.3 / 2.58 = 17.56 = 1.613^6 = (21/13)^2$; $45.3 / 7.17 = 6.317 = (5/2)^3$; $45.3 / 9.56 = 4.74 = 1.68^3 = (5/3)^4$; $55.7 / 7.17 = 7.77 = 1.669^4 = (5/3)^6$; $55.7 / 9.56 = 5.82 = 1.34 = $ cf the 1.325 plastic ratio.

The matches are good for the major peaks. Direct planetary orbits are not present (although Ps = 29.46 versus 28.67). However we noted in other sections that Pj = 11.86 years and 11.86 / 5 = 2.373 which is identical to the Quasi Biennial Oscillation and exactly twice the Chandler Wobble period. Now looking again at the PDO spectrum we see that the 2.38 year PDO period is the QBO and

2.373 x 3 = 7.12 versus PDO period 7.17 ; 2.373 x 4 = 9.5 versus 9.56; 2.373 x 5 = 11.86 , Pj ; 2.373 x 6 = 14.24 versus 14.33 ; 2.373 x 12 = 28.48 versus 28.67 in the PDO. Also note that

2.373 x (3 / 5) = 1.424 versus 1.43 ; 2.373 x $(5/6)^3$ = 1.373 versus 1.37

$11.86 / 7.17 = 1.658 = (5/3)^3$; $11.86 / 2.58 = 4.597 = 1.663^3 = (5/3)$

$11.86 / 14.33 = 0.8276 = (5/6)$; $11.86 / 28.67 = 0.4137 = (5/12)$;

$(11.86 / 3) / 1.43^2 = 2.765 = 1.663^2$ or $(5/3)$.

Jupiter seems to make its presence felt through familiar rational harmonics. What about Saturn? $29.46 / 7.17 = 4.11 = (8/5)^3$; $29.46 / 2.58 = 11.42 = (3/2)^6$; $29.46 / 9.56 = 3.09 = 1.326^4$ cf the 1.325 plastic ratio; $(29.46 / 3) / 1.43 = 6.87 = 1.619^4 = $ phi ; $(29.46 / 3) / 2.38 = 4.126 = 1.603^3 = (8/5)^3$.

Note for Uranus, $84 / 7.17 = 11.71 = 1.504^6 = (3/2)^6$.

The PDO is clearly and strongly linked to solar activity cycles and to planetary orbital period forcing through familiar non-linear system factors.

This is relevant to our climate change problem since some, such as Spencer (21), believe that the PDO could account for up to 2 / 3 of the 'global warming' seen in the 20th century. If so much of that must be solar related given our spectral analysis. He suggests that the response of clouds to the PDO is critical. Clouds again! However Spencer assumes the PDO is a 'natural', intrinsic oscillation of the climate system. Our new analysis suggests it is a 'cycle' driven by solar variation. Between 1998 and 2014 the PDO was in the negative zone and global warming halted. In 2015 it spiked into the positive zone in response to the current strong El Nino in the Pacific.

If the reader still doubts that 'cosmic' influences are in play in the PDO it is instructive to look at the lunar nodal and apsides orbital cycles:

$18.6 / 9.56 = 1.946 = 1.248^3 = (5/4)^3$; $18.6 / 7.17 = 2.594 = 1.611^2 = (21/13)^2$; $18.6 / 2.38 = 7.81 = 1.67^4 = (5/3)^4$; $18.6 / 1.43 = 13$.

Even clearer, for the apsides cycle we have :

$28.67 / 8.85 = 3.24 = 2 \times 1.62 = 2 \times phi$; $14.33 / 8.85 = 1.619 = phi$

$7.17 / 8.85 = 0.8104 = 1.62 / 2 = phi / 2$; $8.85 / 2.58 = 2.27 = 1.505^2 = (3/2)^2$; $8.85 / 1.43 = 6.19 = 1.836^3$ cf Tribonacci constant 1.839.

The harmonic patterns linking PDO, solar activity, planetary and lunar orbital cycles are remarkably precise and consistent and involve well recognised Fibonacci, Tribonacci and other irrational constants such as 1.325, the plastic ratio, $3^{1/3} = 1.442$ and 1.466.

In isolation an observer could be persuaded that lunar cycles determined the PDO! In fact planetary forces modulate both lunar orbital dynamics and solar activity and it is the latter which in turn modulates or drives the PDO …as we have seen for other weather phenomena.

It may be that the PDO also lends support to the Svensmark cloud theory.

Yin & Porporato of Princeton recently compared the Pacific Decadal Oscillation with global cloud coverage for the whole day and day and night periods over the period 2000 to 2017 taking in the global temperature hiatus and the beginning of the long double El Nino period of 2016 to 2019 (21). Their aim was to look for alternatives to the assumption (since disputed) that the 'missing' CO_2 generated heat had 'gone into the oceans' for some reason.

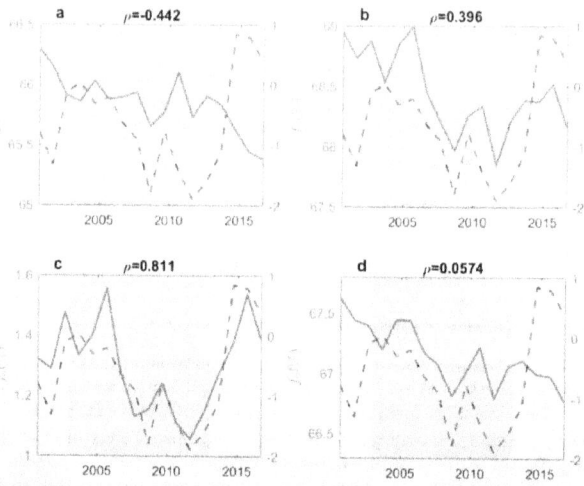

Fig. 1. Comparison of global mean cloud fraction with pacific decadal oscillation. The solid lines are the time series of (a) daytime, (b) nighttime, (c) daily amplitude, and (d) daily mean global cloud fraction; the dash lines are the pacific decadal oscillation (PDO) index. The correlation coefficients (ρ) of the two displayed data sets is reported on top of each panel. The shaded area divides the early 21st century into pre/mid/post-hiatus periods.

They found that the cloud measure representing the full daily amplitude of variation, $A = F_n / 2 - F_d / 2$ was strongly correlated, $R = 0.811$, with the PDO over the full 2000 to 2017 record (Figure 1C above). This included the temperature 'hiatus' from 2000 to 2015 and the following strong El Nino period. They conclude that the daily cloud cycle is correlated strongly with the PDO and thence global surface temperature. This compares with Svensmark's assertion that cloud formation over the tropical ocean areas (largely the vast Pacific) is modified by cosmic ray flux modulated in turn by the variations in the solar magnetic field over the solar activity cycle.
It is interesting that PDO and cloud amplitude in this period has a cycle length of ~11 years.

5.13 Averaged Global Surface Temperature 1860-1990

Vautard et al (25) conducted a comprehensive study of the problems of analysing noisy natural time series generated by 'chaotic' processes. They looked at MEM, MTM and SSA (singular spectral analysis) which we have now met several times. Fortunately one of the series they explored was the averaged global surface temperature compiled from acceptable sources by the IPCC ~1991. They showed how spectral structure was to some extent dependent on technique used and data processing as we have already noted. However the work also confirms the robust presence of certain key frequencies relevant to our hypotheses. An MTM histogram analysis estimating line frequencies gives us : 25, 22.2, 20, 11.8, 10.9, 10, 7.15 years. A separate MEM histogram gives us : 25, 22.5, 20, 12, 10, 7.4, 7.15, 5, 3, 2.8, 2.34. The authors also compute quasi-periodic components for the temperature series with increasing data window length starting with 1860 – 1950 and ending with 1860 – 1990. This gives us some feel for cycle persistency and changes. The periods with a frequent presence include 25, 14.9, 10, 7.4, 6.2, 5.3, 4.7, 3.7 years. Over the period 1980 to 1990 the 14.9 period evolves smoothly to 25 years. Interestingly 25 / 14.9 = 1.676 or ~5 / 3. The picture is consistent and familiar.

25 appears prominently in the sunspot spectrum but is also Csn+.
22.2 – 22.5 is our primary Hale cycle of 22.33 years.
20 is the 19.85 Hale series period but also Cjs- = 19.85.
11.8 – 12 is in the sunspot spectrum as 11.8 but also Pj = 11.86.
10.9 is close to the 11.1 main Wolf sunspot cycle.
10 is simply the second harmonic of Cjs-, 19.85 / 2 = 9.93.
7.4 – 7.15 is the third harmonic of the main Hale period 22.33 / 3 = 7.43
$$2$$
But also 19.85 / (5 / 3) = 7.15 and 17.9 x 2 / 5 = 7.16, reflecting the 3 major Hale spectral peaks and for the lunar nodal cycle 18.6 x 2 / 5 = 7.44.
$$3$$
5.3 – 5 and 22.33 / phi = 5.27, 19.85 / 4 = 4.96.
$$3$$
4.7 is 19.85 / phi = 4.69 and for the nodal cycle, 18.6 / 4 = 4.66.

3.7 is 22.33 / 6 = 11.1 / 3 = 3.72 and 18.6 / 5 = 3.72.

The averaged IPCC global series, from 1860 – 1990, simply confirms what we have seen for regional climate time series before, during and since that period. The Sun's influence is transparently present.

5.14 Nile River Records 622 – 1922 AD

The Cairo Nilometer records of the height of the Nile River may be the oldest instrumental records in the world. Ghil et al (26) have analysed the spectral properties of this time series taking into account annual maxima and minima of the river height and the annual differences. The thought was that the properties might differ. However we will see that the spectral results are very similar for the highs, lows and highs – lows time series. They first considered the period 622 – 1470 AD.

64 years (significant for H, L, H-L series) and $3 \times 22.3 / 2 = 66.9$; $19.85 \times 3 / 2 = 59.6$; $18.6 \times (3 / 2) = 62.8$; $64 = 5 \times 12.8$ from the sunspot spectrum.

32 (H only) $3 \times 11.16 = 33.4$; $19.85 \times phi = 32.1$

18 – 18.3 (L, H-L) 17.9 Hale period; 18.6 lunar nodal period.

12.2 (L, H-L) $19.85 \times 8 / 13 = 12.21$; $18.6 \times 2 / 3 = 12.3$; $Pj = 11.86$.

7.2 – 7.3 (H, H – L) $22.33 / 3 = 7.44$; $18.6 \times 5 / 8 = 7.27$; $19.85 / (8 / 5)^2 = 7.15$; Pj, $11.86 / phi = 7.33$.

6.2 (L) $18.6 / 3 = 6.2$; $19.85 / (2 \times phi) = 6.13$; $25 / 4 = 6.25$; $11.86 / 2 = 5.94$.

3.6 (H) $11.16 / 3 = 3.7$; $18.6 / 5 = 3.72$.

2.9 – 3.0 (H, L, H-L) $11.16 / 4 = 2.79$; lunar apsides, $8.85 / 3 = 2.95$
Pj, $11.86 / 4 = 2.96$; Ps, $29.46 / 10 = 2.95$.

2.2 – 2.3 (L, H) $11.16 / 5 = 2.23$; lunar apsides, $8.85 / 4 = 2.21$;
$17.9 / 8 = 2.24$.

We see familiar Wolf and Hale sunspot periods plus lunar periods and signs of the planets via simple and Fibonacci harmonics.

Now for the period 622 – 1922 AD which involved some infilling of data gaps.

85 years (H) $22.3 \times 4 = 11.16 \times 8 = 89.2$; $19.85 \times (13 / 8)^3 = 85.2$; $Pu = 84$
$11.86 \times 7 = 83.1$; note that $85 \times 3 / 4 = 63.8$, see below.

64 (L, H-L) see 622 – 1470 AD period analysis above.

23.3 (H) 22.33 ?; 11.86 x 2 = 23.7 ; apsides, 8.85 x (13 / 8)^2 = 23.4.

19.7 (L) 19.85 Hale cycle and Cjs.

18.3, 12.2 (H-L), **7.3** (H , H-L), **6.2** (L) see above.

4.2 (H) Ps, 29.46 / 7 = 4.2; Pj, 11.86 x (3 / 5)^2 = 4.27; Cjs,19.85 x (3 / 5)^3 = 4.28; 11.16 / (13 / 8)^2 = 4.22; also 85 / 20 = 4.25.

2.9 – 3.0 (L, H, H-L) and **2.2** (L,H,H-L) see above.

The two data periods are spectrally very similar with the longer series displaying additional 85, 19.7 and 4.2 year cycles. The 19.7 period is clearly Cjs = 19.85, which is also a major Hale solar peak. Also note that 19.85 x (13 / 8)^2 = 85.2 years and 85 x 3 / 4 = 63.8 versus 64 observed over both periods and 85 / 19.85 = 4.27 in the longer period. Some spectral power has shifted in an organised way over the 1470 – 1922 period.

The strong links to the Sun and Moon have some wider significance since the Nile data reflect annual rainfall over central and east-central equatorial Africa. Recall also the strong spectral power at ~60 to 70 years found in the AMO proxy time series.

5.15　East Asia Winter Monsoon.

We began with an analysis of a 2,500 year temperature proxy series for the Tibetan Plateau and showed clear links to solar activity. Here we look at a reconstruction of 8,000 years of winter monsoon rainfall by Xiao et al (27). Xiao used a long sediment core from the Yangtze River mouth and analysed variations in the sedimentation rate for a defined range of fine particles deposited by the 'ECS winter coastal current' which is mainly controlled by the East Asian winter monsoon. They were able to derive a high resolution proxy spectrum for monsoon intensity with ~18 clear peaks. Several of the peaks corresponded to peaks in a well established C14 spectrum from tree rings which they claimed (reasonably) as a solar activity proxy. 'Coherence' between monsoon intensity and the Sun was claimed on this basis. However the spectra repay a closer look applying our usual detailed approach.

Five periods are significant at >95%; seven at 80% - 95%; six at <80%, but close. We will label these as L,M,S. Many of the periods are harmonically related. The periods appear in close pairs, as split peaks. Taking the longest first we have 2463, 1368, 456, 397, 280, 237, 173, 140, 128, 111, 106, 100, 91, 88, 78, 76, 72, 70. Note that 2463 = 14 x 176; 2463 = 19 x 129; 1368 = 3 x 456; 1368 = 8 x 171; 456 = 5 x 91; 4 x 100 = 400 (~397); 280 = 2 x 140; 237 = 3 x 79. Familiar Fibonacci 2 harmonics also occur: 280 / phi = 173; 456 / phi = 281; 456 / phi = 174. We can also see familiar solar / planetary / lunar links :

456 (S)　and 23 x **19.85** = 456.6 but Cjs = 19.85 and this is a major Hale series peak; 18 x **25** = 450.

397 (M)　and 20 x **19.85** = 397 but Cjs = 19.85. Also 22 x **17.9** = 394. 17.9 is also a major Hale peak, as is 25 and 16 x **25** = 400.

280 (S)　and 25 x **11.165** = 279.2; 14 x **19.85** = 278; for the lunar nodal cycle 15 x **18.6** = 279.

237 (S)　and 12 x **19.85** = 238.2; for Pj, 20 x **11.86** = 237.2.

173 (S)　and we have **22.3** x 8 = 178.4; 7 x **25** = 175; 9 x **19.85** = 178.6.

140 (S)　and 7 x **19.85** = 139; 11 x 12.8 = 140.8, Wolf ss peak and Cjn

128 (M)　and 7 x **17.9** = 125.5; 10 x 12.8 = 128, Wolf ss peak and Cjn 5 x **25** = 125.

111 (S) and 10 x **11.16** = 5 x **22.33** = 111.6; 11 x **9.93** = 109.4 and Cjs / 2 = 9.93, also a Wolf ss peak; 6 x **18.6** = 111.6.

106 (M) and 6 x **17.9** = 107.3; 12 x **8.85** = 106.2, lunar apsides cycle.

100 (M) and 9 x **11.16** = 100.4, for the main Wolf ss peak; 5 x **19.85** = 99.3; 4 x **25** = 100.

89-91 (L) and **22.33** x 4 = 89.3; **17.9** x 5 = 89.5; **9.93** x 9 = 89.4; **8.85** x 10 = 88.5.

76-78 (L) and **11.16** x 7 = 78.1; **9.93** x 8 = 79.2

70-72 (L) and **9.93** x 7 = 69.5; 6 x **11.86** = 71.2; 8 x **8.85** = 70.8.

Note also that the large, longest period peak is a simple multiple of the 'multiple' conjunction periods of the four gas giants of 171 – 181 years that we have seen many times : 2463 = 13 x **176**. Also 110 x 22.333 = 2457 and 124 x 19.852 = 2462. The (nearly 95% significant) medium peak at 1368 is also 7 x **171**. It is also 3 x 456 but we noted 456 is closely 23 x **19.85**, the Cjs period which is also a major Hale ss series peak.

Overall each 'monsoon' spectral period is a simple integer sub-harmonic of at least two of : a major Wolf ss peak, a major Hale series peak; a gas giant conjunction period particularly Cjs, the Jupiter orbital period, Pj, or a lunar cycle (all highlighted in **bold** above). This does not include the many Phi / Fibonacci ratio links. For example 22.33 x (5 / 3)⁸ = 1332 cf. 1368 and 19.85 x (8 / 5)⁹ = 1363 and 22.33 x (21 / 13)⁶ = 396.3 cf. 397 and 18.6 x (5 / 3)⁶ = 398.2.

The coherence between the East Asian monsoon and solar activity claimed by the Chinese team seems to be solid. They did not make the lunar and planetary connections for some reason. They note a further link to the Indian monsoon with similar cycles through the Holocene.

5.16 Drought Cycles in the Western USA

One of the earliest papers noted by the author on climate / solar links days from 1976 by Walter Orr Roberts based on a PhD thesis by Marshall (28). Here is Marshall's record of High Plains droughts.

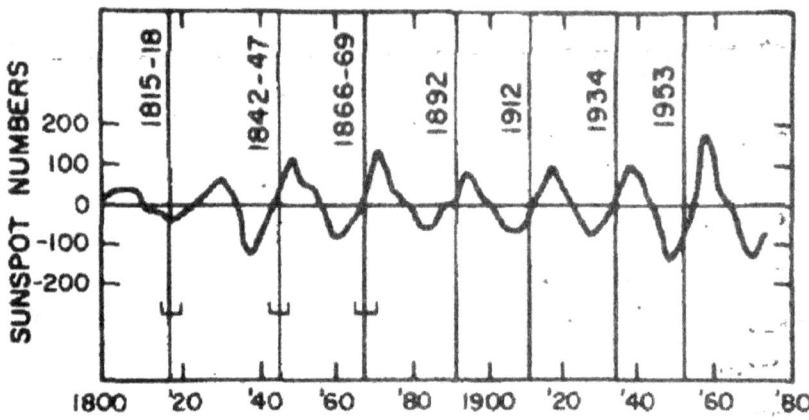

FIGURE 3.—High Plains droughts. This figure is adapted from the Ph.D. thesis of Marshall (1972). The vertical lines correspond to the center dates of all droughts cited by Marshall from rainfall data over the High

We can note a remarkable correspondence from ~1840 to 1960 between the midpoint of historic drought periods and the crossover (minimum) point of the Hale 'magnetic' solar time series. At these points the sun reverses magnetic polarity. Roberts points to variations of the Hale cycle length in the range 20 – 22 years. These should be familiar by now since spectral analysis has shown us the main cycle at 22.3 years and major cycle at 19.85 years which is exactly the Jupiter – Saturn conjunction period. In fact we showed that $1 / 19.85 - 1 / 178.6 = 1 / 22.33$ where 178.6 is the mean of a number of close gas giant planetary conjunctions. In fact all three major Hale spectrum peaks are related since $\sqrt{(22.3 \times 17.84)} = 19.85$, the JS conjunction.

Some years later Mitchell et al examined drought Area Indices based on tree ring data for the entire USA west of the Mississippi River (29). Each DAI was based on 40 to 65 tree ring sites from Canada to Mexico in latitude and from the US west coast to the Great Plains states.

The drought series went back to ~1600 AD. The various DAI time series all yielded a concentration of spectral power near 22 years at statistical significance levels from 0.1% to 5% (relative to a pink noise reference background). The authors were able to confirm **phase locking** between droughts and the Hale cycle at significance levels of 0.1% to 1%. That is the smoking gun of a causative link.

They also found evidence of an amplitude modulation of the 22 year cycle of about 90 years with a good correlation between drought severity and the solar activity cycle amplitude. This cycle is very familiar and corresponds to the often claimed 70 – 100 year Gleissberg cycle. However we see that 90 is just 180 / 2, first harmonic of the cycle of several planetary conjunctions involving 3 and 4 gas giant planets. Note also that 22.33 x 4 = 11.15 x 8= 89.32; 17.9 x 5 = 89.5; 9.925 x 9 = 89.35; 11.86 x 4 = 94.8; 29.46 x 3 = 88.8. The origin of the ~90 year drought modulation is clear. The 90 year cycle was significant at the 1% to 5% level across DAIs.

The authors concluded that the USA drought rhythms are controlled by 'long term solar variability directly or indirectly related to solar magnetic effects.'

We cannot disagree. The variation of the solar magnetic field modulates the cosmic ray flux (at some energies) entering the Earth's atmosphere which in turn affects cloud formation over the oceans, according to Svenmark's heresy. The strange thing is that by the time of the IPCC foundation all this clear, definitive early US work was forgotten or discounted.

5.17 Global Temperatures and Solar Activity Indices

Chinese scientists have carried out correlation studies relating sunspot numbers, total solar irradiance and long global surface temperature reconstructed time series using wavelet and cross coherence analyses (30). In other analyses we have simply pointed to *the matching of periods* in weather variables and solar – planetary time series. In this analysis the authors demonstrate not only period matching but also stable phase relationships and high cross correlations at particular, significant frequencies i.e. coherence. This makes a stronger case for causation.

Sea surface temperature and total solar irradiance show strong links with wavelet coherence = 0.95 at a periodicity of ~ 60 and C > 0.8 for 52 – 70. A second strong peak is at 20 years with coherence = 0.9 and a range of 17 – 23 for C > 0.8. These periods are familiar.

60 $19.85 \times 3 = 59.55$; $17.9 \times (3/2)^3 = 60.4$; $11.86 \times 5 = 59.3$; $29.46 \times 2 = 58.9$; $18.6 \times (3/2)^2 = 62.8$; $22 \times (5/3) = 61.1$; $8.85 \times (8/5)^4 = 58$.

The C > 0.8 band also contains $22.3 \times 3 = 66.9$; $18.6 \times 3 = 55.8$; $12.8 \times 4 = 64$; $13.8 \times 5 = 69$; $8.85 \times 7 = 62$.

20   19.85 is the JS conjunction period; $8.85 \times (3/2)^2 = 19.9$; $(17.9/2) \times (3/2)^2 = 20.1$; $29.46 \times (2/3) = 19.64$; $12.8 \times (3/2) = 19.2$; For the C >0.8 band (17-23) we have 22.3, the Hale solar cycle; $11.86 \times 2 = 23.7$; $8.85 \times 2 = 17.7$.

Surface global temperature on land and sea are clearly linked to total solar irradiance in coherence zones which match major Hale solar cycle periods 22.3, 19.85, 17.9. But 19.85, 12.8, 13.8 are planetary conjunction periods for the gas giants and we also see multiples of the Jupiter and Saturn orbital periods. There is equally good evidence for harmonics of the lunar nodal and apsides cycles of 18.6 and 8.85 years. The authors note that the solar correlations with sea temperature are the strongest suggesting that whatever the solar link is it manifests through the oceans. The Svensmark hypothesis points to low cloud formation over the tropical ocean regions.

It is not surprising that sunspot numbers and total solar irradiance are linked with four peaks above C=0.6, ~81, 51, 32.6 (30-33) C > 0.8), 20 (18 – 22, C >0.7) years.

We see that 19.85 x 4 = 79.4 versus 81; 12.8 x 4 = 51.2; 11.1 x 3 = 33.3; 19.85 is the JS conjunction as before. It is notable that the authors conclude that the correlation of TSI to temperature is **stronger** than the correlation of TSI to sunspot activity. But TSI and sunspots are definitely generated in the same physical system…the sun.

On the basis of a stronger correlation between TSI and Earth surface temperature we can reasonably infer that these phenomena too are physically connected.

The variation in total solar irradiance across the solar activity cycle is small and usually discounted by climate modelers. However the variation of the ultraviolet bands in SI is 6 X larger than the total variation . Some analysts have suggested it is the ultraviolet frequency bands, or some solar effect strongly connected to them, which count in weather modulation.

5.18   Svalbard Temperature Record 1912 – 2010

Svalbard is an Arctic island at 78 degrees N. It is of interest because of the expectation that man made climate change will show up most clearly in high latitudes. Humlum et al (31) have analysed the Svalbard annual, winter and summer season mean surface air temperatures using Fourier and wavelet techniques. The wavelet results appear to have good period resolution. The annual, winter and summer series are similar in frequency composition. However it is striking that the winter season data shows periodic power several times that of the summer season. Whatever is driving these periods annually acts primarily though winter season variations. Humlum mentions that El Nino and the North Atlantic Oscillation are recognized in climate records and that several show periods of around 60 years. The authors also speculate on the possible presence of lunar orbital periods and in this we will see they are correct. They do not recognize however the obvious solar related periods, nor other planetary forcing periods.

A few comments can be made about the Fourier analysis although resolution is limited. They pick out 68.4, 25.7, 16.8 and 2.45 years as significant periods. From other places we note that $22.33 \times 3 = 67$; $19.85 \times (3/2) \times (3/2) \times (3/2) = 67$, where 19.85 is the Jupiter- Saturn conjunction period ; $11.86 \times 3 = 71.1$, where 11.86 is the Jupiter orbital period; $13.8 \times 5 = 69$ where 13.8 is the Jupiter- Neptune conjunction period.
$29.46 \times (3/2)^2 = 66.3$, where 29.46 is the Saturn orbital period. The 25.7 period is 2×12.85, where 12.8 is the Jupiter-Uranus conjunction period; $17.9 \times \sqrt{2} = 25.3$; $8.85 \times 2\sqrt{2} = 25.1$. The 16.8 period is also familiar. $11.15 \times (3/2) = 16.73$; $11.86 \times \sqrt{2} = 16.77$. The short period of 2.45 is also significant. Note that $22.33 / 9 = 2.48$; 2.5 is the major El Nino period; $19.85 / 8 = 2.48$.

The winter wavelet analysis shows strong, persistent frequencies at 83, 62.5, 37.7 moving to 37, 25.6 moving to 26.3, 16.1 moving to 17.2, 12 moving to 11.8, 7.9 moving to 9.2, 5.2 years. We find that solar and planetary harmonics are clear:

83   $83 / 7 = 11.86$, the Jupiter orbital period; $29.46 \times 2\sqrt{2} = 83.3$; $19.85 \times (8/5)^3$
$= 81.4$; $17.9 \times (5/3)^3 = 82.5$; $13.8 \times 6 = 82.8$; $11 \times (5/3)^4 = 84.6$, $12.8 \times (8/5)^4$
$= 83.9$.

62.5 $18.6 \times (3/2)^3 = 62.8$, where 18.6 is the lunar nodal cycle; $7 \times 8.85 = 62$, where 8.85 is the apsides cycle; $22.33 \times (5/3) \times (5/3) = 62$; $12.8 \times 5 = 64$. $11.86 \times (3/2)^4 = 60.1$; $(17.9/2) \times 7 = 62.6$.

37.3 (37.7 -37) $22.33 \times (5/3) = 37.2$; $12.8 \times 3 = 38.4$; $25 \times (3/2) = 37.5$; $18.6 \times 2^3 = 37.2$; $8.85 \times (21/13) = 37.3$.

26 (25.6-26.3) $12.8 \times 2 = 25.6$; conjunction 25; $11.86 \times (3/2)^2 = 26.6$; $11.2 \times (3/2)^2 = 25.2$; $8.85 \times 3 = 26.5$; $18.6 \times \sqrt{2} = 26.3$; $(19.85/2) \times \text{phi}^2 = 26$.

16.64 (16.1-17.2) $11.15 \times (3/2) = 16.72$; $11.86 \times \sqrt{2} = 16.77$; $(19.85/2) \times (5/3) = 16.54$; $25 \times (2/3) = 16.65$.

11.9 (11.8-12) 11.86 is Jupiter orbital period; $29.46 \times (2/5) = 11.78$; $19.85 \times (3/5) = 11.91$; $17.9 \times (2/3) = 11.93$; $(22.33/3) \times (8/5) = 11.91$.

8.53 (9.2-7.9) $22.33 / \text{phi}^2 = 8.53$; $11.86 / \sqrt{2} = 8.39$; $29.46 \times (2/3)^3 = 8.73$; $12.8 \times (2/3) = 8.53$; $17.9/2 = 8.95$; $19.85 \times (2/3)^2 = 8.82$; 8.85 is the apsides cycle; $11.15 / \sqrt{2} = 7.9$.

5.2 $11.86 \times (2/3)^2 = 5.27$; $22.3 / (21/13) = 5.3$; $8.85 / (5/3) = 5.31$; $17.9 / (3/2)^3 = 5.3$.

We do see signs of the lunar nodal and apsides cycles but solar activity spectrum frequencies and planetary orbital and conjunction cycles are even clearer.

It is interesting to note that summer temperatures for three Svalbard lakes have been reconstructed going back ~12 kilo years just after the end of the last ice age(32). Peak warming occurred at ~ 10 ky BP when the temperature was 7 degrees C higher than today. Temperatures then fell between 9.5 & 8 ky responding to freshwater fluxes from melting ice. By ~4 Ky BP temperatures had stabilised near current levels. So for much of the last 10 Ky temperatures were at or well above current levels. Similar observations, based on lake sediments, can be made about conditions in Greenland with peak temperatures 4-7 degrees C above current levels and on average 2.5- 3 degrees C warmer during the Holocene Thermal Maximum (46). For the last 10 Ky ice volume was significantly less than today (6.5 units) with a peak at ~ 9 Ky (6.35 units) and a minimum at 5 Ky (6.2 units). Variation over the last millennium has been ~ +/- 0.04 units.

There would appear to have been no collapse of the Greenland ice cap despite temperatures elevated 4 – 7 degrees C above current conditions for several hundred years. It would seem that fears of global floods have been somewhat exaggerated in the climate change jihad.

5.19 The Indian Ocean Dipole, Monsoons & Solar Activity

The IOD is a long term cycle in surface sea water temperature which is connected to the monsoon patterns over Indonesia and to drought cycles in Australia. It is the Indian Ocean equivalent of El Nino in the Pacific Ocean. Nugraho (33) carried out a high resolution wavelet spectral analysis of the IOD using monthly data for the period April 1978 to January 1996. He analysed separately the complete monthly record; the monsoon wet season (December- February period) data; the monsoon dry season (June-August period) data. The spectra are full of detail but Nugraho merely noted that in the wet season spectrum there was significant power (but not peak) power) at a period of ~11 years while ignoring the higher power peak centred on ~9 years throughout the full 18 year data series. The author noted immediately that this period, 8.84 years was 17.7 / 2. The three main spectral peaks in the Hale (magnetic) solar activity spectrum are at 17.9, 19.86 and 22.33 years. The data series are too short to pick up these long periods but could pick up their shorter harmonics. On this basis the author carefully measured the periods of peak power for each series of Nugraho with interesting results.

Full monthly data series
In 1978 peak power is centred on 4.45 years. After ~1987 peak power period falls but also intensifies into a narrow peak centred on ~3.65 years. This kind of frequency shift in climate data is common and confuses automated interpretation. Note for now that 4.45 x 5 = 22.25 years, the primary Hale solar cycle. Note for now that 3 x 3.65 = 10.95 versus the sunspot primary cycle of 11.16 years (or 22.3 / 2). There appear to be clear harmonics of solar activity variation in the full data series. There is also a moderate annual periodicity as we would expect.

Wet season data series
The wet season spectrum is dominated by a broad band of power between ~7 and ~11 years. However there is a narrower band of intense power centred on 8.84 years throughout the data series. This period immediately suggested 17.9 / 2 = 8.95 or the first harmonic of the strong 17.9 year peak in the Hale solar activity spectrum. However we also note that 8.85 years is the period of the lunar orbital apsides cycle which shows up in other climate series.

Dry season data series
Medium power between ~5.5 years and ~3.5 years dominates until ~1990 but for most of this period there is an intense power peak centred on 4.5 years. After 1990 this peak declines in power and moves to a lower period peak of renewed intensity centred on 3.31 years. We note that 4.5 = 18 / 4 while 17.9 is a major Hale period. Also 3.31 = 9.93 / 3 but 9.93 appears strongly in the Hale spectrum as the first harmonic of the major 19.86 peak giving us 19.86 / 2 = 9.93 years. There is also a narrow, moderate annual peak.

These preliminary results suggest looking at all the low order integer multiples of the periods we have found. These are presented in Table 1. This simple harmonic analysis in this case points directly to prominent periods in the '11 year' sunspot activity spectrum and the '22 year Hale' magnetic solar activity spectrum. We have for the full series, 10.95 (11.15) from the sunspot spectrum and 17.8, 18.2 (17.9) and 22.25 (22.33) from the Hale solar spectrum.

TABLE 1

series	Full series		Dry season series		Wet Season	
Obs. Period	4.45	3.65	4.5	3.31	8.84	
2 X	8.9	7.3	9.0	6.62	**17.7**	
3	13.4	**10.95**	13.5	**9.93**	26.5	
4	**17.8**	14.6	**18.0**	13.24	35.36	
5	**22.25**	**18.2**	22.5	16.55	**44.2**	
6 X	26.7	**21.9**	27.0	**19.86**	53.04	

For dry season we have 9.93 (9.93) from the sunspot spectrum and 18 (17.9) and 19.86 (19.86) and 21.9 (22.33) from the Hale spectrum. Note that 19.86 is also exactly the Jupiter-Saturn conjunction period. For wet season we have 17.7 (17.9) from the Hale spectrum. We note 44.2 = 2 x 22.1 (22.33) from the Hale spectrum.

In other climate systems we have seen phi and Fibonacci ratio harmonics appearing. This seems to be the case here since, for example 1.618 x 4.5 = 7.28 and (2 x 3.65 = 7.3) and 7.3 x 3 = 21.9 years versus Hale 22.3 years. (1.618 x 1.618) x 4.5 = 11.78 but 11.86 years is Jupiter's orbital period.

1.618 x 3.65 = 5.91 but this is 11.82 / 2 and 11.86 is Jupiter's orbital period. 1.618 x 18.2 = 29.44 but Saturn's orbital period is 29.46 years.

Can these phi observations help us with the remaining harmonic results in Table 1? They can. Some of the numbers are familiar as being close to minor peaks in the solar spectra. There are six harmonics with a common mean factor of 13.37 years but 13.37 x 5 / 3 = 22.28 years, the Hale primary activity cycle being 22.33 years. 5 / 3 is a Fibonacci ratio. We also note full series 14.6 years but 14.6 x 3 / 2 = 21.9 versus the Hale cycle. In the dry season we have 16.55 years but 16.55 x 4 / 3 = 22.1 years. This is not a Fibonacci ratio but close to another irrational number, the plastic number,1.325 which also appears in nonlinear dynamic systems.

We can conclude that much of the spectral power in the IOD and hence in monsoon rainfall is related in a nonlinear but transparent way to well known sunspot and Hale magnetic cycle frequencies. Monsoon spectral power over periods of decades shifts between solar activity harmonics but in an organised fashion. The wet season spectrum was stable for ~1980 – 2000.

The dry season and overall series show very similar frequencies. There should be potential in principle for medium term forecasting. It should be noted that there are strong direct hints of the Jupiter-Saturn conjunction period (in the dry season data at 19.86 years) and Jupiter's orbital period. For example in the wet season we have 8.84 x 4 / 3 = 11.78 years versus 11.86 years via the plastic number (i.e. 1.325) while 8.84 x (5 / 3) = 14.73 = 29.46 / 2 versus the Saturn orbital period of 29.46 years (and recall also the full series period 3.65 harmonic multiple at 14.6 and 14.6 = 29.2 / 2 but 2 / 1 is also a Fibonacci ratio). For the full series, 3.65 x 13 / 8 = 5.93 = 11.86 / 2. The ratio 13 / 8 = 1.625 is another Fibonacci ratio close to phi. Also 4.5 x (8 / 5) x (8 / 5) x (8 / 5) x (8 / 5) = 29.5 versus Saturn orbital period 29.46 years. 8 / 5 = 1.6, another Fibonacci ratio. The planetary dynamics matches involving the two largest gas giant planets are surprisingly good.

5.20 Global Surface Temperature Spectral Analysis using SVD

Several analysts have used various techniques to analyse a number of data sources on global surface temperature. This analysis uses the official monthly temperature anomaly time series for 100 years ending in the mid 1990s. The study also used the singular value decomposition technique and Monte Carlo simulation to establish statistical confidence limits. It is a solid piece of work producing results that correspond closely with all the other analyses in this book in terms of key spectral frequencies and it also attempts to link those frequencies to known internal weather cycles such as El Nino and the North Atlantic Oscillation. The authors note a positive secular trend in the temperature anomalies but talk only of 'possible anthropogenic effects on global climate' (34). Their aim it seems is to look at the effect of natural internal cycles around that trend. That is eminently sensible. What makes this study of special interest is that it is co-written by Professor Michael Mann, originator of the infamous Hockey Stick graph (see section 5.24) and denier of the effect of external forces (e.g. the sun) on the earth's weather. Mann is 'Mr. CO_2'.

We will see if Mann's spectral analyses agree with the several other weather series we have examined in highlighting frequencies *easily related* to external forcing sources i.e. solar output variation driven by planetary orbital dynamics and possible direct effects of these on the Earth / Moon system. Finally we examine his appeal to internal 'natural' weather cycles. These of course we have already linked in detail to the sun and planets.

Mann seems well aware that frequencies visible in weather spectra can be subject to slow drift and to modulation (from interaction with longer period variations). He studies the full 100 year data series but also the first 89 years and the last 90 years. This demonstrates in this case that the frequencies are quite stable. We will therefore discuss the full 100 year data results shown below. All the peaks are above 95% significance.

119 years   11.86 x 10 = 118.6; 29.46 x 4 = 117.84; 19.85 x 6 = 119.1;
11x11.1 = 122;
4
17.9 x (8 / 5) = 117.3. This peak is broad and may not be well estimated although it is highly significant. Even so we have simple multiples of the Jupiter and Saturn orbital periods two major sidebands of the Hale solar cycle at 17.9 and 19.85 years. The latter is also the JS conjunction period but also a sub-harmonic of the main 11.1 years SS cycle. Overall, remarkably clear links. .

44.6 This peak is clearly visible on the side of the broad 119 peak.
It is of course 2 x 22.3, or twice the Hale main cycle; 19.85 x $(3/2)^2$ = 44.66; 17.9 x (5 / 2) = 44.75; 8.85 x 5 = 44.25, based on the lunar apsides cycle.

16.4 (15-18) 11.1 x (3 / 2) = 16.65; (19.85 / 2) x (5 / 3) = 16.5; 17.9 is a Hale solar sideband; 22.3 / (3 / 2) = 14.9.

11 11.1 is the primary SS cycle; 17.9 x (8 / 13) = 11.02; 29.46 x $(8/13)^2$ = 11.16 but **12** is probably 11.86, Jupiter's orbital period.

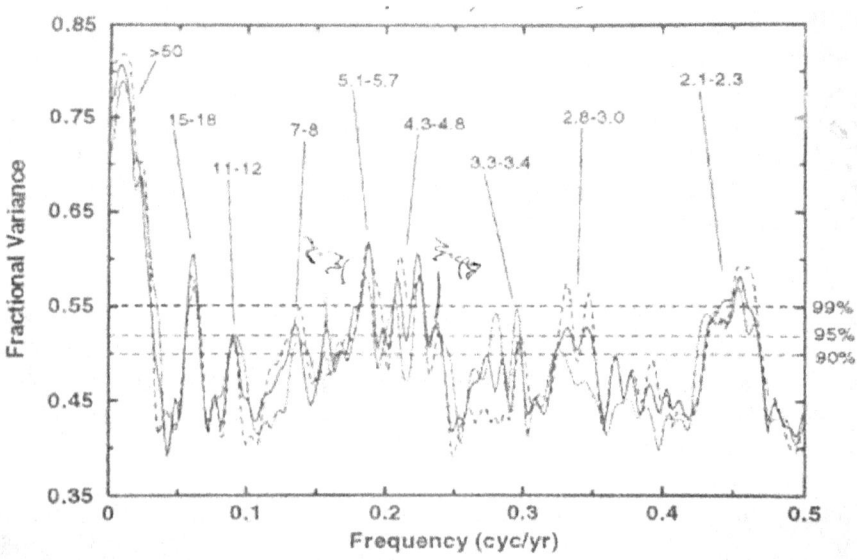

7.46 22.3 / 3 = 7.43; 19.85 x $(8/13)^2$ = 7.51; 11.86 x (5 / 8) = 7.42; 29.46 / 4 = 7.37.

5.4 (5.1-5.7) 22.3 / 4 = 11.15 / 2 = 5.57; 11.86 / $(3/2)^2$ = 5.27; 29.46 / $(3/2)^4$ = 5.8; 11.86 / 3 = 5.93.

4.83 29.46 / 6 = 4.91; 22.3 / $(5/3)^2$ = 4.82; 19.85 / $(8/5)^2$ = 4.85.

4.46 22.3 / 5 = 4.46; 17.9 / 4 = 4.48; 11.86 / $(13/8)^2$ = 4.49; 29.46 / $(8/5)^4$ = 4.49

3.37 $19.85 / 6 = 3.31$; $11.86 / (3/2)^3 = 3.5$; $11.15 / (3/2)^3 = 3.3$;
$17.9 / (3/2)^4 = 3.53$.

2.9 $11.86 / 1.6^3 = 2.84$; $29.46 / 10 = 2.95$; $17.9 / 6 = 2.98$; $19.85 / phi^4 = 2.9$
$22.3 / (5/3)^4 = 2.89$.

2.2 $11.1 / 5 = 2.22$; $19.85 / 9 = 2.2$; $11.86 / 5 = 2.37$.

We see direct evidence of sub-harmonics and harmonics of the 22.3 years main Hale magnetic solar cycle: 44.6, 11.1, 7.43, 5.57, 4.46, 2.22 and also Fibonacci harmonics such as 16.6, 2.89. The solar link could not be clearer. Similarly we see harmonics of the Hale sideband periods at 19.85 and 17.9. Links to planetary orbital periods are also clear in sub-harmonics and harmonics of 11.86 (Jupiter), 29.46 (Saturn) and the JS conjunction period 19.85, yielding the 119 year spectral peak but also 7.43, 4.91, 3.31, 2.84, 2.2. Jupiter and Saturn periods and their conjunction period also show up in shorter temperature periods as Fibonacci harmonics. It is ironic that such a clear result comes from a careful spectral analysis of a standard global surface temperature time series by the arch denier of external influences on Earth's weather and key promoter of manmade climate change, Michael Mann. We must assume that Mann, knowing the sun is irrelevant a priori, because climate change is manmade, has not bothered to check out solar activity periodicities never mind planetary dynamics. Why bother when you know the causation answer already? There is further irony in Mann's assertion that internal cycles such as El Nino, explain the periodicities he finds.

'…significant peaks…within a broad period range from 2.8 to 5.7 years exhibit characteristic El Nino- SO patterns …An inter-decadal mode in the 15-18 year period range appears to represent long term EN-SO variability…This mode has a sizeable projection onto global-average temperature and accounts for much of the anomalous global warmth of the 1980s.'

Well we noted that his $2.9 = 11.86 / 4 = 2.84$ and $29.46 / 10 = 2.95$ and the JS conjunction (and Hale sideband), $17.9 / (10 / phi) = 2.9$ but also $22.3 / 4 = 5.57$, $22.3 / 5 = 4.46$, $19.85 / 6 = 3.31$.

His El Nino 'band' contains several planetary and solar harmonics.

In section 5.10 we also looked at a modern MESA analysis of El Nino / Southern Oscillation Index data for recent decades and noted a major peak at **2.5 years** or $22.3 / 9 = 2.48$, $19.85 / 8 = 2.48$, $17.9 / 7 = 2.55$, $25 / 10 = 2.5$ years with all the main Hale solar peaks present. Mann found peaks at 2.2 and 2.9 but we note that $\sqrt{(2.2 \times 2.91)} = 2.53$. His method split the **2.5 year** peak in the older data. Similarly his 15 to 18 year band is centred on $\sqrt{(15 \times 18)} = 16.43$ which is $11.1 \times (3 / 2) = 16.6$, a simple Fibonacci harmonic of the main SS cycle. Another large El Nino MESA peak is at 13.1 years which is just $19.85 \times (2 / 3) = 13.2$. This is small in Mann's spectrum.

Based on the recent MESA analysis El Nino is *saturated* with solar activity periods and planetary periods. Mann's 'anomalous warming' in the 1980's is strongly related to the sun. The large El Nino event in 1999 and again in 2016-2019 also lifted global temperatures with a long 'hiatus' of no increase in between 1999 and 2015. In early 2021 the hiatus may have returned.

Mann also proposes that the power around 7-8 years and near 2.1-2.3 years is related to the North Atlantic Oscillation. We examined the NAO MESA spectrum in section 5.3 and noted large peaks at 13.3, 7.8, 5.1, 3.5, 2.94, 2.3 years. The latter 5 peaks also occur in Mann's spectrum, though not the largest at 13.3. So he was correct that NAO and global surface temperature share some frequencies.

However we showed that the NAO was strongly linked to solar activity and the planets. So, to take a few examples, $11.86 \times (2 / 3) = 7.91$; $22.3 / 4 = 11 / 2 = 5.5$; $19.85 / 4 = 4.96$; $11.1 / 3 = 3.7$, $11.86 / 6 = 3.31$; $29.46 / 10 = 2.95$; $11.86 / 4 = 2.96$; $11.86 / 5 = 2.37$. Recall that the 2.37 period is the mean value of the ubiquitous tropical Quasi Biennial Oscillation. Again a closer look at Mann's NAO attribution brings us back to strong solar / planetary correlations. Ironic indeed that his work confirms all that we have found in other data analyses.

5.21 Pacific Sea Temperatures since 1600 via Coral D18 O Records

This early spectral analysis (1994) of a long period temperature proxy record has important lessons to teach us. It has been widely referenced because of its relevance to the El Nino – Southern Oscillation debate but strangely the strong solar activity and planetary signals have been ignored(35). The authors examined oxygen isotope ratios and skeletal growth in examples of the massive corals Pavona clavus and P. gigantean on the west coast of Isabela Island in the Galapagos chain. The P. gigantean isotope record from 1961 to 1982 correlates strongly with the instrumental record odd of sea temperatures with R = -0.9. The P. clavus annual D18 O record taken from a 10m diameter colony, covers 1587 to 1953. The authors note that during the Little Ice Age the D18 O variations correlate well with North American tree ring variations but later warming between 1880 and 1940 is not seen at Isobela. That is, the 20th century 'manmade warming' effect is not seen here. The ENSO spectrum has significant periodic signals at 3.3, 4.6, 6, 8, 11, 17, 22, 34 years. The largest peak is at 4.6 years accounting alone for 12% of the total variance. The periods are familiar solar activity and planetary dynamics:

3.3 $11.15 / (3/2)^3 = 3.3$; $22.33 / (8/5)^4 = 3.4$; $19.86 / 6 = 3.31$;
$11.86 / (3/2)^3 = 3.5$; $29.46 / 6 = 3.27$.

4.6 $(22.33 / 3) / phi = 4.6$; $17.9 / 4 = 4.48$; $19.85 / (13/8)^3 = 4.62$;
$11.86 / (8/5)^2 = 4.63$; $29.46 / (8/5)^4 = 4.5$; $18.6 / 4 = 4.65$; $8.85 / 2 = 4.44$; $13.8 / 3 = 4.6$.

6 $19.85 / (3/2)^2 = 5.88$; $17.9 / 3 = 5.97$; $11.86 / 2 = 5.97$; $29.46 / 5 = 5.9$;
$18.6 / 3 = 6.2$; $8.85 / (3/2) = 5.9$.

8 $22.33 / (5/3)^2 = 8.04$; $19.86 / (8/5)^2 = 7.76$; $17.9 / (3/2)^2 = 7.96$;
$12.8 / (8/5) = 8.0$; $11.86 / (3/2) = 7.91$.

11 $22.33 / 2 = 11.15$; $18.6 / (5/3) = 11.15$; $29.46 / (13/8)^2 = 11.15$.

17 $11.15 \times (3/2)^2 = 16.73$; $22.33 / (phi/2) = 17.05$; $11.86 \times \sqrt{2} = 16.77$;
$29.46 / \sqrt{3} = 17.0$; $25 / (3/2) = 16.7$.

22 $22.33 = 2 \times 11.16$; $8.85 \times (8/5)^2 = 22.65$; $18.6 / (5/3) = 11.16 = 22.33 / 2$.

34 $11.16 \times 3 = 33.35$; $11.86 \times 2 \sqrt{2} = 33.55$ or $11.86 \times (5/3)^2 = 32.94$; $(19.85 / 2) \times (3/2)^3 = 33.5$; $29.46 \times (2 / \sqrt{3}) = 34.0$.

The authors do not look at longer cycle periods despite the 347 year long coral record but do remark that the amplitude of the 11 year $D18\,O$ cycle varies with the amplitude of the solar cycle. This suggests the presence of much longer modulation cycles well known from the sunspot and Hale solar time series.

5.22 The Relative Contribution of Anthropogenic Forcing And Natural Climate Change Cycles

This section is based on an extensive paper by Abbot & Marohasy from 2017 (36). It points out that the problem of nailing down the various contributions to alleged climate change in part lies in the short duration of most instrumental climate records, even for temperature. Even the famous Central England Temperature series dates back only to the late 17th century. Wavelet analysis indicates long cycles of ~90, and ~215 and much longer possible amplitude modulation cycles around 420, 700 and ~1,000 years. These cycles are of considerable power which means on shorter time windows like a century they will generate apparent secular trends. Abbott therefore decided it was essential to look at longer proxy temperature records from 750 to 1950 years long, for locations in both hemispheres and phenomena including lake varves, tree rings, pollen, stalagmites, bore holes, and coral, lake and marine sediments. They analysed six data records by estimating power spectra for the period from the start of the various time series until 1880. This identified up to 10 spectral periods for each series. The shortest period considered was ~ 25 years and the longest ~1,200 years. The spectra had familiar frequencies in the higher frequency bands and picked up much longer periods in addition.

We take New Zealand proxy temperature as an example. For short periods we have 51.1, 57.3, 67, 83.8 years. We see immediately that 67 is $3 \times 22.33 = 66.99$, marking the Hale magnetic solar activity cycle. Similarly 57.3 is close to $5 \times 11.86 = 59.2$, and $2 \times 29.46 = 58.9$, $3 \times 19.85 = 59.5$ reflecting the orbital periods of Jupiter and Saturn and their conjunction period. Also 51.1 is $4 \times 12.8 = 51.2$, $2 \times 25 = 50$, marking other gas giant conjunctions. 83.8 is $7 \times 11.86 = 83.1$ from Jupiter's orbital period and also

Uranus' orbital period of 84 years and half Neptune's period, 165 / 2 = 82.5 years. Solar and planetary signals are clear. Most spectral power is concentrated in a group of longer periods, 140, 148.5 and 154.8. These are obviously related since √ (140 x 154.8) = 147.3. Also 8 x 18.6 = 148.8, based on the lunar nodal cycle, and 5 x 29.446 = 147.3, reflecting Saturn's orbital period. 140 comes from 12 x 11.86 = 142.2 and 13 x 11.86 = 154.2.

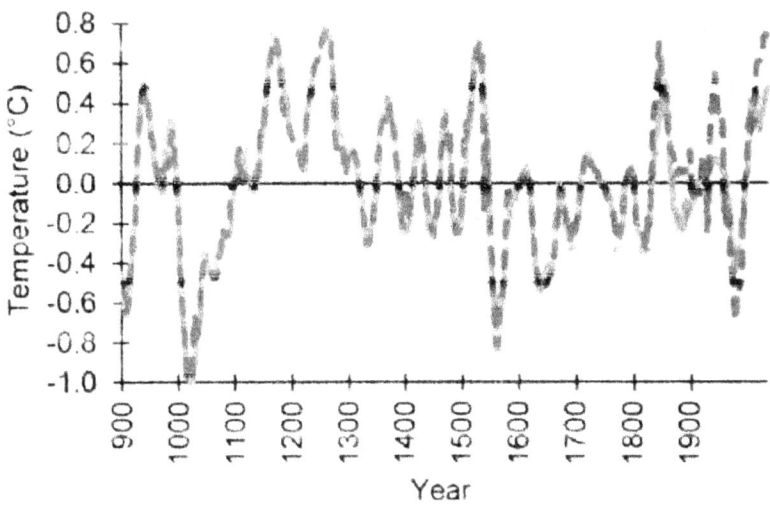

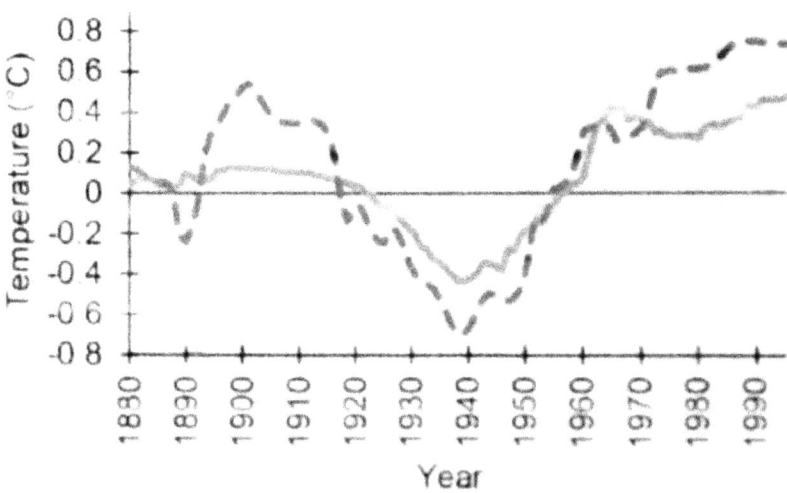

The central period, with the peak power comes from 5 x 29.46 = 147.3, from Saturn's orbital period but also 8 x 18.6 = 148.8, reflecting the lunar nodal cycle.

The periods discussed make up 94% of the spectral power in the time series. The other five temperature proxy series show similar solar activity/planetary links. The authors use the spectral frequencies found for each time series as inputs to an Artificial Neural Net in order to optimize the fit to data training sets up to 1830 using a subset of frequencies and choosing the best phase relationships. For the New Zealand data this process improves correlation from 0.87 in the raw spectrum to 0.98 in the ANN model accounting for at least 80% of the record variance with a few spectral components related to the sun and planet dynamics. This is an impressive fit as can be judged by the graphs above covering the whole data set and the post training period. The ANN model fitted in the pre-industrial period fits the post 1880 data well and the upward trends in temperature. All six ANN models capture the temperature trends of their respective data series in the 20th century which has obvious implications. Below as a further example is southern South America temperature predicted from 1880 to 1995.

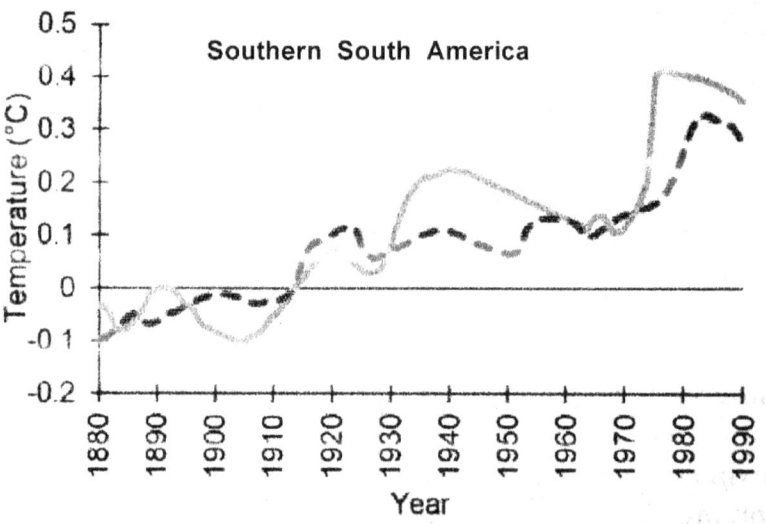

In this case a strong observed trend in temperature is completely explained by the natural cycle ANN model fitted to data *before* 1880. We note again the key periods found : 52, 89.2, 129.2, 141, 216.6, 245.4, 383.3, 745.4 years.

We again find obvious matches to sun and planets: 3 x 17.9 = 53.6; 4 x 22.33 = 89.3; 7 x 18.6 = 130.1; 12 x 11.86 = 142.3; 11 x 19.85 = 218.3; 11 x 22.33 = 245.6; 17 x 22.33 = 380; 4 x 179 = 8 x 89.5 = 32 x 22.33 and 25 x 29.46 = 737.

If these 'natural' cycle ANN models account fully for the post fitting period temperature trends and oscillations through the 20th century this leaves little or nothing for a CO_2 manmade warming effect to explain. By looking at the differences between the actual temperature changes over the post fit periods and the ANN model predictions the authors find deviations of from 0.06 to 0.2 d C. They generously take the largest deviation as a measure of CO_2 warming contribution and for an increase of ~100 ppm for the gas over the period **they estimate an equilibrium climate sensitivity of ~0.6 d C for a doubling of CO_2**. The IPCC have given various mean values and wide ranges for ECS over time but at the time of this paper in 2017, the General Circulation Model set mean was 3.2 d C and the range 2 to 4.5. So the temperature proxy spectral ANN models ECS is < 20% of the IPCC GCM mean estimate. We noted that spectral models for Tibet based on very long proxy series and for central Europe similarly leave no space for a large CO_2 effect. Now we see similar results for the southern hemisphere, plus Canada and Switzerland.

Incidentally the 1200 – 1950 AD spectrum for Switzerland is dominated by a period of ~ 35.5 years but 3 x 11.86 = 35.6 and 2 x 17.9 = 35.8, so we have harmonics of the Jupiter orbital period and a major sideband of the Hale solar activity spectrum. We also note 3 x 22.33 = 67 versus 67.7 and 11.16 x 7 = 78.2 versus 78.7 years in the spectrum. The longest periods, 346 and 722, can be seen as 2 x 179 = 358 and 4 x 179 = 716 years. We noted that ~179 coincides with three, 'three gas giant planet' conjunctions and an approximate four planet conjunction of between 171 and 181 years.

The authors point out that their ECS result is closest to the estimates produced by considering water vapour as the dominant greenhouse gas (unlike the IPCC models) and using spectral analysis results for the atmosphere (and experimental simulated atmospheres) coupled with global hourly measurements of temperature and *relative humidity*. These analyses yield an ECS range of 0.33 to 0.9 d C and a mean of 0.68 d C. Estimates of ECS have fallen significantly over the last two decades but, based on long duration **observed** temperatures proxies from across the Earth, not far enough. The psychological basis for this IPCC lagged response to strong emerging evidence is discussed in Appendix 3. It is an old disease in science.

5.23 The Relationship between CO2 and Global Surface Temperature

The 'consensus' is that there are no external driving forces acting on climate and that small variations in total solar irradiance have a negligible effect: therefore the sun has no effect. The 'consensus' does not extend to the quantitative results of the several dozen models produced so far. These produce forecasts of 'climate sensitivity' to CO2 (delta T for a CO2 doubling) spread evenly across a range of 1 to 6 degrees C. In various documents the IPPC now prefers to trim this range to 2 to 4 degrees C…for the sake of apparent 'consensus' preservation. The models are a poor fit to history and now look to be 'running hot' and particularly so in the tropics. Here is an analysis written in 2021 by Professor Christy of the University of Alabama in Huntsville, and a leading expert on satellite MSU temperature measurement.

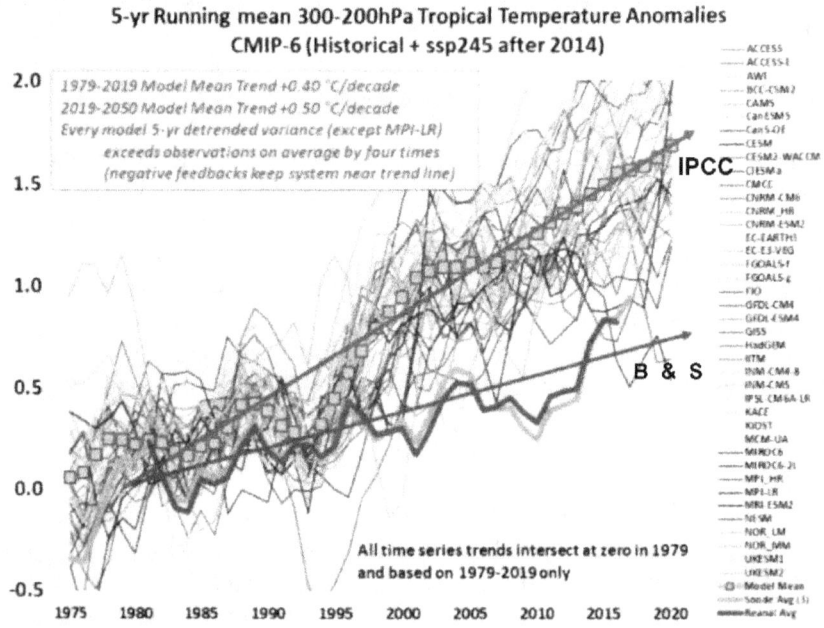

This shows a sample of the latest CMIP-6 'official' models for the period 1979 to 2019 and the line marked IPCC gives the mean trend line. By contrast the trend line from Balloon and Satellite real world records is marked B & S. The reader can decide if the IPPC approved GCMs match actual history and if there is a model 'consensus'.

Even adding in the major 2016 – 2019 double El Nino spike does not take us near the 'consensus' line and now global temperatures are falling. The 2000 – 2015 temperature hiatus in particular has attracted attention since CO_2 continued to rise strongly over that period.

Analysts have studied the hiatus in the hope of understanding its cause and one is of particular interest in that it uncovers the complexity of what is going on along with identifying some familiar forcing frequencies.

Ruzmaikin and Byalko of JPL and the Landau Institute of theoretical Physics in Russia made a detailed study of CO_2 levels and global temperature over the period 1979 to 2014 (40). They found a complex pattern of correlation between detrended CO_2 and temperature series far removed from the simple idea that temperatures rise in response to CO_2 increases which we will examine later. First we will examine the wavelet coherence between CO_2 and temperature in terms of frequency. Coherence looks for periods of phase agreement between variables at particular frequencies. In such periods we can infer causation.

They note that on periods of less than a year the situation is 'chaotic' with only small, scattered islands of coherence. Frequencies around one year occur in two islands over the periods 1986 to 1994 and 1997 to 2007 (these islands record areas of 95% statistical significance or above). At other times annual coherence is weak. CO_2 and global temperature are coherent for only ~50% of the 1979 to 2014 record. The only other stable islands (95%) occur for frequencies of 2 to 2.4 years from ~ 2010 onwards and 2.5 to 3.9 years for the period 1993 to 2000. There is a hint of moderate coherence centred on ~3.0 -3.9 years for the period 1979 to 1983. Moderate coherence occurs centred on ~5.6 years for the periods 1979 to 1984 and 2008 to 2014. That is it. Nevertheless these frequencies are of interest.

5.6 (5.0 – 6.0) 22.33 / 4 = 5.58; 11.86 / 2 = 5.93; 19.85 / 4 = 4.97; 29.46 / 5 =
$$5.89; 9.93 / 4 = 4.97^2; 17.9 / 3 = 5.96; 12.8 / (3 / 2) = 5.7.$$

3.9 19.85 / 5 = 3.97; 11.86 / 3 = 3.95; 11.154 / (5 / 3)^2 = 4; 29.46 / (5 / 3)^4 = 3.83.

3.0 22.3 / 7 = 3.18; 19.85 / 6 = 3.3; 17.9 / 3 = 2.98; 11.86 / 4 = 2.97; 29.46 / 9 = 3.27; 18.6 / 3 = 3.1.

2.4 Quasi Biennial Oscillation is 2.37 years and 11.86 / 5 = 2.37; 22.3 / 9 = 2.47; 19.85 / 8 = 2.48; 29.46 / 12 = 2.45;

2.5 El Nino primary peak at **2.5** years also visible from 1993 to 2000.

2.1 $29.46 / 14 = 2.1$; $18.6 / 9 = 2.07$; $11.15 / (3/2)^4 = 2.2$; $22.3 / 11 = 2.03$; $11.15 / 5 = 2.22$.

Note that some islands are harmonically related in that $\sqrt{(3 \times 3.9)} / \sqrt{(2.1 \times 2.4)} = 1.52$ or $3/2$, a familiar low Fibonacci harmonic. Also $5.6 / \sqrt{(2.1 \times 2.4)} = 2.49$ or $5/2$ and $5.6 / \sqrt{(3 \times 3.9)} = 1.635$ or $\sim 13/8$. The occasional islands of coherence are harmonically related and those harmonics in turn are well matched to simple solar activity harmonics and planetary orbital periods and conjunctions. Opening period and closing period power at ~5.6 years is punctuated in the central period (from 1991 to 2000) by that power transferring to the coherency island centred on ~3.4 years, the two being harmonically related as ~13 / 8. The 5.6 island is $11.15 / 2 = 5.58$, half the main SS cycle and this morphs into a coherency island including the main El Nino peak at 2.5 years ($22.33 / 9 = 2.48$). This not surprising given our other analyses. What is surprising is that the phase relationships for the coherent frequency islands are not what we are told to expect by the 'CO2 does it all advocates'.

Over the period 1979 to 1985 when there is a moderate ~5. 6 year cycle visible CO2 and temperature are **in phase**.

Over the period 1993 to 2000 when the 3.0 to 3.9 cycle is visible **CO2 leads temperature** in phase.

Over the period when the ~5.6 year cycle returns from 2010 to 2014 we see that CO2 and temperature are back **in phase**.

For the island centred on ~2.2 years over the period 2010 to 2014 CO2 and temperature are also **in phase.**

For annual periods the 1986 to 1994 island CO2 and temperature are in anti-phase. In the 1997 to 2007 island 97 to 2002 has CO2 and temperature in anti-phase and from 2002 to 2007 temperature leads CO2.

So, in summary for the multi-year periods of CO2 / temperature coherence from 1979 to 2014 CO2 leads temperature only for the period 1993 to 2000 or just 20% of the full record. For 80% of the record when there is coherence, CO2 and temperature are in phase. This means we cannot infer the direction of causality. Perhaps the lag between the variables is too short to pick up for most cycle periods. Looking again at the cycles close to one year we see that the 1986 to 1994 island shows CO2 and temperature in antiphase.

By 2012 with a new moderate island centred on a one year cycle, CO2 leads temperature again until 2014 and the record end. There is little evidence that CO2 consistently leads temperature in this data set across all islands of coherence. Where CO2 and temperature are in phase (for 80% of the multi-year frequency record) we can reasonably argue that causation could be in either direction or that both variables are driven by some other phenomenon. This result is surprising when we consider the 'consensus' certainty that CO2 causes global temperature change... dooming us all! We have also shown that all the coherent frequencies can be explained by harmonics of well established solar activity and planetary cycles found in other weather phenomena.

There seems to be a general problem among the 'consensus' climate community when it comes to phase relationships. Al Gore of course resorted to the long record of ice age temperature and CO2 variations measured in Antarctica and widely used, including by the US NOAA, to claim that CO2 causes, that is, leads in time, temperature changes. The classic series of recent ice ages is shown below. The reader should examine it closely. First note the sudden rises in temperature at the end of each ice age, and particularly for the last five. The temperature and CO2 levels rise 'simultaneously', taking into account the uncertainties, although it is probable there is still a several hundred year lag between temperature change and a CO2 response.

But look now at the decline of temperatures in the long fall back into glacial conditions after each interglacial temperature peak. It is absolutely clear that temperatures fall away rapidly and CO2 levels *follow*. The evidence tells us that temperature changes on these time scales *lead* CO2 changes. It is important to note that there is universal acceptance that the drivers of the main ice ages are long cycles in the shape parameters of the Earth's orbit and in the tilt of the Earth's axis (see section 5.11).These change the total energy input reaching the Earth's surface from the Sun. Changes in input energy change temperature. The author knows of no climatologist suggesting that changes in input energy from the Sun *directly* modulate CO2. Some analysts also suggest that the Sun itself undergoes long ~100,000 year cycles in energy output also related to planetary dynamics modulation effects (see section 5.11). Recently groups like NOAA have admitted the problem but hedged their bets

'While it might seem simple to determine cause and effect between CO2 and climate *from which change occurs first* ... the determination of cause and effect remains exceedingly difficult.'

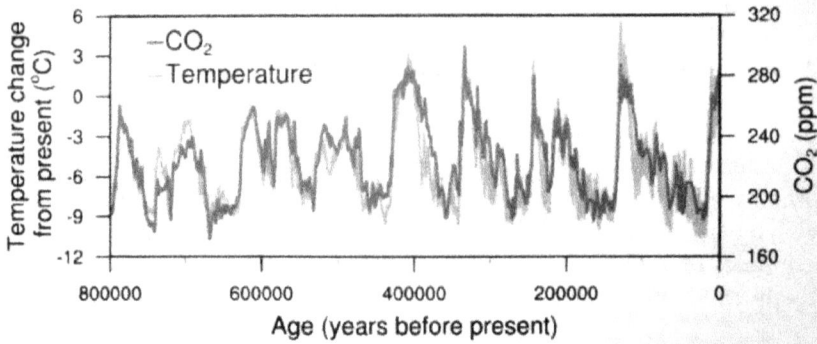

Temperature change (light blue) and carbon dioxide change (dark blue) measured from the EPICA Dome C Ice core in Antarctica (Jouzel et al. 2007; Lüthi et al. 2008).

The age old scientific principle that 'cause **must** precede effect' is being abandoned. Even so this statement is a long way now from the certainty of 'An Inconvenient Truth' and the claims of the IPCC about manmade climate change requiring all countries to declare a 'global climate emergency'. This is not science but some kind of millennial, global mass neurosis.

Other workers have used various means to estimate 'climate sensitivity' instead of the General Circulation Models used exclusively by the IPPC, which give a CS range of 1 to 6+ d C. As for the GCMs the range of independent estimates is wide but some are as low as 0 degree C. To put these numbers in context the basic, undisputed physics calculation tells us that, with no other knock on effects, a doubling of CO2 should yield a CS of ~ 1d Centigrade. The issue is whether there are positive feedback effects in the real climate system or negative feedback effects. By definition the IPCC models appear to favour positive feedback, significantly magnifying the basic CO2 sensitivity. Looking back to the comparison of the IPPC models with historical reality we started with it seems a lot of positive feedback is assumed and little negative feedback. One such feedback is the Svensmark proposal about the effect of low clouds over the ocean.

Kauppinen and Malmi have looked again at the IPCC evidence (39). Kauppinen worked as an expert reviewer of the IPCC AR5 report and challenged the IPCC about the lack of *observational* evidence for the claims of high climate sensitivity values. The response was that the Technical Summary of AR5 contained it. The authors looked critically at that 'evidence'.

They attempted to reproduce the components of temperature change in natural factors and manmade green house gas factors.

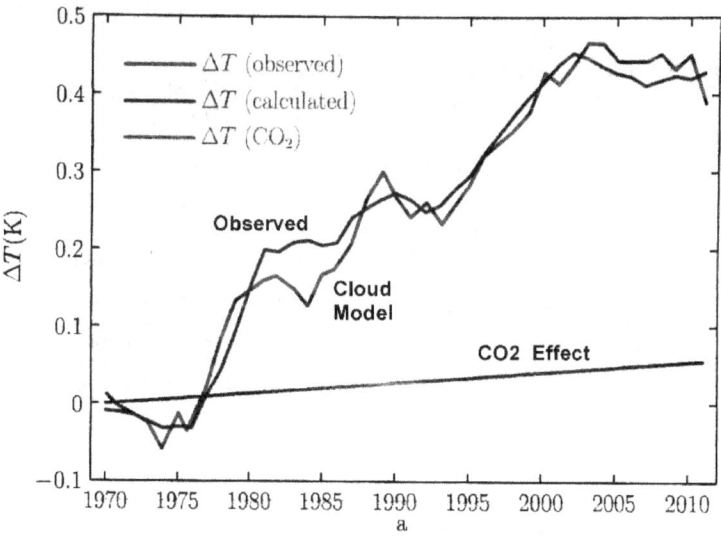

FIGURE 4. [2] Observed global mean temperature anomaly (red), calculated anomaly (blue), which is the sum of the natural and carbon dioxide contributions. The green line is the CO2 contribution merely. The natural component is derived using the observed changes in the relative humidity. The time resolution is one year.

They examined the data in Figure TS.12 of the Technical Summary of the IPCC Fifth Assessment Report which presented 15 regional temperature series covering periods from 1908 to 2008 and global total and land and ocean temperatures.

Also presented were the IPCC *model* means using only alleged natural forcings and using natural and anthropomorphic forcings. The authors point out *that the models do not count as experimental evidence of anything*. They assert that failing to include the feedback effects of low clouds means the GC models are biased in their calculation of the green house gas effects. These have to be increased to account (poorly) for the observed temperature history. Others have pointed to the same problem.

They then compare the global temperature anomaly series and observed global low cloud cover changes over a period from 1983 to 2008 for which the latter variable was available. These are *real observations* which can be used to estimate that a 1% increase in cloud cover decreases temperature by 0.11 d C. For the same period they back calculate the contribution of CO2 using

$$DT = (CS \ln C / C0) / \ln(2) - 0.11 \cdot Dc$$

Where DT is the observed global temperature change, CS is the climate sensitivity, C0 the starting CO2 level and C the closing CO2 level. Dc is the change in observed cloud cover.

They develop a simple model using cloud records to estimate the variation in low cloud cover over the oceans and hence its feedback effect in recent decades (back to 1970) and add that to the estimated CO2 trend effect. They conclude from joint model estimation that the real contributions to the global temperature anomaly series are as shown in Figure 4 above. The fit of the estimated cloud cover component and negative feedback to the global temperature anomalies is excellent, **capturing periods of low and high rates of change** almost exactly. The correlation is very high. The corollary is that there is little for the CO2 growth to explain. This gives a low estimate of 0.24 d C for 'climate sensitivity' compared with the theoretical 'no feedbacks', 1 d C and the up to 6 d C and more of the GCMs of the IPCC. The authors conclude that 'natural cycle derived' low cloud variations account for much of the global mean temperature variation over the modeled period. That is a very powerful result deserving of confirmation or refutation over longer periods.

Of course we have shown that many of the 'natural' decadal and shorter scale weather cycles over the oceans such as El Nino-Southern Oscillation , AMO, NAO, PDO and the QBO show clear spectral signals of solar variation driven in turn by planetary dynamics. On that basis much of the global temperature variation of the last century is explained by astronomical forcing.... just like the ice age cycles on a much longer time scale which are widely accepted. Why then should this shorter term result be a surprise? The Sun is King as every early agrarian civilization on this planet recognized. Somehow we have lost touch with the real world.

5.24 Hockey Sticks & Ski Jumps

We could say that the Great Climate Panic was set in motion by a single graph which grabbed the imagination of politicians and environmentalists on the make and BBC 'science' reporters. It is the infamous Hockey Stick of Michael Mann who we met in section 5 X. The HS dominated the IPCC's Third Assessment Report (2001) and attitudes and assumptions in later reports.

Mann's Hockey Stick

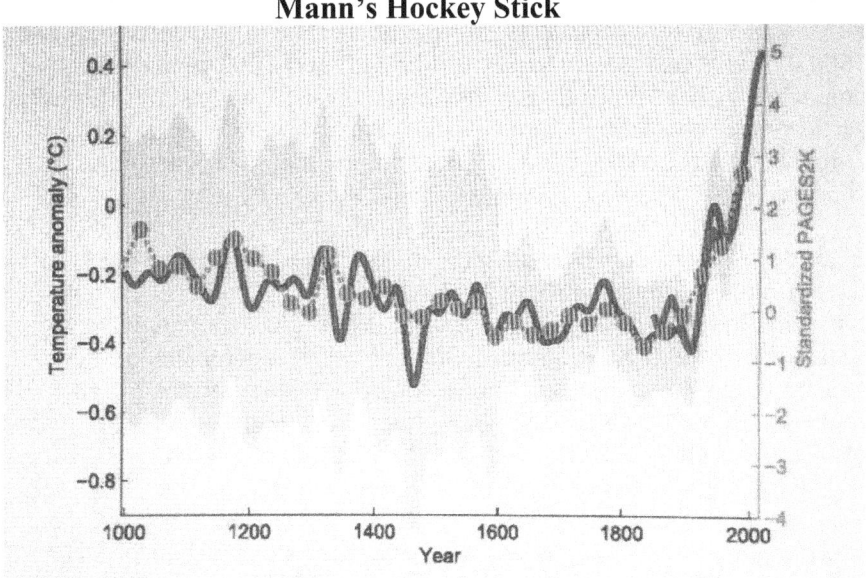

It shows global temperature *reconstructed* from many proxy time series such as tree rings from 1000 AD to the mid 20th century. The last, huge increase to 2000 AD is based on a 'spliced in' recent instrumental temperature series. Many papers have looked at the reconstructions critically. It has been claimed that some tree ring series were based on few, selected trees and so on. Readers may consult the vast literature on the HS debate for themselves.

The above diagram tells us many things of interest. The first is the high uncertainty in the temperatures shown. The pale gray area indicates the confidence limits on the various reconstructions. The band of uncertainty is large. Note that the period 1600 to 1900 AD is essentially flat. There is no sign of the Little Ice Age. Yet looking at many respected geophysical proxy series we have seen the prominence of the LIA in the climate record.

The HS supporters have also argued that the LIA if it existed, occurred only in Europe and parts of north America. We will test this below. Of course eliminating the LIA leaves no room for the argument that the Grand Solar Minimum of the 17th century could have influenced global scale climate!

Notice also that reconstructed temperatures fall slightly until around 1900 when, suddenly, wicked human industry kicks in and temperature rockets in response to CO2, instantaneously bending the Hockey Stick upwards. In fact CO2 was already growing though the 19th century. However the most interesting feature is that the Medieval Climate Maximum, centred around 1000 AD, with peak temperatures comparable to the 20th century, has gone! Yet we have noted that respected geophysical proxy series show the MCM repeatedly. Truly, Mann's Hockey Stick is a curve for all seasons …if you are an advocate for or religious follower of man-made climate change.

Since 2001 others have used a wider range of proxies to create what I call the Ski Jump. Ljungqvist used 120 proxies including ice-cores, pollen types, marine sediments, tree ring widths, lake sediments (varves), speleotherms and historical records from diverse geographical locations (Asia, Europe, N America, the tropics, and the high latitudes north and south) (41). In 2012 Christiansen and Ljungqvist published a further study of 32 proxies going back to 1 AD. The MWP peaked between 950 – 1050 AD at around 0.6 d C warmer than the calibration base of 1880-1960 AD. The LIA of the 17th century was ~1.0 d C below the 1880- 1960 AD datum. The proxy graphs are summarised overleaf. They all show a much steeper fall from ~1000 AD to ~1900 AD than the HS, of the order of 2 d C compared with 0.2 d C in the HS. That is a tenfold difference so either Mann or several other analysts are wrong. Note that the recent proxy studies also have much narrower confidence intervals than Mann's study which gives us confidence in the trends and shape.

The following graph from Loehle in 2008 takes us back 2000 years with 18 non-tree ring proxies. It shows the central estimate curve and the 95% confidence limits (42). The data confirm that the large, long term temperature swings are real. It shows a clear maximum between ~800 AD and ~1100 AD. It shows temperature stabilizing for a few centuries and then the steep decline into the late 16th century and 17th century, LIA. The temperature span is ~0.9 d C. Since the end of the LIA temperatures recovered to ~1800 AD halted until ~1900 and then increased rapidly through the 20th century. Temperatures today are comparable with the MCM peak temperatures.

Ljungqvist's 120 Temperature Proxy Times Series

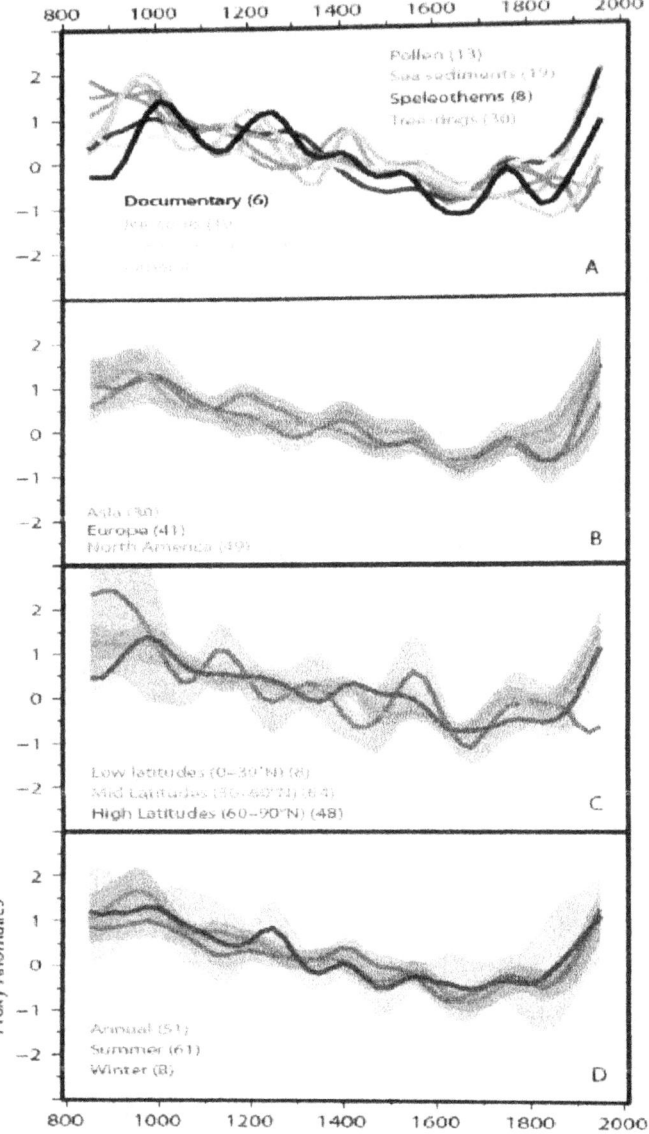

Ljungqvist et al 2012 Fig. 4 Mean time-series of centennial proxy anomalies separated by: (A) data type, (B) continents, (C) latitude, (D) seasonality of signal. The curves in (B-D) show the mean and moving block bootstrap confidence intervals (±2 standard error) (Wilks, 1997).

Loehle's Ski Jump

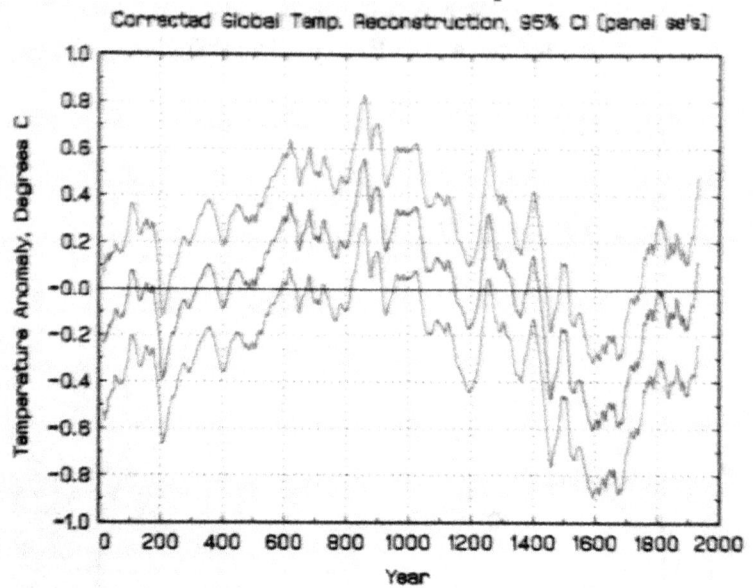

China temperatures from 1 AD

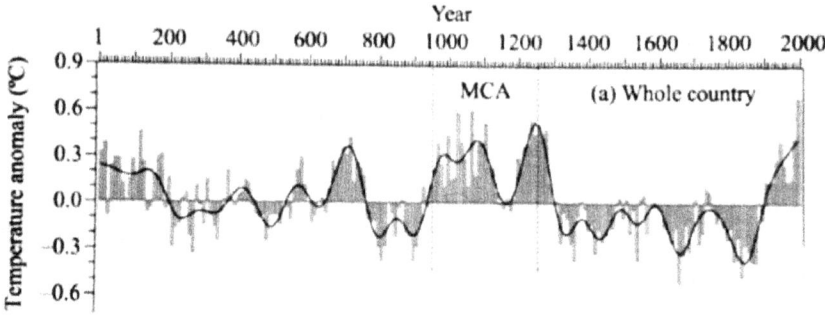

Source: **Hao et al., (2020)**

Who should we believe Mann or all the others? Well in the last few decades Chinese climate scientists have been busy filling in the historical climate map for East Asia which adds confirmation. Zhixin Hao et al (43) seem inclined to take the data as it comes. Hao studied four regions of China: northeast, northwest, central-east and the Tibetan Plateau using many proxies checked against detailed, local historical records. They used an 'ensemble empirical mode decomposition' method to analyze the times series.

We see immediately that there was a prominent double peak MCM between ~1000 AD and ~1300 AD. They note some variation in phasing of warm periods within this period from region to region but the high temperatures overall are clear. They are comparable with late 20th century temperatures. A similar narrower peak occurred around 700 AD in all regions and such high, but intermittent, temperatures also occurred between 1 AD and 200 AD. The Chinese have no doubt their 'Medieval Climate Anomaly' is real. It seems to mirror the MCM in Europe and other regions of the world. Their reconstruction also shows significantly lower temperatures around 1600 for several decades and a similar cool period centred on 1800 AD. This is compatible with the LIA in other regions.

The same reconstruction was used by Quansheng Ge et al to look in detail at periodicities and rates (44). They note the rapid temperature climb in the period 1870 to 2000 AD at 0.56 d C per 100 years but find similar rates of warming for 981 – 1100 AD and 1201 – 1270 AD. The team identified periodic signals in the time series at 50 – 70, 100-120 and 200- 250 years, periods which are simply harmonically related. The authors note that the large temperature excursions correlate with solar activity variations. Looking in more detail we see familiar patterns:

50 – 70 mean 60 years : $3 \times 19.85 = 59.6$; $22.33 \times (5/3)^2 = 61.5$; $11.86 \times 5 = 59.3$; $29.46 \times 2 = 58.9$; $3 \times 22.33 = 67$; $3 \times 17.9 = 53.7$.

100 - 120 mean 109.5 : $10 \times 11.16 = 111.6$; $11 \times 9.925 = 109.2$; $9 \times 11.86 = 106.7$; $4 \times 29.46 = 117.8$; $6 \times 18.6 = 111.6$; $12 \times 8.85 = 106.2$; $6 \times 19.85 = 119.1$.

200 - 250 mean 223.6 : $10 \times 22.33 = 223.3$; $11 \times 19.85 = 218.4$; $12 \times 17.9 = 214.8$; $12 \times 18.6 = 223.2$; $25 \times 8.85 = 221.3$. Notice that $22.33 / 8.85 = 5/2$, another Fibonacci harmonic and $18.6 / 11.165 = 1.666 = 5/3$.

We see major solar periods, lunar orbital periods and Jupiter – Saturn orbital / conjunction periods seen in many other climate series.

It is clear that the details differ between the various reconstructions reported in this book, including China, *but* the common large natural swings in temperature are clearly real. We can see this by looking at recent China temperature history which shows the same pattern of movement as the rest of the world but with interesting variations in magnitude and timing related to winds and cloud cover (45) which takes us back to the Svensmark Heresy to end our review of climate time series in this book.

The figure below shows temperature variations for the whole of China and for locations on the Tibetan Plateau above and below 4000 m. The dark dashed line shows global temperatures for comparison.

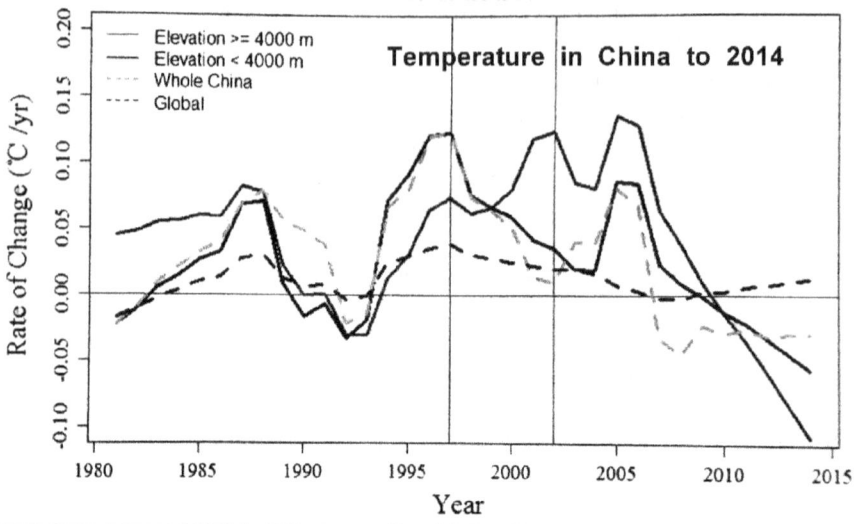

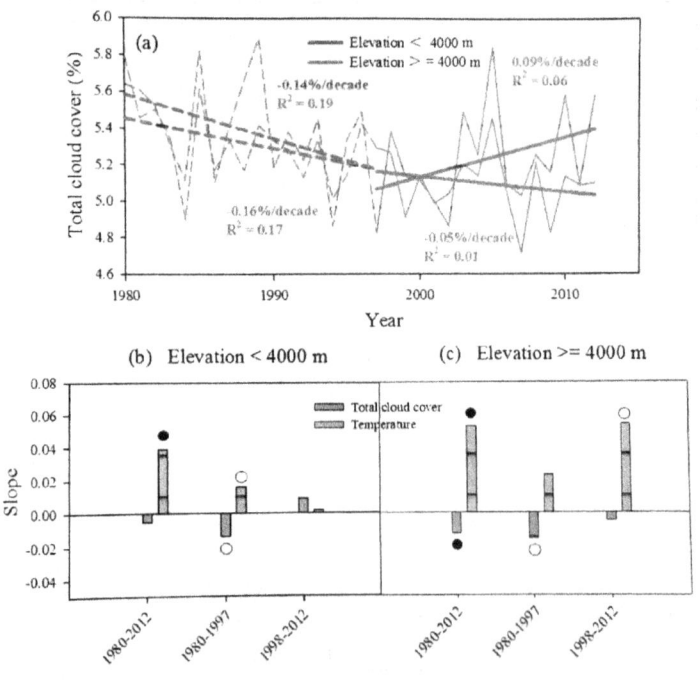

It is interesting that variations in China overall since 1980 showed strongly rising temperatures until the late 1990s (as occurred globally but even larger) and then a strongly falling temperature until ~2010 followed by a stabilization. The Chinese temperature hiatus was stronger than elsewhere.

On the Tibetan Plateau below 4000 m, temperature also peaked at ~ 1998 And then fell away as for the rest of China. On the Tibetan Plateau at elevations above 4000 m temperature continued to rise until 2005 and then fell strongly as far as 2014. The authors are interested in these differences in pattern and discuss the role of regional winds in detail. However what is of note is their analysis of the history of cloud cover from 1980 to 2014 in the two elevation zones with a mean fall rate of -0.15% per decade. At the lower and higher levels Cloud cover fell until ~1998. At higher levels cloud then stabilized to a small fall rate of -0.05% per decade. At lower levels cloud cover increased again at 0.09% per decade. The lower table shows comparisons of temperature trends and cloud cover trends. Overall temperature and cloud cover for both elevation ranges are negatively correlated and the trends are significant in the range 90% to 95%. Looking at similar studies the authors cautiously suggest cloud cover and wind speeds affect TP temperature variations. This is another modest regional result in favour of Svensmark's Heresy.

Section 5 References

1. Liu Y; '2,485 years of tree ring data for the central-eastern Tibetan plateau', China Science Bulletin, Vol. 56, No. 28-29, October 2011.
1A. 'Tibetan Plateau warming and precipitation changes in East Asia', Geophysical Research Letters, Vol. 35, 2008.
2. Murphy et al; 'Regression model for the 22 year Hale sunspot Cycle derived from high altitude tree ring data', Proc. Astronomical Society of Australia 11(2), pp157-163, 1994.
3. Landscheidt T; 'Solar eruptions linked to North Atlantic Oscillation', Schroter Institute for Research in Solar Activity, Belle Cote, Nova Scotia.
4. 'World will cool for the next decade', New Scientist, 12th September 2009, p10.
5. McCarthy G D et al; 'Ocean impact on decadal Atlantic climate variation revealed by sea level observations', Nature, 521, pp508-510, 28th May 2015.
6. Chiessi et al; 'Past South American Precipitation & AMO', Geophys. Res. Letters, Vol 36, L21707, 2009.

7. Knudsen M F et al; 'Tracking the AMO through the last 8000 years', Nature Comms. 2, No. 178.
8. Kane R P et al; 'Maximum entropy spectral analysis of total ozone', Pure & Applied Geophysics, Vol. 122, Issue 5, pp747-762, September 1984.
9. Schonwiese C D; 'Some statistical aspects of regional and global Climate change within the instrumental period', in 'Climate & Societies', Eds. Yoshino M et al.
10. Baldwin M P et al; 'The Quasi-Biennial Oscillation', Reviews of Geophysics, Vol. 39, Issue 2, pp179-229., May 2001.
11. Kane R P; 'Quasi-biennial and quasi-triennial oscillations in the rainfall of Northeast Brazil', Revista Brasileira de Geofisico, Vol. 16, No. 1, Sao Paulo March 1998.
12. Landscheidt T; 'Solar Activity controls El Nino & La Nina', Proceedings of The 1st Solar & Space Weather Euro-conference, Tenerife, 25-29 September 2000.
13. Sharma M; 'Variations in solar magnetic activity during the last 200,000 years: is there a Sun-climate connection?', Earth & Planetary Science Letters Vol 199, issues 3-4.
14. Park J et al; 'Plio-Pleistocene time evolution of the 100 Kyr cycle in marine records', J Geophysics Res. 98; B1, pp447-461.
15. Huybers; 'Obliquity pacing of the late Pleistocene glacial terminations', Nature, 434, pp491-4, March 2005.
16. Rutherford et al; 'Early onset and tropical forcing of the 100 Kyr Pleistocene glacial cycles', Nature, 408, pp72-75, November 2000.
17. Rial J A; 'Earth's orbital eccentricity and the rhythm of the Pleistocene ice ages : the concealed pacemaker', Global & Planetary Change, 41, pp81-93, 2004.
18. Mitsui T et al; 'Bifurcations and strange nonchaotic attractors in a phase oscillator model of glacial-interglacial cycles', submitted to Physica D : Nonlinear Phenomena, June 16, 2015.
19. Ditlevesen P D; 'The bifurcation structure and noise assisted Transitions in the Pleistocene glacial cycles', Neils Bohr Institute, University of Copenhagen, December 2013.
20. Aban J E et al; 'Characterisation of times series spectra from NOAA – AVHRR data', Centre for Environmental Remote Sensing, Chiba University, Japan; www. academia.edu
21. Spencer R W; 'Global warming as a natural response to cloud changes associated with the PDO', October 20th 2008, www.drroyspencer.com/research-articles/global-warming-as-a-natural response/
22. Ludecke H J; 'Multiperiodic climate dynamics: spectral analyses of long term instrumental and proxy temperature records',

Climate Past 9, p447-452, 2013.
23. Krahenbuhl D S; 'Investigating a solar influence on cloud cover using North American Regional Reanalysis data', J. Space Weather Space Clim. 5, A11 (2015).
24. Knudesn M F et al; 'Evidence for external forcing of the AMO since termination of the Little Ice Age', Nature Comms. 5, Article No. 3323, 25th February 2014.
25. Vautard R et al; 'Singular Spectral Analysis : a toolkit for short, noisy, chaotic signals', Physica D 58 (1992), pp 95 – 126.
26. Ghil M et al; 'Nile River Records Revisited: how good were Jacob's predictions?', Advanced spectral methods for time series analysis, Lecture V, Nelder fellow Lectures, UCLA, 3rd March 2014.
27. Xiao S et al; 'Coherence between solar activity and the East Asian winter monsoon variability in the past 8000 years from Yangtze River-derived mud in the East China Sea', Palaeogeography, Palaeoclimatology, Palaeoecology, 237, (2006), pp 293-304.
28. Roberts W O; 'Relationships between Solar Activity & Climate Change'; University Corporation for Atmospheric Research.
29. Mitchell J M et al; 'Evidence of a 22 year Rhythm of Drought in the Western US related to the hale Solar Cycle since the 17th century'; Proc. Sym. Ohio State Universty, 24-28, August 1978.
30. Zhao X H et al; 'Periodicities of Solar Activity and Surface Temperature variation of the Earth and their Correlations'; Chin. Sci. Bull 2014, 59: 1284.
31. Humlum O et al; 'Spectral Analysis of the Svalbard Temperature record 1912-2010'; Advances in Meteorology, Vol 2011, Article ID 175296.
32. Willem G M et al; 'Early Holocene temperature oscillations exceed amplitude of observed and projected warming in Svalbard lakes'; Geophysical Research Letters, 03 December 2019.
33. Nugraho J T; 'Appearance of solar activity signals in Indian Ocean Dipole phenomena and monsoon climate pattern over Indonesia'; Bulletin of the Astronomical Society of India, December 2007.
34. Mann M E & Park J; 'Global-scale modes of surface temperature variability on interannual to century timescales'; Jour. of Geophysical Research, Vol. 99, No. D12, pp 25,819-25,833, December 20th, 1994.
35. Dunbar R B et al; 'Eastern Pacific Sea Surface Temperature since 1600 AD : the D18 O record of climate variability in Galapagos corals'; Paleoceanography, Vol 9, Issue 2, pp 291-315, April 1994.
36. Abbott J & Marohasy J; 'The Application of machine learning for evaluating anthropogenic versus natural climate change'; GeoResJ, 5th August 2017.

37. Ogurtsov M G; 'On the possibility of long term forecasting of solar activity using radiocarbon sunspot proxy'; Multi-Wavelength Investigations of Solar Activity, Proc. IAU Symposium No. 223, 2004.
38. Patterson J S; 'DSP analysis of global temperature data series suggest global cooling ahead'; Watts Up With That website, September 11th 2013.
39. Kauppinen J & Malmi P; 'No Experimental Evidence for Significant Anthropogenic Climate Change'; arXiv: 1907.00165v1 [physics.gen-ph] 29 June 2019.
40. Ruzmaikin A & Byalko A; ' American Journal of Climate Change, Vol. 4 pp181-186, 18th May 2015.
41. Lungjqvist F; 'A Regional Approach to the Medieval warm Period & the Little Ice Age'; published August 2010, DOI 10.5772 / 9798.
42. Loehle C & McCulloch J H; 'Correction to A 2000 year Global Temperature Record based on non-tree ring Proxies'; Energy & Environment, 1st January 2008.
43. Hao Z et al; 'Multi-scale Temperature Variations and their Regional Differences in China during the Medieval Climate Anomaly'; Journal of Geophysical Sciences', January 6th 2020.
44. Quansheng G et al; 'Characteristics of temperature Change over China over the last 2000 years and Spatial Patterns of Dryness / Wetness During Cold & Warm Periods'; Advances in Atmospheric Sciences, August 2017, Vol. 34, Issue 8, pp 941-951.
45. Schweinsberg A D et al;' Multiple independent records of local glacier variability on Nuussuaq, West Greenland, during the Holocene'; Quaternary Science Reviews, 215, 2019.

Section 6   Climate Futures: Ice or Fire?

A day like today I realize what I've told you a hundred times: that there is nothing wrong with the world. What's wrong is our way of looking at it.

Henry Miller; A Devil in Paradise

The simple answer to the above question based on our review of climate history is: both. The more complex answer depends on the time scale we consider, on 'natural' cycles, on our future actions or failure to act and, frankly, on luck. The luck element partly relates to the many 'cosmic' influences on the Earth which from time to time, impact in major ways on climate and life: star burst events as we cross galactic spiral arms or pass through the dusty galactic plane; dwarf galaxy collisions; random gamma ray bursts and supernovae; comet showers from the Oort Cloud; collisions with one of the many thousands of NEOs, near earth asteroids, in the 100 mt diameter and above class, along with the occasional 'Dinosaur Killer'.

The other primary luck factor concerns the restless Earth itself. We live on thin plates of slag drifting on a sub-surface ocean of magma. As we saw, there are mantle 'hot spots', occasional 'flood basalt' events and more frequent 'super volcano' eruptions in the Yellowstone class. Such events charge the atmosphere with SO_2 and aerosols. In our climate review we also saw that the drifting continents play a major role in determining climate for periods of many millions of years. Like giant chess pieces their positions on the great game board of the Earth's surface control the ocean currents and the winds. If a continent or supercontinent occupies one or both poles an ice epoch will follow. Also a polar ocean surrounded by land is cut off from warming ocean currents and growing ice and snow will reflect the Sun's heat back into space and enhance cooling.

Mountain ranges and plateaus thrown up by plate collisions will change the winds and rainfall patterns and the level of chemical weathering, sometimes with global effects. These threats are all too real and as yet, largely beyond our technical powers to mitigate. Perhaps in a few millennia, if we survive, we will no longer have all our eggs in one basket, the Earth. Now at least, we have the Spacewatch networks mapping the NEO orbits and in a few decades, given early collision forecasts, we will have the power to deflect the smaller rocks. That will remove one of the 'short term' threats to climate and life.

In our historical walk through climate we also came to suspect that life has also played, and still plays, a major role in influencing climate on many time scales. Life and climate have co-evolved from the beginning when microbial life transformed the atmosphere more than once, pumping out gases like methane and later toxic oxygen as photosynthesizers became dominant, creating the 'Great Oxygen Catastrophe'. Huge quantities of carbon were and are exchanged between atmosphere and life forms on land and in the oceans where there are massive stores of carbon in biomass and in the soil, sediments and sometimes permafrost layers. Sometimes sea level changes caused by plate movements and ice epochs disturb the equilibrium and flooding of the continental margins releases vast quantities of buried CO_2, or worse, deposits of clathrates, methane hydrate. In the long game, deep buried, life generated, carbonates are carried into plate subduction zones where they change the composition of the magma and 'lubricate' plate motion.

We also saw that ocean life forms are the source of vast flows of sulphur into the atmosphere via dimethyl sulphide and that this is the primary source of SO_2 and sulphuric acid droplets which create cloud condensation nuclei. Volcanic eruptions and human industry are further sources of SO_2. What is less certain, but plausible, are the views of Lovelock and others, that plankton productivity, sulphur transport into the atmosphere and cloud formation are parts of an evolved cybernetic system which affects albedo and the Earth's energy balance ...and so keeps conditions in the 'optimal' range for the ocean plankton. Perhaps.

It is interesting that in looking at the Earth's energy balance we noted the views of eminent scientists like Professor Hoyle that the efficiency of water vapour and CO_2 radiation traps in the atmosphere were 'insensitive' to changes in the levels of those gases over large ranges. Hoyle and others point out that the key factor in the energy balance is albedo, the reflectivity of the Earth, and that is controlled by the relative areas covered by cloud, snow, ice and to some degree, vegetation. We also saw that the effect of cloud variation is highly significant. Perhaps Lovelock has a point about the scope for self-regulation. It is fascinating that our chief heretic, Svensmark, also focussed on clouds and albedo. Svensmark took the role of sulphur compounds as a given and concentrated on another aspect of the process: the catalysing role of electrons derived from collisions of secondary cosmic rays with atmospheric molecules. His CLOUD experiments at CERN appear to have shown the validity of his model. Svensmark and others like Shaviv have argued that the variation of galactic cosmic ray flux caused by several 'cosmic' factors, on many time scales, has strongly impacted our climate since the beginning.

One cannot criticise these 'heretics' for lack of vision. Just about every other aspect of their heresy has been hounded unmercifully by the majority 'global warming tendency'. But in science the rule is: big ideas, particularly if they contradict the current consensus, require big evidence. Conflict was inevitable since Svensmark has claimed that his cloud forcing model and experiments lead to the same climate sensitivity parameter as that claimed for CO_2, around 1.3-1.4 Watts per square metre.

Is Svensmark correct? We do not know yet. All I can say to the reader is consider what we found when we analysed the spectral 'fingerprints' of solar activity variation in several forms. Such variables included the strength of the Sun's magnetic field, sunspots and the solar wind on time scales of years to centuries. The solar wind definitely modulates the cosmic ray flux reaching the Earth. We then took the spectral 'fingerprints' of a range of climate records in some cases, going back thousands and millions of years. In all cases we found clear links between these 'fingerprints' …provided that we allowed the possible existence of classical non-linear behaviour in our climate systems. This is not an unreasonable proviso since the very equations which describe heat and mass transfer in atmosphere and ocean are non-linear and capable of rich behaviour (as are the equations of the proposed solar dynamo). We discovered that very often harmonics, sub-harmonics and interactions between prominent activity periods in the Sun showed up strongly in the climate 'fingerprints'. We saw other work which pointed to other effects like frequency and phase modulation and that some climate sub-systems do exist on the 'edge of chaos'. This behaviour showed up frequently through Fibonacci ratio harmonics in the Sun and climate records. Let's just summarise here the geophysical and climate records we have analysed :

1. Orbital parameters of the Earth-Moon system.

2. Global earthquake energy release and the Chandler Wobble (window : 1904 – 65) & Volcanic Activity.

3. Tibetan tree ring records (window : present – 2,485 BP)

4. Western USA tree ring records (window : 1673 – 1986 AD)

5. The North Atlantic Oscillation. (window : 1825 – 2000 AD)

6. The Atlantic Meridional (multi-decadal) Oscillation (window : present – 19,000 BP)

7. Central England Temperature. (window : 1659 – 2007 AD)

8. Total ozone variation. (window : 1930 – 1982 AD)

9. Northern hemisphere mean surface air temperature. (window : 1871 – 1980)

10. QBO & Brazilian rainfall. (window: 1951 – 1990 AD)

11. El Nino and the SOI.

12. Oxygen isotope records for the North Atlantic. (window : 0.8 – 2.77 million years BP)

13. The Pacific Decadal Oscillation (window : 1900 – 2010)

14. HadCRUT3 global surface temperatures (window: 1850-2012)

15. IPCC averaged global surface temperature (window: 1860 – 1990)

16. Long term River Nile level records (window : 622 – 1922 AD)

17. East Asia winter monsoon (window : present – 8000 BP)

18. Drought Cycles in the Western USA.

19. Global Temperature & Solar Activity Indices.

20. Svalbard Temperature Record 1912 – 2010.

21. Indian Ocean Dipole, Monssons & Solar Activity.

22. Global Surface Temperature Spectrum Analysis using SVD.

23. Pacific Sea Temperature since 1600 AD via Coral D18 O records.

24. The Relative Contributions of Anthropogenic Forcing & Natural Climate Change Cycles.

25. The Relationship between CO_2 & Global Surface Temperature.

26. Temperature Hockey Sticks & Ski Jumps.

Many aspects of climate, on many times scales, are clearly modulated by solar 'influences' which probably include: modification of the cosmic ray flux by the solar magnetic field and wind; variation in ultraviolet radiation now known to be much larger than thought; occasional 'grand minima' in solar activity, slow occasional variations in the total output of the Sun due to intrinsic instabilities and, perhaps, to planetary gravitational forcing.

We noted the association between the Maunder Minimum in solar activity and the Little Ice Age in the 17th century, now known to be a global, not just a USA / Europe phenomenon. There have been other such associations in the past which are suggestive of a causative relationship. The 'warming' of the last century has been associated with very high solar activity, not just sun spots, which some have called a 'Grand Maximum'. The consensus has done everything to dismiss the GM by 'adjusting' sunspot numbers to allow for 'bad data'. We showed that actually solar behaviour based on the suspect data has been consistent since at least 1650 by looking at evolutive spectra. Other activity proxy data show clearly that 20th century activity was exceptional. Looking at long tree ring records along with C14 levels (1) we find that solar activity has not been at 20th century levels since ~11,000 years ago, just after the ice age ended, although a few periods have come close.

We also note that the low point of the Maunder Minimum occurred at the end of a long slow fall in solar activity since ~2,500 BP. The last similar slow fall grand minimum occurred around 7,500 years ago. We now know from the study of other middle aged sun-like stars that solar activity cycles are the norm and that such stars spend ~20% of their lives in grand minima. For example HD9562, which is close to the Sun's mass and a slow rotator, has shown no stellar surface activity since 1966; HD3651 (54 Piscium), a K0 star slightly smaller than the Sun, has cycles which have strongly decreased in amplitude since the 1960s.

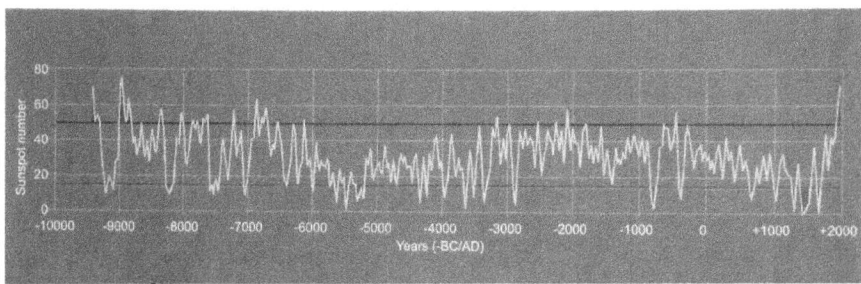

The most recent maximum was lower than the minima of the 1970s: in effect 54 Piscium has been caught entering a grand minimum.

Younger stars show a much lower occurrence of grand minima (5% ?) which emphasises that stable main sequence stars like our own do evolve significantly, with consequences for any life on their planets.

Taking this evidence for G-K class stellar populations in conjunction with our spectral analyses showing strong solar-climate links on a remarkable range of time scales, from years to millennia, tells us we should take such links very seriously. The fact that the physical mechanisms involved have not been confirmed is insufficient reason to dismiss the mass of empirical evidence now available. The author is reminded of the 'continental drift' controversy of the last century. Alfred Wegener pointed out the remarkably close fit between the coastlines of North and South America, Africa and Eurasia if pushed together on a globe. Similar fits were noted for Australia, Antarctica, India and South Africa (see Figure 3, section 1). Also he and others noted the identical geological and biological fossil formations on lands now separated by oceans. The evidence was dismissed as obvious nonsense or famously explained by an embarrassing host of now sunken 'land bridges'. Wegener and his few supporters were mocked and ridiculed as lunatics. After all there was no *theoretical* basis for continents ploughing their way through solid rock...until, that is, the discovery of mantle convection driven by the vast heat of radioactive decay along with the tectonic plates, the edges of which were and are, clearly defined by earthquake clusters.

We have seen how important the movements of those plates have been in determining climate for at least 2.5 billion years. Perhaps Svensmark and the other heretics, likewise, will one day be vindicated. If the Sun plays a major role in modulating climate, what about the arch villain CO_2? It is remarkable in the author's experience how many non-scientists believe that CO_2 is some kind of toxic pollutant. But we saw that CO_2 is a key component in the cycle of photosynthesis which has preserved life on Earth for at least 3 billion years. Normally the carbon / oxygen cycle chugs along in a relatively stable way. But we saw that occasionally in the history of climate, as in the Palaeocene Hyperthermal Events of 55 million years BP, disturbances to vast 'buried' carbon deposits can cause major CO_2 and temperature excursions lasting for hundreds of years. We can think of this sequence of hyperthermal events as 'negative ice ages'. However even in such extreme events the climate system somehow responded and temperatures returned to their original levels. There was no cataclysmic runaway even though CO_2 levels at the time were at least 2000 ppm (5X higher than today). There was no Palaeocene +2 degree C 'tipping point' or doomsday temperature runaway. This is the key issue as we consider the future impact of increasing man made CO_2

in the atmosphere. CO_2 was as low as 180 ppm in the last ice age, rising rapidly in a few hundred years at its end, to ~300 ppm and equally as quickly stabilising. In the last few industrial centuries we have raised CO_2 to ~400 ppm. It should be recalled that an estimated 57% of human generated CO_2 has already gone 'somewhere': into land and ocean biomass, soils and sediments. Experiments in the 1990s (2) on 156 plant species showed that doubling CO_2 (as is predicted for this century) increased plant growth by an average of 37%.

What is happening in the real world? Well analysis of satellite imagery at the Commonwealth Science & Industrial Research organisation showed that plant growth across the planet was increasing and especially in marginal 'desert' areas such as the USA south west, East Africa, the Middle East and Australia's red centre. Ground cover increased and transitioned from dry grassland to trees and shrubs (3). From 1982 – 2010 CO_2 increased ~14 % and foliage by 11%.Remember that typically 10 to 40 times more CO_2 at equilibrium is sequestered in biomass and soils than in the atmosphere. Are we therefore noticing already, part of the biosphere stabilising mechanism in play? If the ocean plankton respond similarly will increasing dimethyl sulphide emissions transported to the cloud forming levels increase low cloud cover? Remember changes in albedo can have a potentially larger effect than changes in the radiation traps. CO_2 levels would need to change drastically to alter the trap.

As Professor Hoyle reminded us: the CO_2 heating effect is logarithmic and the higher its concentration becomes the smaller is the effect of further increases. We also noted that for the last four ice ages, going back over half a million years, the global climate has bounced between remarkably fixed, upper and lower temperature levels. We saw that this can be explained in terms of a non-linear response of the climate system to modest cycles in the Earth's orbital and axis dynamics (and possibly direct solar changes) with periods of 23 to ~100 Kyrs. Powerful feedback mechanisms are needed to account for this limit cycle behaviour. Will the rise from 300 ppm to above 400 ppm of CO_2 trigger the usual feedbacks? Perhaps it already has. There is some intriguing evidence from the non-linear dynamic properties of global tropospheric temperature and solar irradiance. Karner (17) studied the period of recent rapid growth from 1979 – 2001 and found both series displayed antipersistency of the same scale. That is daily temperature can be described by a fractional Brownian motion process and shows a long term correlation to the present …a negative correlation. This suggests a cumulative negative feedback in the troposphere and this same property is found in solar irradiance. The Hurst exponent measures persistency and antipersistency in a time series.

For the tropospheric temperature $H_t = 0.27$ and for solar irradiance $H = 0.23$. For the stratosphere $H_s = 0.57$. The stratosphere behaves very differently, as a random walk, well modelled by an ARIMA model. It is the troposphere which responds to the Sun…on the time scales of this study. Karner concludes, provocatively, that CO_2 induced climate change 'had not yet started' up to year 2000. After 2000 of course the rate of temperature increase fell dramatically…by some measures to zero, as CO_2 continued to rise steadily.

Here are a few observations about CO_2 forcing which appear to be largely unknown to the general public, thanks to the supine media. The IPCC approved climate models attempt to predict headline, average global surface temperature. The models in effect **assumed** CO_2 forcing (including feedbacks) of anything from 1 to 6 degrees C associated with a doubling of CO_2 level. In formal reports the IPCC usually stuck to the middle range, 2– 4 degrees C …to appear judicious. This huge range in 'guesstimates' tells us how uncertain the game is even though the 'consensus' holds to the simple 'scientific fact' that CO_2 is a greenhouse gas and **must** cause warming (see Appendix 3 on the 'sociology' of consensus). The actual 'scientific fact', again largely unknown to the public, is that the Earth did not warm from 2000 to 2015. Following a double El Nino event and twin temperature spikes, in late 2020 a La Nina cooling event set in and by March 2021 global temperatures were back to 2001 levels. The figure below shows temperature from (unfiddled) RSS satellite records for three important layers in the atmosphere since 1979. The UAH satellite series is very similar. For the lower troposphere (below) there has been no warming trend since 1998.

There has been a significant cooling in the important troposphere / stratosphere boundary zone of -0.27 degrees C per decade since 1980 (see graph below). The various approved models would predict temperature increases for the period 1980 to 2015 of +0.4 to +1.3 degrees C (based on 90 published CMIP5 climate models). The model average was +0.8 degrees C. Over the same period the actual RSS satellite temperature for the lower troposphere went up by ~0.4 degrees C. The real troposphere / stratosphere boundary went down by ~1 degree C. Across the three atmospheric layers the real average fall was -0.5 degrees C compared with the model average increase of +0.8 degrees C. This is not an impressive performance I suggest. The middle stratospheric (C13 channel) cooling is even stronger at a rate of -0.5 degrees C per decade. There has been no significant warming in the upper troposphere (below) since 1986. How could this happen?

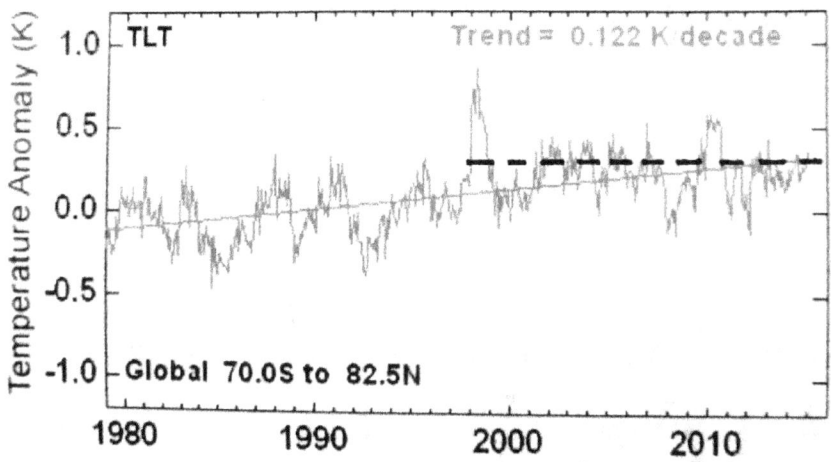

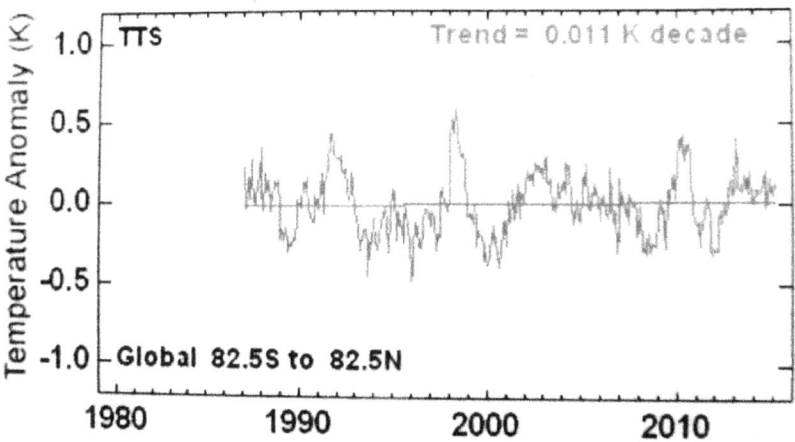

Well now we are told that the climate models *did not* take into account 'natural cycles' like El Nino, the NAO and the AMO which we analysed in section 5. Nor were we told, or the modellers do not know (or care), that these cycles show the clear spectral 'fingerprints' of the Sun as we demonstrated. These naughty cycles halted the upward temperature trend due to CO_2 …apparently. So some process originating in the Sun halted the warming trend. But then some of the 'high' warming trend intervals in the 20th century must also have been boosted by these naughty natural cycles. But this must mean that the underlying, supposedly CO_2 forced, trend has been significantly over 'guesstimated'.

This 'simple fact' has not exactly been shouted from the IPCC rooftops. Rather awkward, yes?

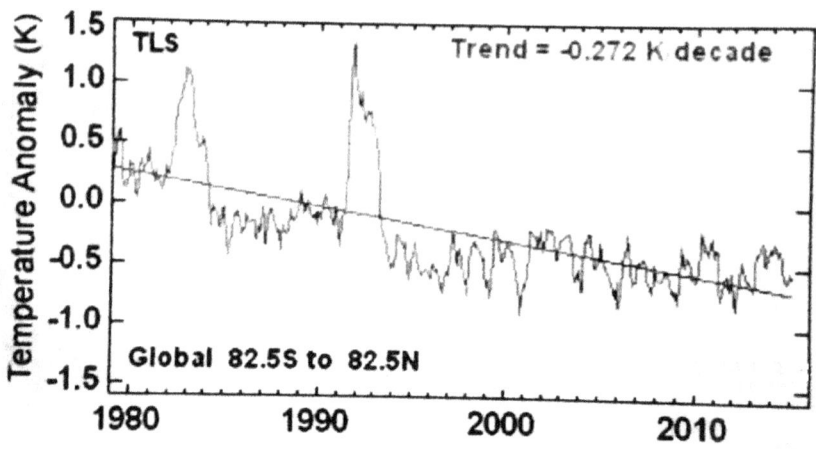

Very recently the story from some 'warmist' quarters has changed tack. Apparently using climate models and CERES satellite data over 15 years it was found that radiative cooling in response to global warming, the well known Plank Effect, and new strong local effects, are sufficient to stabilise global temperatures. This result is used to argue that therefore external forcing of temperature must be occurring. What could this be? Why, man made greenhouse gases of course. Apparently this means that 'denialist' claims that 'natural cycles' explain global warming can be dismissed. Of course these conclusions require the reader to ignore other possible external forcing processes and ignore the halt in global temperature rise since ~2000 (while CO_2 emissions have accelerated). Who are the real 'denialists?

I ask the reader this: the models forecast a range of temperature increase in this century from 1.0 to 6.0 degrees C and every possible increase in between. With such a range how can the IPCC and their fellow travellers speak of a 'scientific consensus' on climate change? Their claim is patent nonsense. Whatever the excuses about natural cycles, if CO_2 really did cause the predicted warming (as some still insist) and we do not see it, it must have gone somewhere, just like the missing CO_2 emissions. Where did it go? Well the surface layers of the oceans did increase in temperature in line with the land temperatures through the 20th century until the great early 21st century halt. Ah, ha said the climate modellers the heat must have somehow transferred into the lower ocean layers, yes that's it!

Wrong. Willis et al (4) of JPL, in 2014 examined NASA satellite data and direct temperature measurements of the surface and deep ocean below 2000 mts. They then computed sea level rise due to thermal expansion and ice melting. With all this information they could calculate how much the deep oceans must have warmed, by difference. Unfortunately there was not any warming over the period 2005 to 2013. Awkward. Also in 2014 a paper by Wunsch and Heimbach, eminent oceanographers both, in the Journal of Physical Oceanography (5) concluded this:

'[shifts in deep ocean properties] may indeed be so slight that their neglect in discussion of heat uptake and sea level change is justified.'

In fact they estimated a cooling of the deep ocean below 2000 mts of 0.01 degree C over the last 19 years. So where is the missing heat generated by the CO_2 rise? Nobody knows: perhaps because it does not exist. By the way, some areas of the global ocean are consistently cooling including, notably the Great Southern Ocean, which has played such a role in maintaining the necessary conditions for the present ice epoch. From 1985 to 2015 the cooling was -0.23 degrees C…by no means negligible. While we are looking at ocean changes some recent work by our old friends Shaviv and Svensmark may be of interest (12). They looked at twenty years of satellite altimetry data which gives highly accurate global sea level measurements. From these it is possible, as above, to calculate thermal expansion and hence the amount of heating and cooling of the oceans. Shaviv had earlier looked at tide gauge data in the same way. They found that the variations in sea level could be very closely explained by fitting a model with solar harmonic components and an El Nino / Southern Oscillation contribution. The dotted line below is the de-trended sea level data and the continuous line is the model. The fit is remarkably good and explains about 71% of the observed variance in sea level change. The model represents accurately, variations on yearly to decadal time scales. But there is more. We saw earlier that the El Nino / Southern Oscillation is *totally dominated* by spectral peaks related to solar activity variation. This means most of the variation in global sea level and ocean heating is explained by the Sun. Shaviv uses the model to calculate that solar forcing is currently 1.33 +/- 0.34 Watts / Square metre…very close to the IPCC claimed CO_2 forcing parameter.

The climate farce never ends and we should be proud of the role our country plays in it. The British Meteorological Office, despite the, by then, failure of global surface temperature to rise since 1998, put out the decadal forecast illustrated above in 2011 (6). Up to 2020 we have an even faster rise than was predicted than in the 1990s!

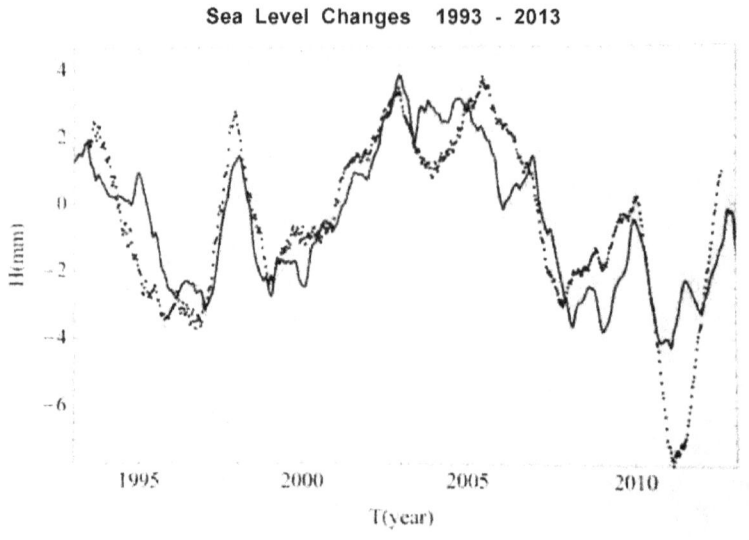

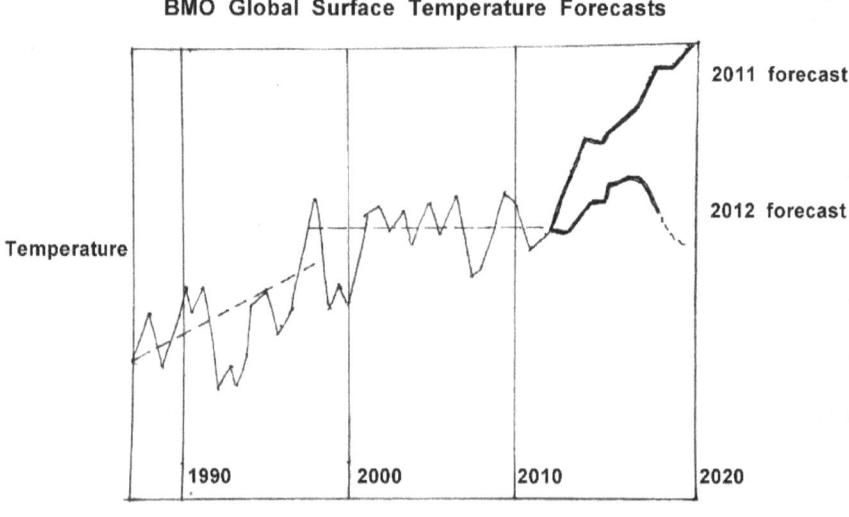

Somebody must have noticed the, er, obvious deviation from reality. On Christmas Eve 2012 when the world was properly distracted, the BMO quietly put out the new forecast.

As you can see after a modest rise (presumably due to anticipation of the current double El Nino period) temperatures are forecast to fall back towards the average level of the 2000s. **In this book update we can now say in early 2021 that this is happening (see below).**

Let us consider the naughty 'natural cycles' which have cancelled out the alleged CO_2 effect. In section 5 we looked in detail at the Pacific El Nino / Southern Oscillation Index, the NAO and various Atlantic Meridional Oscillation proxies. In the SOI we found the major spectral peak at 2.5 years was a higher harmonic of several solar activity periods simultaneously including : 22.3 / 9 = 7.43 / 3 = 2.48 ; 9.93 / 4 = 19.85 / 8 = 2.48 ; 17.9 / 7 = 2.55 ; 25.5 / 10 = 2.55 ...Large spectral peaks at 13.1, 6.7, 4.9, 3.6, 2.1 years were also simple solar harmonics such as 19.85 x 2 / 3 = 13.2; 19.85 / 3 = 6.62 ; 19.85 / 4 = 4.96 ; 11.16 / 4 = 2.79, etc, etc. The solar forcing presence is very clear.

For the AMO we analysed proxy records from 7 sites around the Atlantic boundary and found large spectral peaks at ~60, 66, 78, 88 and 100 years. These are obvious solar activity spectrum sub-harmonics including 19.85 x 3 = 59.6 and 6,7,8,9 times sub-harmonics of the primary 11.16 years sunspot cycle. The various data records covered the range from the present back to 19,000 years BP.

The story was similar for the NAO spectrum where we found all its periods were related to solar activity periods such as 22.3, 19.85 and 8.3 years via simple or Fibonacci ratio harmonics. The patterns are clear and consistent. Also clear was the strong appearance of planetary pairwise and triple conjunction periods, via yet more simple harmonics, in the NAO spectrum. Much of the spectral power in these supposedly 'intrinsic natural cycles' is solar related. Considering also the spectral 'fingerprint' match to a dozen geophysical and climate variables there can be no doubt about the role of the Sun in forcing climate on many time scales.

We also saw in sections 3, 4 and 5 that climate, some geophysical phenomena, Earth-Moon system dynamics and the Sun's variations all carry the spectral 'fingerprints' of the orbital dynamics of the planets. The author was surprised just how clear the links are given the supposedly 'minute' tidal and torque effects in play. But then as we saw with the 'tiny' changes in the longer Croll-Milankovitch orbital cycles, somehow the climate responds strongly to them. In both cases we are dealing with non-linear dynamic systems where such counterintuitive effects occur naturally. We noted in many cases of 'unlikely' spectral links that Fibonacci ratio and phi harmonics showed up strongly, which is a classic indicator of quasi-chaotic non-linear behaviour.

In considering earthquakes we also discussed Per Bak's theory of critically poised systems where natural systems self-tune to a point where small forces can cause a rupture or reconfiguration. In principle the patterns of planetary gravitational 'forcing' are predictable for centuries ahead so surely they can be used to predict solar variations and via the Sun the effects on our climate...after all, the signals are clear in the spectra. Unfortunately we saw in Appendix 1 that the Sun itself in some aspects is described by a strange attractor and taken at face value is chaotic.

That tells us not to be surprised at phenomena like the Grand Minima in solar activity. It also tells us that long term forecasting of the Sun's moods may not be possible. Perhaps when we understand the solar dynamo better (now we know there is a double meridional circulation loop in each hemisphere), medium range forecasts will be possible based on planetary forcing. But that is not the end of our problems. The climate system is also non-linear and quasi-chaotic at least. Until we know more and have better physical and biological models, forecasting capability will remain limited.

If the Sun is important on time scales of decades and centuries we should sensibly ask what is in store for it and us, at least on modest time scales. This is a difficult question to answer given the above analysis. The occurrence of grand minima is not regular. However solar activity is now in decline following the peak in the 1960s – 1980s.

If we look again at the 11,000 year solar activity series we can see several occasions when the fall from a grand maximum to a grand minimum typically took 50 to 200 years. The slow fall from the last maximum to the recent (Little Ice Age) minimum was exceptional.

We may have other useful clues. Below is a graph of the magnetic field strength in the centre of sunspots from 1998 to 2011, straddling cycles 23 and 24 (Penn et al). The trend is downwards. By ~2024 if it continued, the local fields will be too weak for sunspots to form. This was the situation during the Maunder Minimum and other grand minima. However Penn found that for the period 2011- 2015 during the rise phase of cycle 24 the field strength was constant. That slows any overall fall in field strength significantly. Penn's earlier forecast of a cycle 25 SS peak of ~7 may be low. Recently Hill et al (of the US National Solar Observatory) reported helio-seismology results (7) which showed that the usual polar 'jet streams' (which migrate equatorwards, generating and conveying sun spots) for the upcoming cycle 25 had not formed by 2011, some 3 years later than in previous cycles.

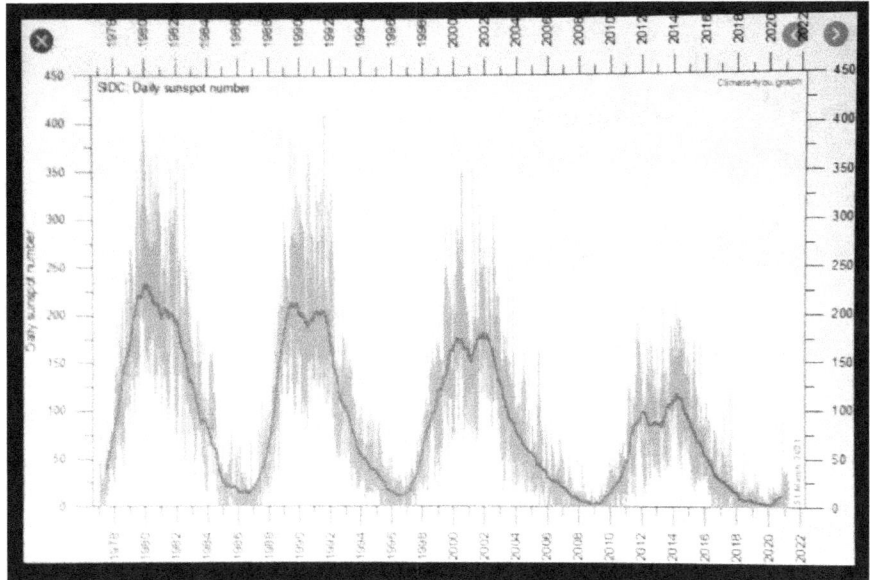

See the figure below. Hill believes cycle 25 maximum due around 2024 will not happen. This is in line with the declining field strength forecast. Altrock (7), an observer of the solar corona, reported that the usual shift of the corona towards the poles of the Sun, associated with the forming jet streams below had also not happened (see figure below). The Sun is changing. Why this is happening has emerged only in the last few years …possibly.

At the Royal Astronomical Society sponsored National Astronomy Meeting in 2015, Professor Zharkova presented a new duplex model of the solar dynamo (8). She used principal component analysis to dissect the long 1976-2008 Wilcox Solar Observatory record of magnetic field observations and found that the Sun acts as though it has two dynamos : one at the base of the convection zone and one near the surface. The new model implies two cyclical components both of about 11 years duration but with a phase lag. It is interesting to recall that many solar activity peaks show evidence of bimodality. The model fitted the magnetic and sun spot records well. Zharkova found that the two cycles were moving out of phase and by 2030 will interfere strongly and cancel out producing another solar grand minimum. See the figure below. This model is not in accord with conventional solar theory. Should we believe it? Well the author did a little searching around and found a 2013 paper by Junwei Zhao et al (15). The assumption has always been that each solar hemisphere holds a single meridional circulation cell.

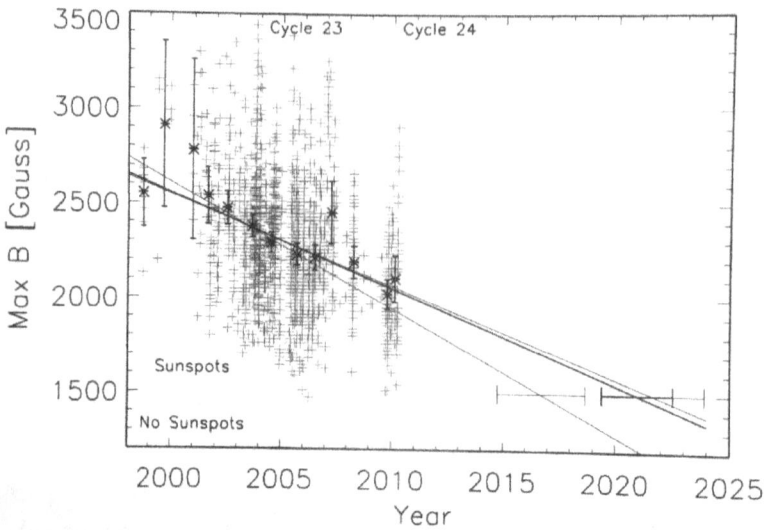

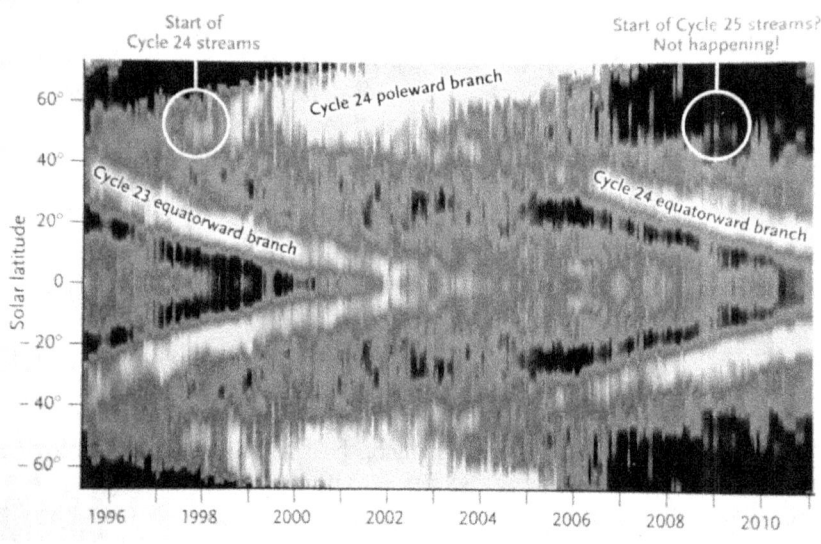

Material flows along the solar surface from equator to pole and then sinks deep into the Sun for a return to the equator. The slow return loop was supposed to account for the 11 year cycle. Now Zhao has found evidence for *two* meridional cells one above the other. This result has been described as 'catastrophic' for current theories but the two cells perhaps explain Zharkova's double dynamo result.

Such results show us how little we know about our home star and should perhaps open minds to hypotheses like planetary torque modulation of solar activity, which empirically exists as we saw, whether we have a complete physical theory or not.

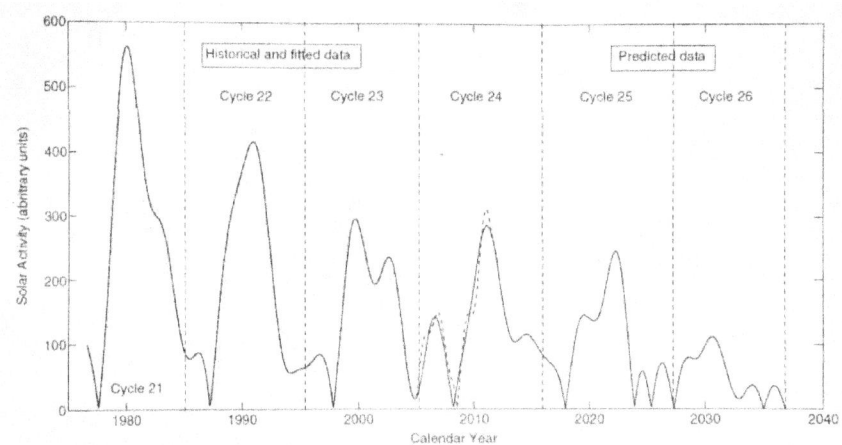

The author had a sense of déjà vu in looking at all this work. In the 1980s my statistical model of solar activity used paired non-linear oscillators and predicted a long period of solar activity decay …which was a decade premature! (section 3, ref. 13 and Appendix 2). However the Zharkova model is more physically grounded and may prove accurate given the other recent evidence. My coupled oscillators were essentially a Lotka-Volterra system which originated in studies of prey / predator dynamics. These were chosen because of links between a specific observed response of the Earth's magnetic field and *later* solar activity. The Earth was picking up a precursor signal in some coupled solar activity generator. The author recently discovered others had later thought along similar lines. Some astrophysicists have used equations of the form

$dX / dt = k1 \, X - k2 \, XY$ and $dY / dt = - k3 + k2 \, XY$

where X represents the poloidal dynamo field and Y the toroidal field. The equations produce coupled, out of phase oscillations of similar magnitude. In discussing such equations Consolini et al also interpreted them as describing a *double dynamo* in the Sun, one at the base of the convection zone (tachocline) and the other a shallow sub-surface dynamo (24). It will be recalled that in discussing how small planetary torque effects could modulate solar activity, the proposers pinpointed effects at the tachocline layer as the scene of the crime. 3-D MHD simulations of the Sun have been able to reproduce double dynamo type results.

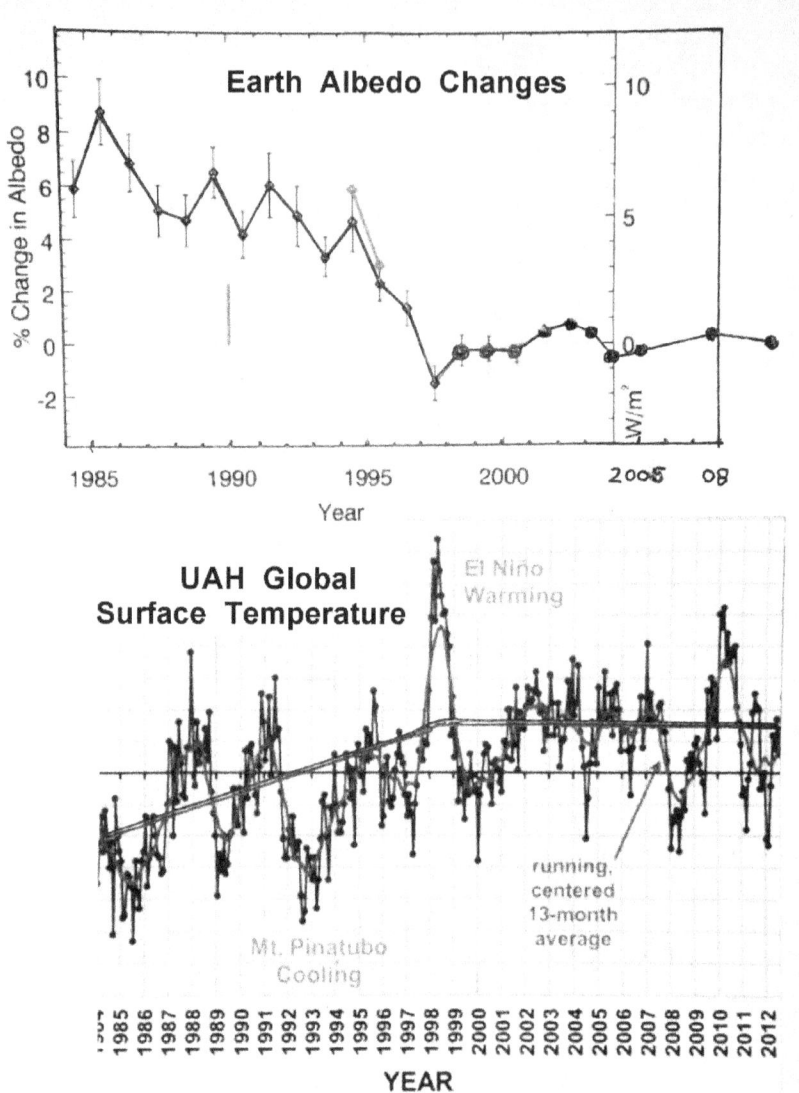

So if solar activity declines significantly (or completely) over the next few decades what should we expect? We have seen that many climate phenomena have a strong spectral link to solar activity variation. If we accept the historical correlation of the Maunder Minimum with the Little Ice Age and the other observational evidence, a cooling of the Earth would be expected. Well we have seen for ~18 years a halt in global surface temperature rise (from the satellite data) despite the continued rise in CO_2.

What else do we know from actual observations? We have whole Earth albedo measurements based on Earthlight reflected from the Moon since 1984 (9). These measurements show a slow decline in albedo from 1984 to ~1996 followed by a very rapid decline to ~1998 (see figure above).

Since then albedo has changed little. The constancy from 2000 onwards is confirmed by CERES satellite and MODIS cloud fraction records. The panel above shows the UAH satellite global surface temperature record from 1984 until 2012. Temperatures increased slowly until the major El Nino event in 1997-1999 and then stabilised. The RSS satellite data series is very similar. The temperature curve is the inverse of the albedo curve. The sudden discontinuity in temperature and albedo trends coincide around 1997 and the El Nino event. As the Earth dimmed in the 1980s and 1990s, presumably as ice, snow and cloud cover declined under the hyper-active Sun and temperatures increased. After El Nino, for at least 12 years, albedo remained constant and so did surface temperature levels. Coincidence? Perhaps, but this qualitative behaviour is what we would expect from Svensmark's 'cloud' forcing model. If he is correct future decline in solar activity should increase cosmic ray flux leading to more low clouds, higher albedo and a cooling Earth...all other things being equal.

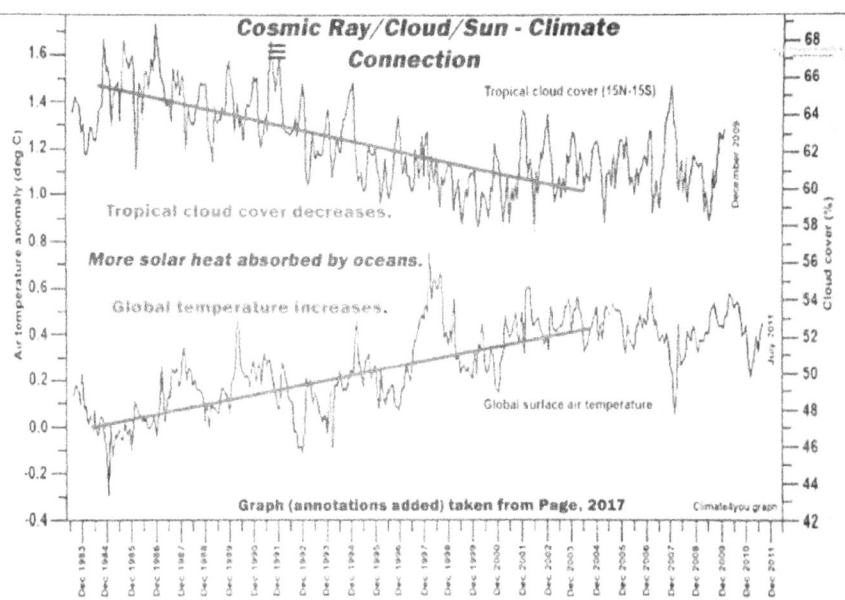

In 2017 Stozkhov made some interesting observations concerning cosmic rays and cloud cover (22). Above he shows the opposed trends in tropical cloud cover and HadCrut3 global surface temperature.

The cloud cover downtrend ends in the late 1990s as does the down trend in albedo. Stozhkov provides a forecast of temperature based on a spectral decomposition model from to 1880 to 2015. Like several statistical models we have examined in section 5 it predicts a turnaround in temperature after ~2014 and in this case a fall back to 1950s levels over the next twenty years. The model does not predict the large, long double El Nino spike of 2016 – 2019. Is the long El Nino a sign of CO_2 influence? CO_2 continues to rise. The climate forcing parameter assumed for CO_2 is ~1.4 Watts / square metre and for all the greenhouse gases, ~ 2.3 Watts / square metre. However the Project Earthshine estimate of albedo forcing appears to be significantly higher than that.

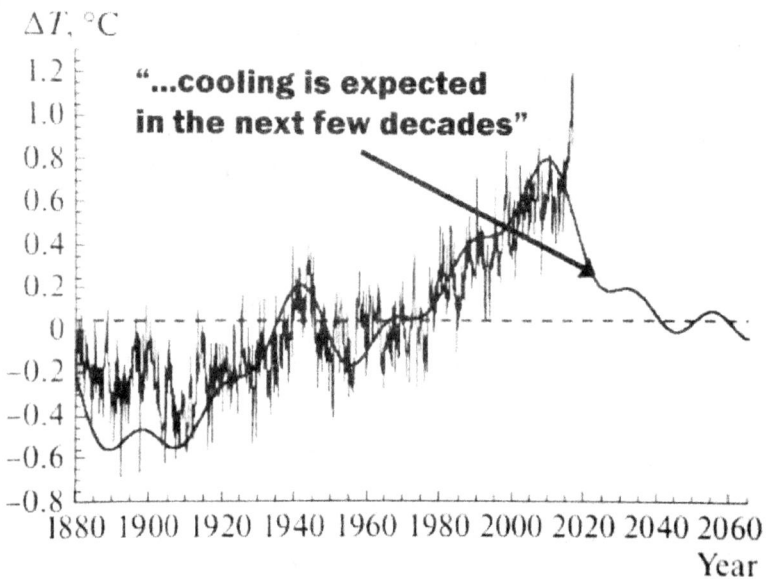

Will albedo effects swamp CO_2 forcing? Also will CO_2 be 'allowed' to accumulate much more in the atmosphere? Over 50% of what we have pumped out so far has already gone 'somewhere' as it did at the end of the last 4 ice ages as CO_2 exceeded ~300 ppm. Will biomass sequestering increase in compensation as we discussed earlier? Will the ocean plankton increase output of dimethyl sulphide further increasing cloud cover? We do not know as yet, but the climate modellers, to be taken seriously, need to understand temperature and CO_2 stabilisation at the beginning of recent interglacial periods and say why that, so far unconsidered, stabilisation mechanism has gone away. We have been told by the IPCC climate modellers that the recent global warming 'pause' is due to 'natural cycles'. Very well let us examine the implications of this.

The author was impressed by the model of the very long term, 2,485 year, Tibetan Plateau tree ring growth series we examined in detail, which predicts a sixty year fall, in regional temperatures at least, reflecting again the expected decline in solar activity.

A similar fall is predicted from the spectral model of Central European temperature based on ~2,000 years of proxy data and the other very long temperature proxy series going back several thousand years in some cases. These stable models fit the long data series remarkably well **even in the 20th century.** If CO_2 induced warming over the last century was large we should see a growing deviation between the spectral model fits and forecasts (based on 'natural cycles') and the observed recent temperature records. We see no such systematic deviations. This strongly suggests that any man made greenhouse gas induced warming is small. If this absence is confirmed by long term spectral models for other regions the question can be settled definitively. What can we say for now?

Below is the most recent satellite data from UAH. By the autumn of 2020 the double El Nino event was over and a modest La Nina set in. In March 2021 the global lower atmosphere temperature had returned to the mid 2020s level experienced during the long 'warming hiatus'.

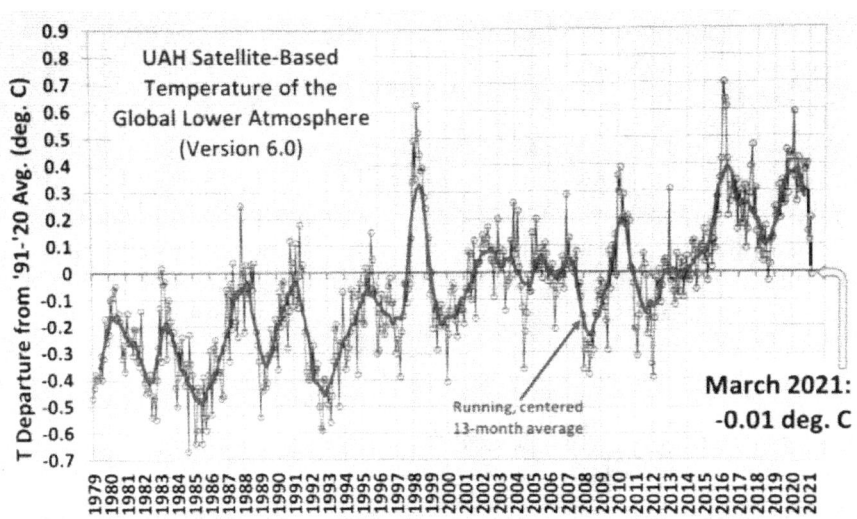

The various long term spectral models we examined predicted a fall back in temperature and solar activity has fallen sharply in the last cycle 24. Despite accelerating CO_2 growth it appears the solar / planetary cycles are dominating the CO_2 warming effect, at least for now.

If a grand solar minimum *is* coming we may get further cooling back to early 19th century conditions and possibly further. Could we find ourselves in another 17th century Little Ice Age? Looking at the best solar activity reconstructions available it seems unlikely anything so severe will happen so soon after the last one. Looking back over 8,000 years only one comparable event (in terms of duration and depth) occurred in ~750 BC. Notice also that the recent high sunspot peak activity has been matched at several points over the full record and the high peaks are often adjacent to sudden drops of significant size, even if not as large as in the LIA. Increasing CO_2 may yet moderate the temperature effects of a decline in solar activity but not balance it.

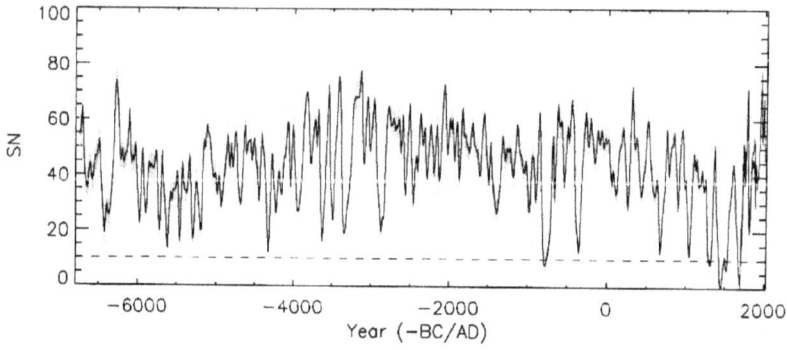

Fig. 14. Reconstructed sunspot number along with its 68% confidence interval (gray shading). This series is available in the ancillary data. The red line depicts the decadally resampled international sunspot number (version 2, scaled by 0.6) from Clette et al. (2014). The dashed line denotes the level of SN=10.

To go further it is necessary to become more ambitious in building mathematical models which test all potential causative links, not just CO_2 effects, and which consider all available data. The author has examined the small number of attempts to build multivariate (linear) ARIMA models linking temperature, CO_2 and 'natural' climate cycles. None are totally convincing for technical reasons which illustrate the problems statistical *and* physical climate modellers face.

It turns out unsurprisingly that *variations* in temperature and CO_2 can be linked although the relationship is complex since increasing temperatures, from whatever source, reduce the solubility of CO_2 in sea water for example. What we really wish to establish is whether the long term upward *trend* in manmade CO_2 can be *causally* linked to the long term growth of global temperature over the last century or so. Allowing for the 60-90 years cycles in temperature (often attributed to AMO, etc) there is an upward trend. So there *is* a correlation between the temperature trend and the CO_2 trend.

However we cannot say that the correlation is causal. Any two data sets with trends will be correlated. Famously after WWII it was shown that the European birth rate increased steadily at the same time as the stork population increased. Did the increase in storks cause the birth rate rise? No. This is the problem of 'spurious regression'. There are techniques to compensate in principle but in the case of temperature and CO_2 the author has not found many convincing results.

Perhaps the most interesting study is that of Wilson (20) in 2010. He built models relating HadCRUT3 global and northern hemisphere temperatures, CO_2 levels and the annual SOI (El Nino) from 1960 to 2010. From spectral analyses he found significant lags of -2, -1, 0, +1 in a model to predict temperature from CO2 and lags 0, +1 in a model to predict CO_2 from temperature ...after trend removal. Lagged SOI values also affected de-trended temperature. He also found a direct effect of CO_2 level on the level of temperature *but* he obtained an equally good overall model by replacing CO_2 with a simple linear trend term. Such a trend of course could be due to other external forcings or a section of a long period cycle in some other intrinsic climate variable (such as AMO, etc). Wilson was confident with the model structures *but* questioned their interpretation himself.

What is most interesting in the light of our earlier climate series analyses is the set of coherency graphs Wilson derived for the trend corrected series. Coherency measures correlation at given frequencies between pairs of spectra. For temperature and CO_2 two zones of significant coherency were centred on peaks at 7.25 and 3.3 years for temperature / CO_2. The significant ranges were 10 – 6.2 and 3.7 – 2.9 years. It is interesting that we have seen 'natural' cycles in climate variables of 7 – 7.5 years and of course 7.4 is the Hale third harmonic 22.3 / 3. 7.17 is also a major PDO period. It is interesting that 3.3 is Cjs and Hale peak harmonic, 19.85 / 6 = 3.31. Considering the coherency ranges 10 = 19.85 / 2, 6.2 = 18.6 / 3 (lunar nodal cycle / 3). Considering the 2nd range 3.7 is 11.16 / 3 and 11.86 / 5 and 11.16 / 4 = 2.8 and 11.86 / 4 = 2.96.Temperature *and* CO_2 appear to share a preference for simple solar, lunar and planetary period harmonics no doubt because temperature affects CO_2, but is this result also showing us that the biosphere is responding to solar variations which partially drive CO_2 variations? Temperature and SOI are maximally coherent at periods of 3.3 (see above) and 2.5 years. Coherency is significant in the range 4.2 to 2.3 years. Recall that SOI has a major peak at 2.5 years and PDO at 2.58 years. We noted earlier that considering major Hale solar periods, 22.3 / 9 = 2.48 and 25 / 10 = 2.5 and 19.85 / 8 = 2.48 and 17.9 / 7 = 2.55. CO_2 and SOI are coherent from 5 to 2.2 years with maxima at 3.3 and 2.5 again. It turns out lagged values of SOI also 'drive' CO_2 variations.

The interconnections here are complex and hard to unravel but it is highly significant that this web of relationships exhibits climate signals which are clearly related to solar activity variation (and probably planetary forcing). (On a personal note: I knew Dr. Wilson when he worked for Professor Gwilym Jenkins, co-originator of ARIMA modelling, at Lancaster University and he is a highly competent statistical modeller: hence my confident use of his work herein.) Overall, multivariate modelling, especially given our well founded concerns about non-linear and time varying behaviour, remains difficult. However, occasionally an analyst spots a pattern which anyone can see. Leamon et al (23) looked very carefully at several measures of solar activity, sunspots, radio flux, neutron flux and sunspot polarity patterns (the 'butterfly' diagram) and the Oceanic NINO Index from 1960 to 2020 (see below).They found a correlation between the start of new sunspot cycles and the termination Hale magnetic polarity cycles, and major positive peaks (and subsequent negative peaks) in the Ocean NINO Index of the Pacific. The vertical lines in the adjacent graphs mark these termination events. The correlation is not perfect but then we have seen the Pacific El Nino phenomena contain many (solar related) spectral peaks, not just a simple ~22 year cycle. However the authors use Monte Carlo simulation to show that the probability of getting this pattern by chance is $p = 0.0034$. This work is by respectable scientists and funded in part by the National Science Foundation.

What is the IPCC saying now? Well the near term global surface forecast for 2016 – 2035 is in the range + 0.3 to +0.7 degrees C. This is 0.1 to 0.2 degrees per decade …or what has actually been *observed* over the last 25 years, and far from what their models predicted (16). They now favour the lower range. The ~0.1 degree trend is that from reliable, un-fiddled, satellite data from 1979 on. In 1990 the IPCC predicted a mean trend of 0.3 degrees per decade. At the very least we can see that the supposedly 'dangerous' +2 degree tipping point will be delayed for at least a century. So after all that work, the IPCC has now done what a school kid could do with a ruler: put a line though the real, historical data. What a triumph for climate science and the 'expert consensus'. But it *still* ignores the decade or more global warming halt after the early 2000s. Nevertheless the 'official' estimates of CO_2 induced climate forcing have been falling over the years. Using the latest IPCC forcing estimates for several 'external' candidates and historical temperature analyses Lewis and Curry (18) derived a median transient climate sensitivity (TCR) of ~1.3 degrees C for a doubling of CO_2. This is a long way from where the IPCC started this global panic. But the forcing for solar effects is **still** set a priori as a tiny fraction of that for greenhouse gases and at 1/4 of that assumed for volcanic effects. The large recent reduction in IPCC climate sensitivity estimates apparently :

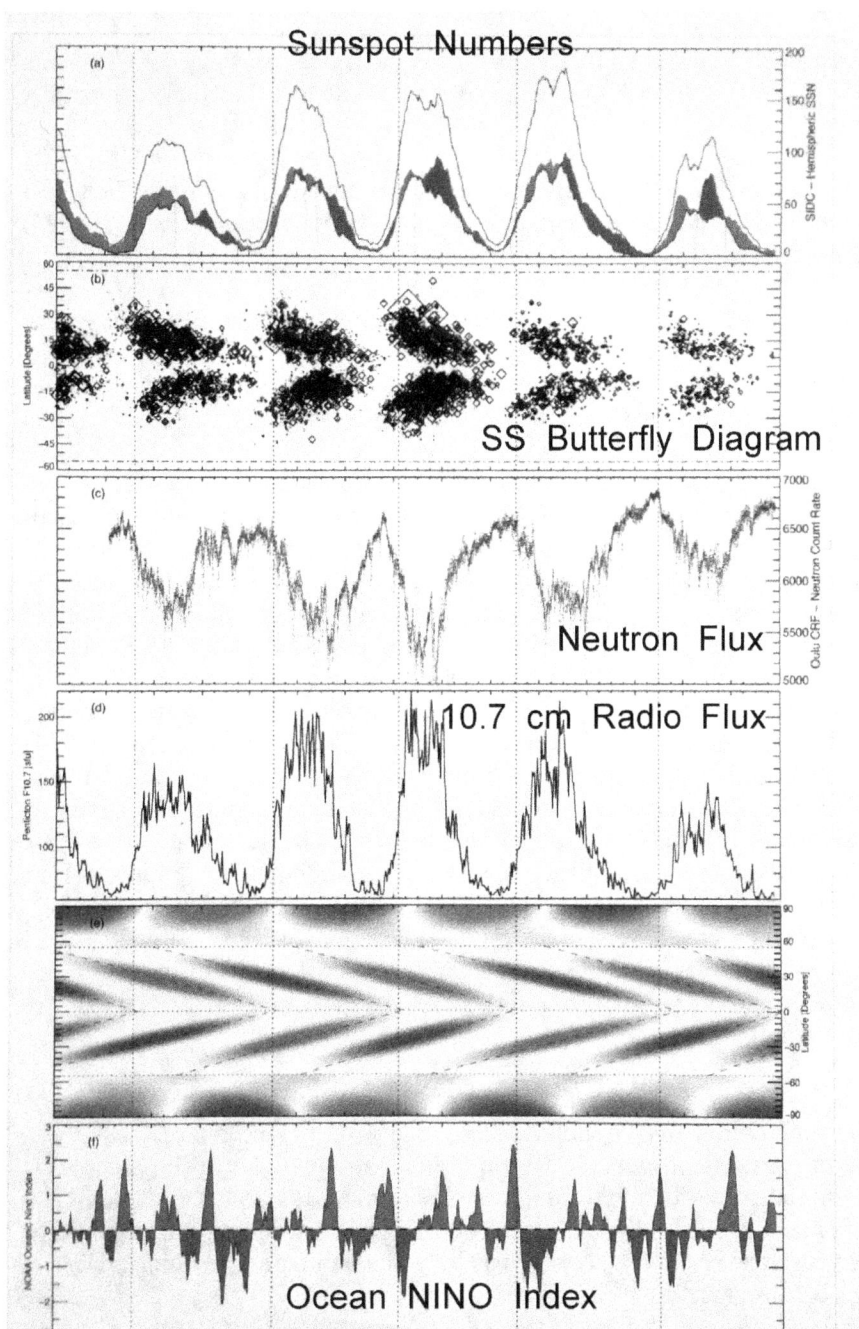

'reflects the evidence from new studies of observed temperature change using the extended records in atmosphere and ocean.'

Translated this means: there was little or no global warming for ~15 years and we can't ignore it any longer. An AR5 note to policy makers now also reads :

'No best estimate for equilibrium climate sensitivity can now be given *because of a lack of agreement on values across assessed lines of evidence and studies.'*

This is after ~25 years of mass climate modelling on the IPCC gravy train and a supposed, and still widely claimed, 'scientific consensus'. *Please, please, remember: caveat emptor.* Bearing the above statement in mind about emerging 'lines of evidence' I note one last paper published in 2011 by Harde (19). Since it deals with the complex spectroscopy of the atmosphere I cannot judge its veracity but its message is important if correct. Harde used the HITRAN 2008 data base which provides detailed measurements of the radiation absorbance of the gases, water, carbon dioxide and methane in the atmosphere. He claims warming saturation effects with increasing CO_2 related to strong overlaps of CO_2 lines with water and methane lines. He uses a 3D model of the Earth with 32 plane surfaces assigned to three climate zones and including surface and atmospheric elements. Cloud scattering is also included. Using the model and the new spectroscopic absorbance data, Harde computes estimates of climate sensitivity for the three climate zones. He finds an overall value of 0.45 degrees C compared with the IPCC (2011) mean value of 3.2. His value is $1/7^{th}$ the IPCC estimate. Is this why global temperatures have 'paused'…as long established 'natural' cycles continue to dominate? We noted that over the decades the ECS values from independent (non-IPPC model) measures have steadily fallen and several are now < 1 d C.
This implies that some physical negative feedback mechanisms *are* compensating for the CO_2 / CH_4 green house warming effect. Gaia is awakening.

But just suppose the above analysis is wrong? Suppose global warming continues? Perhaps we should listen, finally, to the wisdom of a last heretic: Professor Freeman Dyson of the Institute of Advanced Studies at Princeton, a co-founder of military Operational Research and godfather of Quantum Electrodynamics, the bedrock of modern quantum mechanics. He is a very bright cookie who has long had an interest in climate. Here is his opinion (2):

'This [climate business] is a contentious subject …the science is inextricably mixed up with politics. My heresy is that all the fuss about global warming is grossly exaggerated.

Here I am opposing the Holy Brotherhood of climate modelling 'experts' and the crowd of deluded citizens who believe their numbers...They [the models] do a very poor job of describing the clouds, the dust, the chemistry and the biology of fields and forests. *They do not begin to describe the real world we live in'*

Dyson accepts, as do all rational people, that there has been warming over the last century... and that there may be more. He sees the issue as one of management of CO_2 rather than trying, unsuccessfully, to stifle economic growth and fossil fuel use. China, India, Brazil, et al will continue to burn coal and the rest of us oil and gas, for the foreseeable future despite politician derived fantasies.

So what can we do? Dyson does not rely on Lovelock's self-regulating biosphere (which nevertheless may be in action) but points out that humans can take a surprisingly active role themselves. He goes back to the 'simple fact' we have noted repeatedly: the mass of carbon in land and ocean biomass exceeds by 10 - 40 X that in atmospheric CO_2. He calculates that to stop CO_2 rising at the current rate of output we need to increase the average depth of (10% biomass content) soil across half the land surface by **one tenth of an inch.** By changing to 'no till' farming and avoiding deep ploughing he estimates that soil biomass could be increased by at least this amount. Remember that about 10% of the CO_2 in the atmosphere is converted into biomass every summer.

He further points to something else surprising. Growing plants in a high CO_2 greenhouse not only increases growth but changes the root-to-shoot ratio, ie, more growth is diverted from leaves and stems into the roots. Trapping more carbon in the roots leads to a longer residence time of carbon in the soil before decay returns the CO_2 to the atmosphere. Is this one reason why CO_2 appears to be disappearing? Remember that some plants, such as the forest trees, live for many centuries and decay slowly after death. Here is another reason, to preserve the forests and prevent topsoil erosion besides the potential, so far largely untapped, of biological and medicinal gifts hiding in the plants. In 2016 the journal Nature Climate Change published a major paper by 32 authors called **'The Greening of the Earth & its Drivers'.** It was based on 33 years of American satellite records from Modis and AVHRR instruments which can assess vegetation types and densities. The records showed a 25% to 50% increase in vegetation area which represents a huge uptake of CO_2. Just 4% of vegetated land had suffered from plant loss. The plant bloom is attributed to climate & temperature change (+8%), increased nitrogen (fertiliser) in the environment (+9%), shifts in land management (+4%), and finally human derived CO_2 (or as we used to say before it became a pollutant

: plant food) contributing +70%. The lead author Dr. Zaichun Zhu of Beijing University concluded

'The greening reported in this study has the ability to fundamentally change the cycling of water and carbon in the climate system.'

Even so other authors worried that the results undermined the climate emergency 'consensus'. Having demonstrated a greening planet this was dealt with by pointing to other negative consequences of climate change such as sea level rise. But if most CO_2 goes into vegetation or ocean plankton the model projected temperatures will *not* rise strongly and neither will the sea levels. The second argument was that plants will get used to higher CO_2 levels and the 'fertilisation' effect will diminish over time. This is an interesting theory. The author's understanding is that in evolution plants and animals compete by utilising available food to the maximum degree. CO_2 is basic plant food not a 'fertiliser'. As we noted in section 1, in the current ice age epoch CO_2 has been a limiting factor on plant growth and even in the interglacial periods CO_2 is lower by a several fold factor than it has been for most of geological history. Yet somehow plants today will apparently decline a feast of food. Those that do will no doubt quickly become extinct while those that accept the feast will survive and multiply. It is fair to say, as some of the authors do, that in places other inputs such as water or soil nutrients may limit plant growth. So far these other limits do not seem to be in play overall given the huge increase in CO_2 uptake by vegetation.

Dyson also notes, on the basis of the last several ice ages, that we are now nearing the end of our 11,000 year warm interglacial period. He does not know if our interference with the atmosphere will prevent or enhance (by destabilising ocean circulation) the descent into the next ice age. If we return to Professor Hoyle's views for a moment he was deeply concerned that the top layers of the oceans held only a 10 year reserve of heat. Hoyle feared a cooling shock today could precipitate us into that next ice age. He recommended a planetary engineering project to pump (slowly) deep ocean water (currently at ~3 degrees C) to the surface at a location in the tropics to allow the Sun to raise it to perhaps 20 degrees C. In several centuries even with current technologies we could increase the heat reserve from 10 to 20 years. Now of course the continuing rise in CO_2 may or may not do the job of Hoyle's planetary engineering. Or maybe Lovelock's cybernetic biosphere, Gaia, will not allow it.

Dyson proposed one further climate 'heresy' which is worth considering. Seven thousand years ago the temperature was a few degrees warmer than today.

Deciduous forests grew in northern Europe where now there are only conifers. The mountain valleys of the Alps hosted forests where glaciers, despite recent warming, now stand. Most interesting to Dyson, and the author as a student of palaeo-anthropology, the whole Sahara was a well watered grassland where humans hunted antelope and giraffe and the many rivers supported hippos and crocodiles. We know this from the magnificent rock paintings of the northern Sahara at Tassili and other sites. Dyson muses whether the recovery of the First Eden might not be preferable to a new ice age.

Recent work suggests a warmer climate may be beneficial to life overall ...by *some* measures (10). Consider the oceans. Marine fossil records covering the recent 540 million years suggest increased biodiversity in warmer climatic conditions. As usual the picture is not simple: extinction rates, or species turnover, increases, allowing in new species and more species overall. Contrast this with the implied higher productivity (but lower diversity?) of the Earth under ice age conditions. We have to define very carefully what we mean by better.

Dyson does not mention that the IPCC model warming may bring a metre of sea level rise, sufficient to swamp some low lying Pacific atolls and coastal areas like Bangladesh. On the other hand these unlikely things are as nothing compared with the destruction of most of North America and Eurasia if the ice returns. That, depending on speed of cooling, would displace billions and kill many, many millions.

So what should we wish for? Perhaps the preservation of the status quo? To the author that status quo must include the preservation of current civilisation (as the flawed, but only game in town) and indeed the extension of improved living conditions to the billions still in poverty and want. Probably the status quo means living with the natural cycles of climate which we have identified and which have existed for millennia and longer. We should learn those cycles and use them as standards against which we can assess the *true impact* of any changes we make to the planet, deliberately or unintentionally.

Perhaps the planet can look after itself, perhaps not. We need to stop fiddling with toy climate models and find out. That involves looking at some of the key issues and areas of ignorance touched on in this book ...and above all the collation of what we do know **from observation** into an integrated picture of the planet's many dynamics. The Earth cares nothing for our academic micro-categorisation of its parts and processes. Nor, as Hoyle and Dyson contend, are we helpless in the face of any environmental challenge.

We have incredible technology and power at our disposal which we used to wield with confidence.

Six thousand years ago a Neolithic people driven by natural climate change settled in a great river valley. By four and a half thousand years ago the civilisation they built had constructed the Giza Pyramids: still the greatest stone constructs on the planet. They did it for no pressing reason. Surely we can do as well if necessary?

That, dear reader, is that. If you ask why I have not told you explicitly what will happen to climate (unlike the monstrous regiments of climate modellers and politicians) it is because I do not know; nobody knows. We have seen how complex the climate situation is, even to the extent of fierce disagreement about which forcing variables should be included in the forecasting models. We have seen that the climate modellers have made a wide range of 'guesstimates', because guesses they are, about climate sensitivity to CO_2 variation. On recent evidence reported in this book, most of the guesses were wrong. We also saw that on a range of time scales the Sun, somehow, has a big impact on many important climate variables but we lack the physical understanding to build a forecasting model just yet.

If you feel cheated having soldiered through the book, and press me for my own forecast I will reluctantly exercise my cloudy crystal ball. Here goes. We *will* soon enter a period of *reduced solar activity,* perhaps a 'grand minimum' of some kind, for some decades. The PDO is in a negative, cooling phase while AMO is at its maximum and will soon turn downwards. The regional spectral models we looked at also suggested cooling. Minor further warming from rising CO_2 will soon be exceeded by the rising albedo and related cooling of the Earth. We may cool for several decades. Eventually the Sun will recover ...hopefully. What then? That depends on the natural response of the biosphere to raised CO_2 and Lovelock's cybernetic Gaia and Dyson's soil mechanics. If I have to bet, I trust Gaia: the Earth will continue to 'green' until CO_2 is driven down, global temperatures will stabilise, perhaps below mid 20th century conditions, and probably cooler, towards the end of the 21st century...
if we remember the message of the Tibetan Plateau, Central Europe and other spectral model forecasts. **I do not fear the fire.** All this assumes no additional, major, random shock to the climate system, which as we have seen is *always* possible (e.g. a period of strong global, planetary driven vulcanism).

So back to what we should do at the end of the 21st century. By then one can hope that fusion power and cheap surface and space based solar power will be with us and burning fossil fuels will be seen as an unnecessary and wasteful misuse of valuable materials. At that point we will have to pause and consider carefully. Remember we are 11,000 years into an interglacial period. Global temperature reconstructions from ice cores show the durations of interglacials (measured as temperatures above or at the 1960 – 1990 baseline) going backwards from the current one, were 11,000, 2,100, 8,500 and 18,000 years. The mean duration was 9,900 years. By the way the last interglacial had temperatures at up to 3 d C above today. The others had peaks of ~2 d C, 3 d C and 2 d C above today. Earlier in this current interglacial temperatures reached +2 d C for short periods (of a few centuries). Curiously there was no runaway temperature catastrophe in any of these interglacial periods.

In a few thousand years we should be falling rapidly back towards global glacial conditions. That has been the pattern for half a million years. You could try asking the climate modellers and the IPCC what has changed to stop that? If further research on ice age dynamics tells us that a new ice age *is* on the way we may wish to implement something like Hoyle's deep ocean pumping system to increase the heat reserves of the upper ocean. Or perhaps new knowledge will suggest other planetary engineering interventions, the most obvious being genetic engineering of plant and ocean life.

No doubt the windmill tendency would try to stop that (but with cheap, sub-50$ gene editing already here, they will fail.) Similarly in the very unlikely event of 'overheating', supplying mineral nutrients, such as iron dust in tiny quantities and so on, to the nutrient poor tropical oceans could greatly boost plankton productivity and the sequestering of CO_2. In the event of strong cooling it may be that *nothing we can do*, rising CO_2 or no, will prevent the coming of the Frost Giants again. In which case, my long term forecast is

ICE!

Postscript

> The Sun also rises and the Sun goes down
> And returns to the place where he arose.
> All the rivers run into the sea, yet the sea is not full.
> To the place where the rivers rise, there they return again
> And there is no new thing under the Sun,
> But the Earth abides forever.
>
> Ecclesiastes

The first CC book was completed in the late spring of 2016. The author, despite the many temptations, resisted the urge to chronicle the misrepresentation of climate data by the warming lobby and even some scientists. The reader is directed to the many internet sites that specialise in this, often to good effect, and to Booker's excellent book on the history of the great climate panic (10). Let the curious reader type 'Climategate' into his or her favourite search engine and enjoy the extensive and amusing results! (See also Appendix 3 for a discussion of the wider consequences of bad science practice).

One is always supposed to 'play the ball' in science but it is amazing how often the purveyors of heretical ideas get personally hammered. In that spirit the author cannot resist one, small kick back on their behalf. Much of the silliness in the output of the IPCC through the 2000s was enthusiastically promoted by the chairman, Dr. Rajendra Pachauri, the eminent railway engineer, appointed in 2002. We saw earlier that by 2010 some real climatologists (not to mention Royal Society fellows) were, to put it politely, fuming about 'exaggerated' IPCC official reports which contained much non-peer reviewed 'scientific' material and unsubstantiated claims of doom from climate activist groups such as Friends of the Earth and the WWF. Some of the content was even, occasionally, ridiculed. Dr. Pachauri only finally stepped down in 2015 to return to his day job at the Tata Energy Research Institute, whose head he was for thirty five years. Following extensive police investigations of claims of sexual harassment made by three women at Teri, he was first suspended as Director General…then promoted to a new 'executive *vice* chairman' position. Following wide protests it was reported that Teri had decided to 'sever' association with him but in April 2016, he jumped! (11).You could not make it up. Considering the whole IPCC, global circus, the author is reminded of that wonderful comedy show: 'It Ain't Half Hot Mum!'

Apparently all this was big news in India but not widely taken up in Europe for some reason. I wonder why? After all, the IPCC was already in a state of disrepute with many people in the 'scientific know'...but not the public, nor the supine media of course. And too many Western leaders had publically joined the IPCC cult...we would not want them to be embarrassed would we? As western leaders throttle energy supplies but allow 'developing' economies a free for all until 2030, Chinese, Indian and Brazilian leaders must be laughing their heads off. However the fiasco goes on; at the end of 2015 the word went out to the media again that this year was the 'hottest on record'. That was so if one consulted various global surface temperature records favoured by the 'warmists' and notorious for convoluted 'collations', 'infills' and 'technical adjustments' which always seemed to boost current temperatures and decrease historic temperatures. Yet if we examine the same satellite temperature records we looked at earlier, 2015 was cooler than several recent years. However temperatures *were* increasing more sharply at the end of 2015.

In spring 2016 the cry went out again as global January and February temperatures (in the favoured time series) shot up sharply, to match the previous 1998 record. The January to March spikes were added onto the long time series giving the impression that 2016, which had not happened yet, was now the 'hottest year ever'. This is rather naughty. Even worse we noted earlier that the excuse for the global warming halt in the 2000s, after the 1998 temperature spike, was El Nino, the NAO, or other 'natural cycles'. But in the early 2016, gleeful outcry of doom, there was no mention that the January-March spike came during a very strong El Nino period. In fact from data on the NOAA site this El Nino had a warming peak equal in magnitude to that of 1998 (also strong) but had lasted for longer. The 1998 spike was followed by a strong La Nina cooling. The spring 2016 spike showed up in the UAH and RSS satellite data for the lower troposphere but not in the higher layers. In the first CC book postscript I then said this:

'If the pattern of the 1998 El Nino spike repeats, global temperatures should remain high for a few months and then fall back, as in the BMO 2012 forecast, and as we enter a normal neutral and then a La Nina cooling phase in the Pacific. After the 2016-2017 La Nina, will temperatures be back at the year 2000 level or lower? Time will tell. One swallow does not a summer make.'

This forecast was wrong. As in 1998 temperatures did fall back as expected until 2018 / 19, reaching the early 2000s level **but** a second 'El Nino' period followed (which happens sometimes) and a second temperature peak developed.

Only in autumn 2020 did we get a mild La Nina phase setting in. Even so by March 2021 global temperature fell swiftly back to the early 2000s level based on the UAH MSU satellite data. Perhaps the quiet Sun was beginning to bite beyond its normal El Nino / La Nina contribution?

Finally what about the politicians and the scientific consensus? It is good to report that some are concerned about the climate hysteria of recent years. In 2019 the General Secretary of the World Meteorological Organisation finally kicked back, criticising the media in particular for the alarmist narrative. Dr Taalas condemned green extremism which had 'gone off the rails'.

'While climate scepticism has become less of an issue, now we are being challenged from the other side. Climate experts have been attacked by these people and they claim that we should be much more radical. They are doomsters and extremists; they make threats.'

He called for a broader range of opinions to be heard. Even some in the EU have rebelled! Also in 2019 the European Conservatives & Reformist Group of the EU Parliament formally accepted The European Climate Declaration formulated by over 700 prominent scientists which said simply

'There Is NO Climate Emergency'
It also said

'Current climate policies pointlessly and grievously undermine the economic system putting lives at risk in countries denied access to affordable , reliable electrical energy...[current] climate models are unfit... We urge you to follow climate policy based on sound science...'

Signatories included Nobel Laureates Professor Ivar Giaever and Professor Freeman Dyson who we quoted earlier and several senior ex NASA science folks. The shock was too much for some in the Parliament and Irina Von Wiese of the Lib Dems swooned, wept and then *attacked IPCC statements*, quoted in the Declaration, that in fact there were NO global up trends in extreme weather events !!! The declaration made six simple points:

Nature as well as anthropogenic factors cause warming.

Warming is far slower than predicted.

Climate policy relies on inadequate models.

Carbon Dioxide is plant food, the basis of all life on Earth.

Global warming has not increased natural disasters.

Climate policy must respect scientific and economic realities.

The author hopes that the evidence in this book supports these islands of sanity in an ocean of neurotic, post-industrial insanity. If the reader thinks I am exaggerating, he / she/ whatever, might consider the public response to the Covid 19 Pandemic, still running wild as I write. Are we living in rational times? On the one hand we have had a massive disruption to our lives and economic wellbeing while the natural death rate seems to be roughly 180 in 100,000 or 0.18%. Not quite the Black Death. Even so it is a risk still well worth taking every effort to avoid, given Long Covid effects even on the young. On the other hand we have sub-populations, influenced by social media hysterical nonsense, or innate irrational fears about vaccines, or a belief that their human rights are being attacked, or some vague, primitive religious conviction, or a belief that their particular god(s) will protect them, who are refusing to accept a vaccine. There are many ignoring 'the science' in the face of overwhelming, published evidence. As a student of climate change hysteria I cannot be surprised. Fortunately our government, in the Covid 19 case, more or less followed the science. If only they would do the same on climate issues.

DPG Spring 2021

Section 6 References

1. 'Sunspots at all time high', Astronomy Now, June 2012, pp24-25.

2. Dyson F; 'Many Coloured Glass: Reflections on the Place of Life in the Universe', University of Virginia Press, 2007.

3. Donahue R et al; 'Impact of CO2 fertilisation on maximum foliage cover across the globe's warm, arid environment', Geophysical Research Letters, 19th June, 2013.

4. www. science.nasa.gov/secinece-news/science-at-nasa/2014/06oct_abyss

5. Wunsch C & Heimbach P; 'Bi-decadal changes in the abyssal ocean', Journal of Physical Oceanography, 2014, 44, 2013-20130 (pdf).

6. 'A Forecast the Met Office hoped you would not see', Christopher Booker, Daily Telegraph, 12th January 2013, p24.

7. 'What's down with the Sun? major drop in solar activity predicted', report on the American Astronomical Society Solar Physics Division meeting 2011, www. nso.edu/press/Solar.Activity.Drop

8. Royal Astronomical Society: National astronomy meeting Llandudno: 'Solar activity predicted to fall 60% in 2030s, to 'mini' ice age levels : Sun driven by double dynamo', Science Daily, 9th July 2015.

9. Palle 2008; www. bbso.njit.edu/Research/Earthshine/literature/ Palle_etal_2008_JGR.pdf

10. Booker C; 'The Real Global Warming Disaster', Continuum, 2009.

11. The Times of India, April 21st, 2016.

12. www. sott.net/article/300044-Prof-Nir-Shaviv-on-Sunspots-number-recalibration-irrelevant

13. Mills T C; 'Statistical Forecasting : How fast will future warming be?', Professor Terence Mills, Applied Statistics, Loughborough University, 2016.

14. Chopra A & Lineweaver C; 'The case for a Gaian Bottleneck: the biology of habitability', Astrobiology, January 2016; 16(1): 7-22.

15. Zhao J et al; Astrophysical Journal Letters, September 10 th 2013.

16 Ridley M, 'Climate Change will not be dangerous for a long time' Scientific American 27.11.15.

17. Karner O; 'On nonstationarity and antipersistency in global temperature series', Jo. Of Geophysical Research, Vol. 107, D20, pp ACL1-1, 1-11, 27th October 2002.

18. Lewis N and Curry J A; 'The implications for climate sensitivity of AR5 forcing and heat uptake estimates', Climate Dynamics, August 2015, Vol. 45, Issue 3, pp 1009-1023.

19. Harde H; 'How much CO_2 really contributes to global warming: Spectroscopic studies and modelling of the influence of H_2O, CO_2, and CH_4 on our climate', Geophysical Research Abstracts, Vol. 13, EGU2011-4505-1, 2001.

20. Wilson G T; 'Atmospheric CO_2 and global temperatures: the strength and nature of their dependence', Department of Mathematics & Statistics, Lancaster University, 2010.

21. Yin J et al; 'Reinforcement of Climate Hiatus by Decadal Modulation of the daily Cloud Cycle'; Dept. Of Civil & Environmental Engineering, Princeton University, 2018.

22. Stozhkov Y I et al; 'Cosmic rays, Solar Activity and Changes in the Earth's Climate'; Bulletin of the Russian Academy of Sciences, March, 2017; Principia Science Int. (Physics).

23. Leamnon R J et al; 'Termination of Solar Cycles & Correlated Tropospheric Variability'; Earth & Space Science, Vol. 8, Issue 4.

24. Consolini G;' Complexity and Criticality of the Magnetosphere Dynamics'; Memorie della Societa Astronomica Italia, 72 , 605.

Appendix 1 The Behaviour and Analysis of Dynamic Systems

This is a huge topic supported by a vast literature and here we only give a bare outline of some basic issues such as the classes of dynamic systems, their behavioural properties and how these may be explored, for example, by spectrum analysis techniques. Everything in this short book is concerned with how the outputs of physical systems change over time. This includes the solar activity outputs of the Sun in various forms and outputs of the Earth's climate system such as temperature and rainfall. We ask how those outputs can be explored and characterised, perhaps over many time scales. We might also suspect that the outputs of other systems act as forcing inputs to the systems we wish to study. For example, our suspicion that solar activity may be modulated in some way by the physical movements of the planets around the Sun. We might expect that if inputs and outputs have similar quantitative forms then there *may* be a causal relationship between them. In simple dynamic systems and where the inputs and outputs have simple forms this analysis should be easy. Even in complex systems if we can build a plausible physical model of the system a priori, from theory, proving the input – output link may still be fairly easy by using statistical parameter estimation techniques on real data.

Unfortunately the systems we are interested in, solar dynamics and weather dynamics, are very complex and not fully understood and the outputs and possible inputs to these systems are structurally complex in their own right.
This is why despite decades of effort, climate models are still failing to make reliable predictions of global temperature. Modelling
is a wonderful thing but if you leave out important, possible forcing inputs and you insist that you know certain inputs 'must be' critical, from 'simple physical theory', the models, however sophisticated and however powerful your computers, will fail. It may be necessary to go back to basics and look at what the raw data is telling us before moving forward again. That is the simple aim of this book.

So why are the climate modellers having so much trouble and why is it so difficult and contentious to prove that other inputs to the climate system play an important role? The answer lies to a large extent with non-linearity and time varying system parameters. Let us look at this in a little more detail. There are several mathematical classes of dynamical system which may be described successfully by ordinary or partial differential equations.
 In ordinary differential equations we are concerned with one or more variables changing only over time. The time rate of change is written as dx/dt.

This might be velocity for example: the time rate of change of distance x, in a mechanical system. The time rate of change of the time rate of change is written as d^2x/dt^2 : this could be acceleration for example, the time rate of change of velocity. In general we have:

$$dy/dt = C.y + D.x \qquad d^2y/dt^2 = A.dy/dt + B.y + K.x$$

The first equation is a 1st order ODE and the second equation is a 2nd order ODE. Y is the output variable of interest and x an input variable. Often higher order systems may be approximated as 2nd order. The idea is to solve such differential equations by integration to find a final, formal equation for y over time. With linear systems this is often easy.

In partial differential equations the variables are also spatially distributed across two or more space dimensions. For example in modern climate models sets of partial differential equations may model atmospheric flows (winds) in three dimensions using the theoretical equations of fluid dynamics. The models are then run forward routinely to provide local or regional weather forecasts. As experience tells us these forecasts are now reasonably good …for a few days ahead. The attempts to look decades ahead are arguably overambitious, to be polite, given current performance and for all the technical reasons we will examine in this book. For the Sun, partial differential equations are used to model the supposed solar dynamo responsible for the Sun's magnetic field and the convective flows and differential rotation which bring sunspots to the solar surface. So far the dynamo is not fully understood or modelled. We will see in the book that a complete rethink may now be necessary.

For the study of large scale history (in time and space) spatially distributed variables are frequently 'lumped' or averaged across spatial regions to reduce partial differential equations to ODE descriptions. This can be successful even, sometimes, in the study of significantly non-linear systems such as those described by strange attractors. The climatologists happily talk about trends in North Atlantic rainfall or global temperature on this basis and we will also study such 'lumped' variables …cautiously.

Let us look at the common families of ODEs using these descriptors : y is the system output of interest; P is a fixed parameter associated with y; x is some external forcing input; Q is a fixed parameter of x; F (y) is some non-linear function of y; P(t) and Q(t) are time varying parameters.

ODE form	System class
$dy/dt = P \cdot y$	linear, time invariant, homogenous
$dy/dt = P \cdot y + Q \cdot x$	linear, time invariant, non-homogenous
$dy/dt = P(t) \cdot y + Q(t) \cdot x$	linear, time varying, non-homogenous
$dy/dt = F(y)$	non-linear, time invariant, homogenous
$dy/dt = F(y, x, t)$	non-linear, time varying, non-homogenous

Unfortunately the systems we are interested in fall into the last class of ODE other than on very short times scales. Such systems may be anything from mildly non-linear to strongly non-linear when very rich dynamic behaviour is possible, including deterministic chaos. We describe the effects of increasing non-linearity in dynamic systems in more detail in section 3.7. In moderately non-linear systems linear approximations may be valid under small input forcing. Since we are interested particularly in systems which are intrinsically oscillatory, responding to oscillatory inputs a good place to start our understanding is the simple pendulum: the classical example of a 'simple harmonic oscillator'. We have a bob of mass m, suspended from a light string of length L from a frictionless pivot in a vacuum (figure F1).

Displacing the bob along an arc of length S0 and releasing it produces a swinging pendulum which will persist in a regular 'near' sinusoidal motion given no damping. The 'near' is because the sinusoid only approaches a simple sine wave for small values of S0. How does this work? From Newton's second law of motion we have

$$\text{Force} = \text{mass} \times \text{acceleration}$$

In the pendulum the force along the arc, s is $F = -m \cdot g \cdot \sin(\theta) = m \cdot \text{acceleration}$ where theta is the angle of the string to the vertical through the pivot. So acceleration

$d^2s/dt^2 = -g \cdot \sin(\theta)$ Notice that the bob mass m has been cancelled out. The pendulum motion does not depend on its mass. This is a non-linear 2nd order ODE but if the angle theta0 and thence S0 are small we know that sine (theta) is ~ theta in radian measure or s/L. So we have

F 1

The Simple Pendulum

Period T is independent of mass M and displacement s for small so.

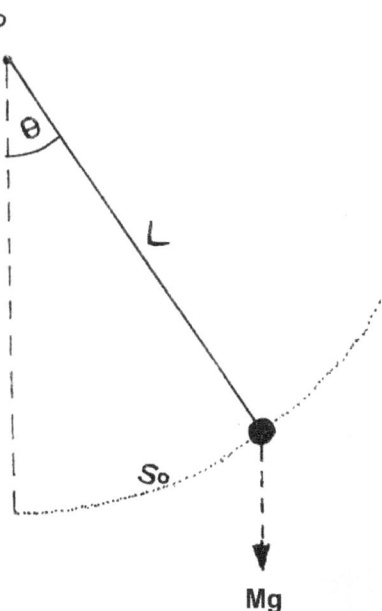

$d^2s / dt^2 = -g.(s/L)$. The integration of this now linear ODE gives the famous result that the period of the simple pendulum is

$$T = 2.\text{Pi} \sqrt{(L/g)}$$

For small s, T is independent of s. However in reality as s and theta increases beyond several degrees, T increases significantly. For theta = 23° the error is 1% but then increases rapidly. The full non-linear solution for T is

$$T = 2.\text{Pi} \sqrt{(L/g)}.[1 + (\text{sine}(\text{theta}/2))^2/4 + (3/4).(\text{sine}(\text{theta}/2))^4/4 +.....]$$

The ...indicates many more terms in sine (theta / 2). By moving the pivot vertically and cyclically all kinds of non-linear behaviour can be produced. In the strangest case an inverted rigid simple pendulum can be stabilised above the pivot in the ultimate magic trick! This perhaps makes the point about other non-linear oscillatory systems of interest: we can expect surprises. In figure F2 we look at how a linear oscillatory system responds to a periodic input, in this case a pure sine wave.

F 2 Linear System Responses

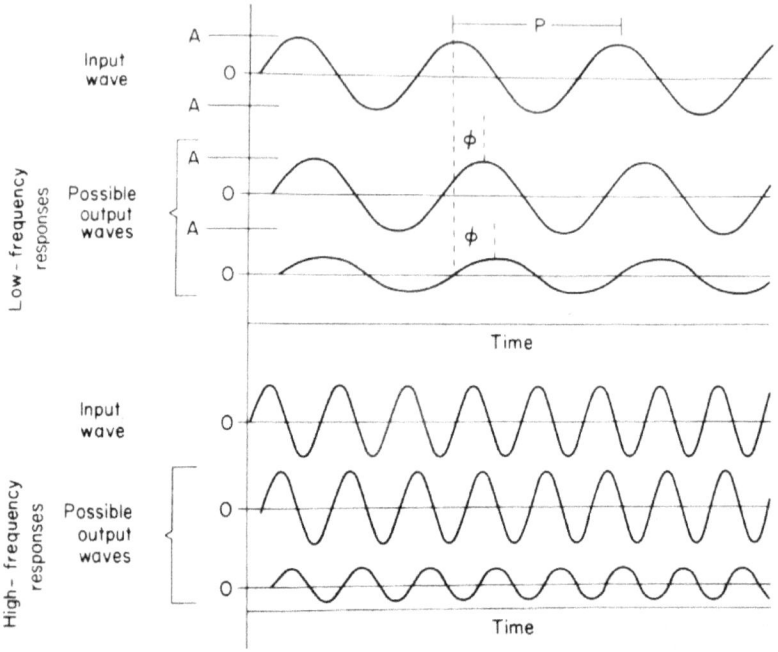

The system may be able to respond to both low and high input frequencies and the output will have that same frequency. The output will be shifted in time: there will be a phase lag between input and output. Also the amplitude of the output wave may be equal to the amplitude of the input or smaller and this will depend on the input frequency. Another useful way of understanding a dynamic system is to make a step change in a steady input. The step response of the system output is shown in figure F3. We show a fast and a slow response system. We see that the output may climb rapidly and exceed the amplitude of the input step in an overshoot but then fall back below the step. We see a damped sine wave which decays eventually to the level of the input step. This is another feature of real physical systems: any oscillation may be subject to resistance usually related to the velocity of the system variable, dx/dt. In the case of the real pendulum this is air resistance. F3 shows the output response for various levels of the damping parameter.

All this behaviour can be summarised in a normalised set of response curves as shown in F4. The horizontal axis shows the frequency (1 / period) of the input signal relative to the natural frequency of oscillation of the system itself, F_n.

F 3 Linear System Step Responses

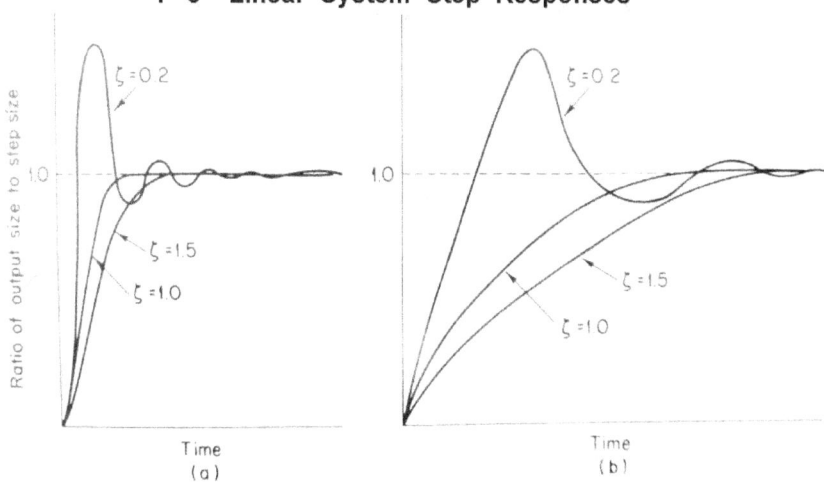

The vertical axis shows the gain of the system on a logarithmic scale (so that 0 marks log (1), a gain of 1). We see that as we increase the frequency of the input signal the gain (and output amplitude) increases as we approach the natural frequency of the system, peaks at that frequency, and then falls away to smaller and smaller gains <1, beyond. The set of curves are for different levels of damping parameter. As damping falls towards zero the peak response at Fn grows without limit. This is the 'resonance' condition which in a real physical system, such as a wind forced suspension bridge, can lead to catastrophic failure. In a non-linear higher order dynamic system the same concepts apply but behaviour may be much more complex. Here is a simple nonlinear system driven by a single periodic signal, Ot

$$d^2x/dt^2 + B(dx/dt) + C(x) = K \text{ cosine (Ot)}.$$

Now the parameters B and C represent non-linear functions of dx / dt and x. This is still a very simple system. In general the linear gain / frequency response curve we looked at in F4 becomes invalid. Instead of a high gain only for input signals near the natural frequency of the system we may also see significant system responses at harmonics or sub-harmonics of the periodic input signal. So if that period is T we may get outputs with periods 2T, 4T, 6T…and 3T, 5T, 7T….and T/2, T/3, T/4, T/5….If the system is strongly non-linear and operating near to 'the edge of chaos' we may see output periodicities which are related to the main periodic responses by Fibonacci ratio harmonics such as : 5 / 3 , 8 / 5, 13 / 8, 21 / 13 or by phi itself, phi = 1.61803…

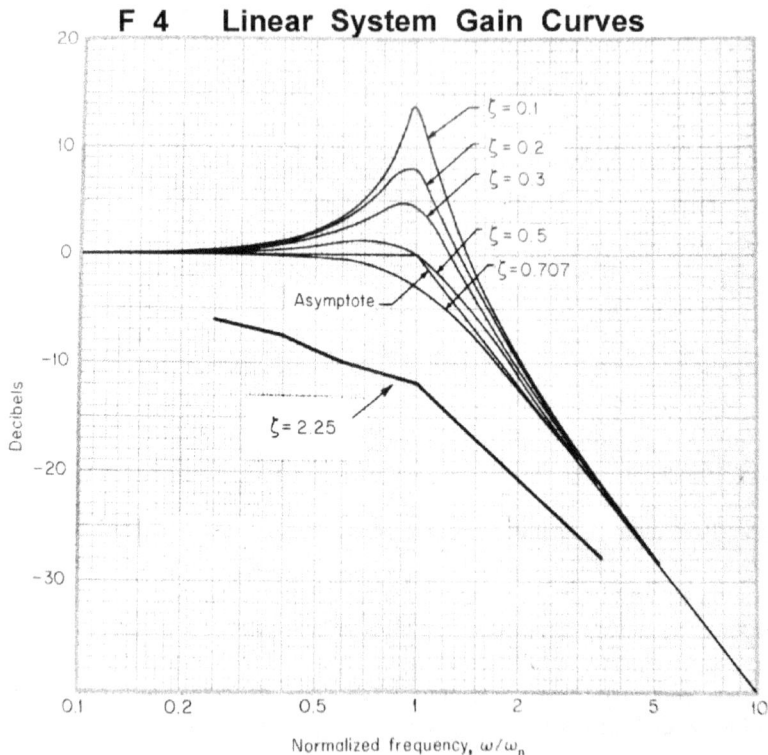

This is discussed in detail in section 3.7. In looking at solar activity and weather variables like temperature we will see that this kind of behaviour is very common.

The primary tool for exploring such dynamic systems forced by multiple periodic input signals, or any complex signal, is spectrum analysis and a brief description of the basic ideas will be valuable to the reader. In the 19th century the mathematician, Fourier, realised that any apparently irregular or complex signal or time series could be decomposed into a series of periodic components: a series of sine or cosine waves of different frequencies and amplitudes. If this seems unlikely consider the waveform in figure F 5 which approximates a sharp saw tooth like wave. In fact this nasty looking curve is simply the sum of smaller amplitude waves which are all simple harmonics of a main period with particular phase relationships to that period. This decomposition is shown in F 6. Such a form of curve is of relevance to our analysis of both solar activity and weather variables.

If we string our (reversed) saw tooth wave forms into a series we get a time record like that in F7. Such a shape is what we see in the solar activity cycle where the growth in sunspot activity is much quicker than the decay, giving an asymmetrical waveform.

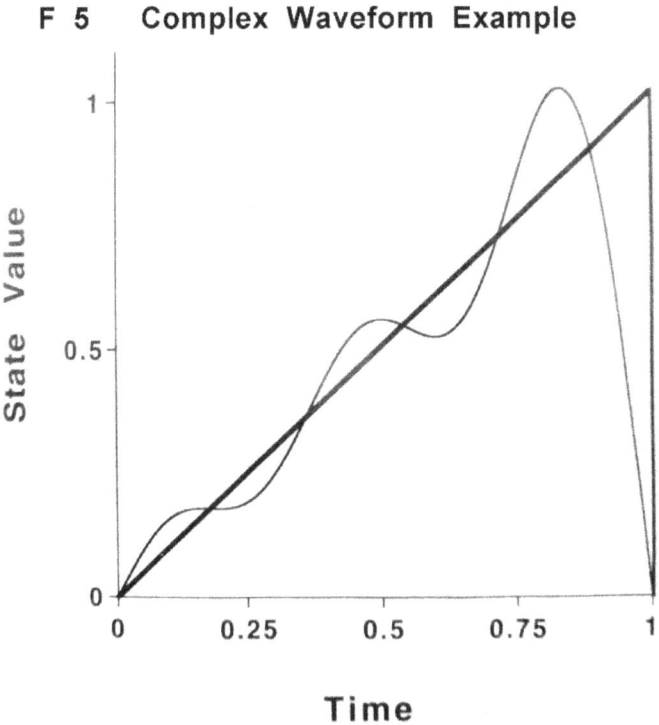

F 5   Complex Waveform Example

Curiously this waveform, with some variation, is also what we see when we look at the history of ice ages over the last half million years: a very rapid, sudden recovery of temperature from the glacial condition, a brief interglacial period (usually ~10,000 years), followed by a slow decline and cooling back to the glacial condition over several tens of thousands of years. It is interesting that like a simple waveform the long term global and regional temperatures move between fixed upper and lower limits as we will explore in section 5.11. We should not therefore be surprised that the sunspot activity spectrum and the spectra of global and regional temperatures over the last million years show harmonically related periodicities. To get at those periodicities we can apply spectral analysis techniques of various forms. For our simple purposes here we feed our system times series of interest into a mathematical process and the output is a curve describing the amplitude or spectral power as a function of frequency across a wide range of frequencies (limited by the sampling period of the data at the high frequency end and the length of the data

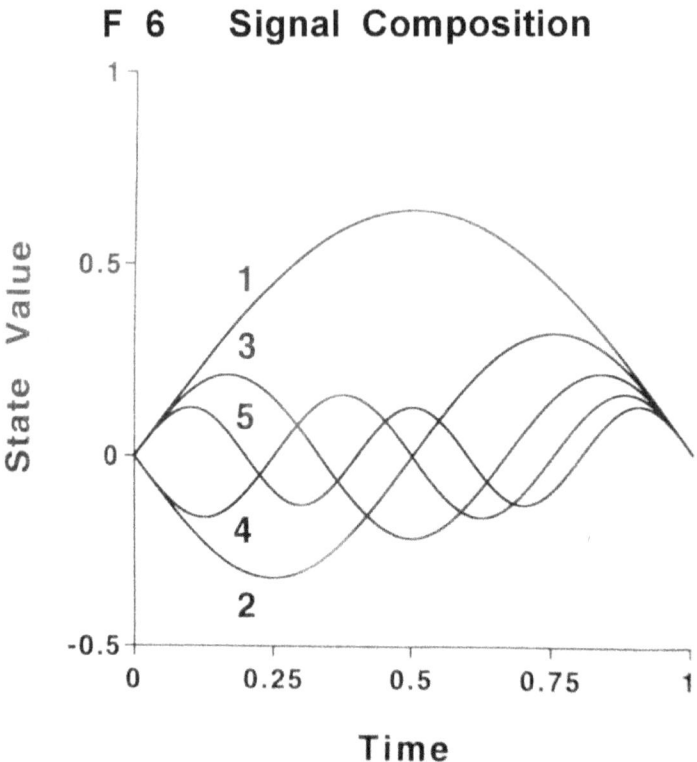

record at the low frequency end of the range). Spectral analysis has the property that a pure periodic, sinusoidal time series shows up as a narrow, sharp, spike in spectral amplitude or power. The taller the spike the greater the amplitude of the signal.

If a complex times series, like our saw tooth curve, is composed of several periodic components, in the spectrum each component will ideally appear as a narrow spike. The relative height of the spikes gives us the relative amplitudes of each periodic component. The spectrum is just like a fingerprint. If we see two spectra with the same set of peaks for the variables of two different physical systems, which could interact in principle, we have reasonable grounds to believe that those systems are influencing each other directly or are each influenced by some third system. This appears to be the situation we will explore when we consider planetary dynamics, solar activity and terrestrial weather variables. There are of course many complications but in essence that is spectral analysis.

F 7   A Complex Periodic Signal

Compare: 1. 3 11 year cycles of the sunspot time series

2. global temperatures over the last three glacial and interglacial periods

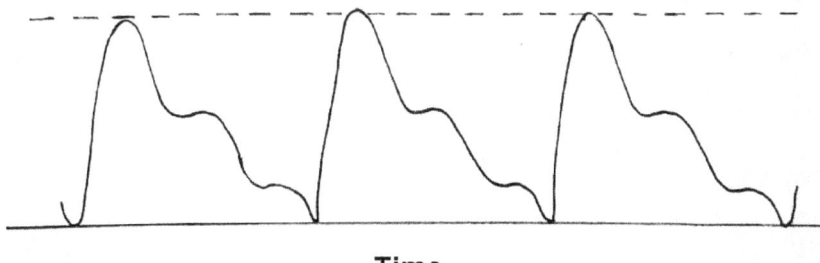

Time

Since Fourier's day much has been learned about how to improve the resolution of spectra, reducing the effects of noise in the time series and in establishing the statistical significance of features in the spectrum. The maximum entropy spectral analysis technique, MESA, is of particular importance since it has great power to reject spurious spectral features and give sharp, narrow peaks which is the ideal result we seek.

The other problem we discussed earlier in complex dynamic systems is that of time varying system properties. Such variation will mean that the system response to the same input signal may change over some time scale from significant to undetectable. This has happened in the past when analysts have detected a strong solar signal in a weather variable for perhaps decades, only to have it disappear. Was the correlation a chance coincidence or was it real before the system physically changed? One way to explore this is to use a high resolution technique like MESA, on successive overlapping data windows, each some sensible fraction of the full data record. These 'evolutive' spectra then give spectral snapshots over time and any systematic change in the behaviour of the system will show up. Of course because each data window is short we have reduced information and less resolving power. It is a trade off. We will look at this approach in considering the long term solar activity records. More recently wavelet spectral analysis has become available. Instead of assuming that a particular period is present to the same extent at all times in the data record, the method allows for the presence of wave packets of fixed period but varying amplitude over some interval.

Outside that interval the period does not exist. This is extremely powerful where we believe the periodic content of a signal may be changing over time. Wavelet analysis will find it. We will look at wavelet analysis as applied to the Sun and to a long term English temperature record.

The alert reader will realise that in a nonlinear system, non-linearity may show up as apparent time variation in some of the parameters of the system and in how it responds to inputs. This is important if we set out to try to build a formal model of a dynamic system using real input – output data. Unless the model structure is flexible enough a priori we may end up misinterpreting what is going on. The ideal model form would allow the data to define the structure. In building linear times series models the most flexible form is the ARIMA, Autoregressive – Integrated-Moving Average class of models originated by Professors Box and Jenkins. The author was once a student of the latter. The basic model form replaces differential equations with difference equations and the sampled system data which is what we have in practice anyway.

In ARIMA modelling the data is processed to produce 'fingerprints' such as the autocorrelation function, the partial autocorrelation function and the spectrum which give clues to the ideal structure of the model for that data. The data determines the model form with no pre-assumptions about trends or any other structures. This is a very powerful procedure. That data derived model form is then statistically fitted and the parameters estimated. Billings (1) extended this model building philosophy of letting the data tell us the ideal model form, to non-linear systems with exogenous variables, the NARMAX model class. In general it has the form

$$y_t = F[\, y_{t-1}, y_{t-2}, y_{t-3}, .. y_{t-d1};\ x_t, x_{t-1}, x_{t-3}, \ldots x_{t-d2};\ e_{t-1}, e_{t-2}, e_{t-3}, \ldots e_{t-d3}\,] + e_t$$

where F is a non-linear function of time delayed values of the system output, y, the exogenous variable (or variables) x, and the noise (error) series e. F may have a number of forms. The simplest would combine a non-linear gain 'box' dependent on y and x current levels say, coupled with linear, time delayed dynamics. Many more general forms are possible. For example time delayed y and x variables could define a set of low order polynomials involving quadratic or cubic or product terms. We will see below that models with quadratic or product terms can reproduce a wide range of non-linear behaviours. The difficulty now is that a very large number of possible model structures exist.

The new methods attempt to generalise autocorrelation, cross correlation and spectral analysis to identify the best non-linear model form before data fitting. We have a stepwise process which gradually identifies the important terms and eliminates others, leading to a sparse and appropriate model structure. The author unfortunately has not so far identified any studies which use NARMAX models to link planetary dynamics, solar activity and the Earth's weather. For this reason the spectra discussed in the book are based on ordinary MESA and wavelet analyses now available in the literature.

It should be noted that these methods will not pick up all the behaviours of a non-linear system. However they still provide much very helpful information if read carefully. For example a careful reading of the spectra can tease out hidden links to forcing signals. In the book we look at the apparent forcing of climate variables such as ice volume and temperature by changes in the Earth's orbit such as eccentricity and rotation axis tilt, the so-called Milankovitch effects. The spectral peaks in the climate time series can often be related to the orbital variables by looking for simple harmonics of the cycles in the forcing signal (see section 5.11). But sometimes the spectra appear to change over time and some known forcing periods do not show up in the record or the spectra. It turns out that the climate system is also susceptible to frequency and phase modulation whereby the amplitude change in a signal is translated into a frequency change or a phase shift. So we look for a period and do not see it but instead, perhaps, a cluster of periods produced by phase modulation. In wavelet analysis some of this may show up. With care and working through possible non-linear models for phase modulation, the mess can be untangled. Remarkably this has been done with apparent success in the case of the Croll-Milankovitch cycles as reported in Science as early as 1999 (2). Using a phase modulation model the author successfully reproduces the climate spectra and the time series of the irregular, semi-periodic ice age cycles of the last million years from the delta oxygen 18 (proxy temperature) records at several sites.

In the book we also look at non-linear dynamic systems of an extreme form: those that conform to a 'strange attractor'. The solar activity cycle may be generated by such an attractor at least if we consider the activity time series without regard to possible external forcing signals. We can really only say that somewhere along the pathway to varying solar output whether it lies in the Sun, or forces modulating the Sun, 'strange attractor' like behaviour emerges. It is worth having a quick look at these extreme non-linear systems.

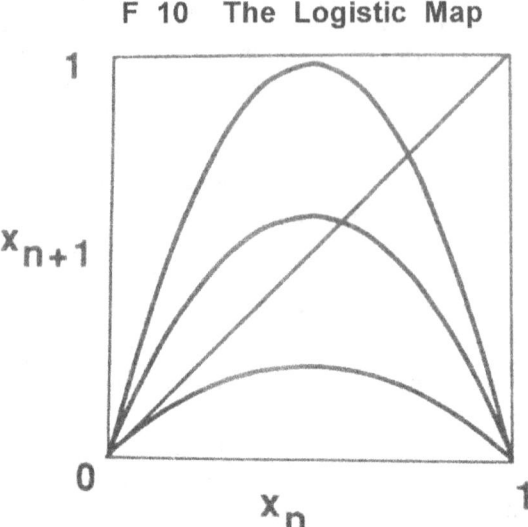

F 10 The Logistic Map

It is curious that there are many relatively simple non-linear systems in physics and biology that can produce limit cycle and even chaotic behaviour. First consider the famous logistic equation which approximates the growth of biological populations in a system with a limiting carrying capacity...in reality all such systems. In its simplest form :

$dN/dt = S.N.(1 - N/K)$ where N is the population;
S is the natural growth rate;
K is a measure of the carrying capacity of the local environment. In the initial case where N0 is small we have $dN/dt = S.N$ and unlimited growth. Rearranging and integrating gives us

$dN/N = S.dt$ and $N = N0.e^{St}$

For small N initial population growth is exponential but as N tends to K growth declines to zero. This seems to be a straight forward if non-linear situation. However the logistic can also be written as a difference equation which has interesting properties.

$X(t+1) = S.X(t).(1 - X(t))$ We can represent this equation graphically which aids interpretation.

The figure F10 shows the quadratic logistic map for various values of S, the growth rate.

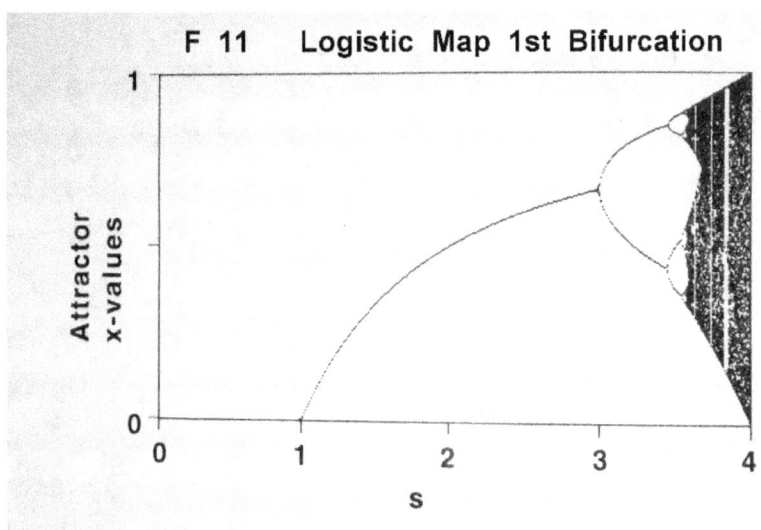

F 11   Logistic Map 1st Bifurcation

When S is < 1 the system will decay to X = 0. We say that 0 is an 'attractor' of the system. For some values of S we find a second fixed point attractor by solving $Xf = S \cdot Xf \cdot (1 - Xf)$. Dividing by Xf gives $1 = S \cdot (1 - Xf)$ and so $Xf = 1 - 1/S$. This holds for $S > 1$. In this case there are fixed points of attraction for the dynamics at 0 and $1 - 1/S$. For $S < 1$ the one fixed point is at 0. If a time series is generated by a logistic map it is always possible to reconstruct the map by plotting time delayed values of X in two or three dimensions. This is shown in Figure 10A. Reconstructing non-linear phase space maps is a powerful tool as we will see.

So far so simple. But the logistic difference equation holds much more as we increase S. Consider F 11. Beyond S = 3 the non-zero fixed point splits into two points. X now oscillates between two values and we have a 2 period limit cycle generated purely by the internal system structure. There is no external forcing signal. As S increases we can see more complex behaviour. This is shown in detail in F 12.

Beyond S = ~3.5 we encounter a cyclic regime which visits 4 values of X repeatedly. Then near S = 3.55 the attractor splits into an 8 level, 8 period cycle. By S = 3.6 and beyond towards S = 4 the behaviour is chaotic with all values of X visited. Even so we can see structure in the chaos. There are breaks in the chaos as near S = 3.83 where we encounter 3 and 6 cycle behaviour… but other windows have 5, 7, 9…cycles. Approaching S = 4 all periods are encountered in principle but their windows of existence are by then very narrow.

F 10A Reconstructed Logistic Maps

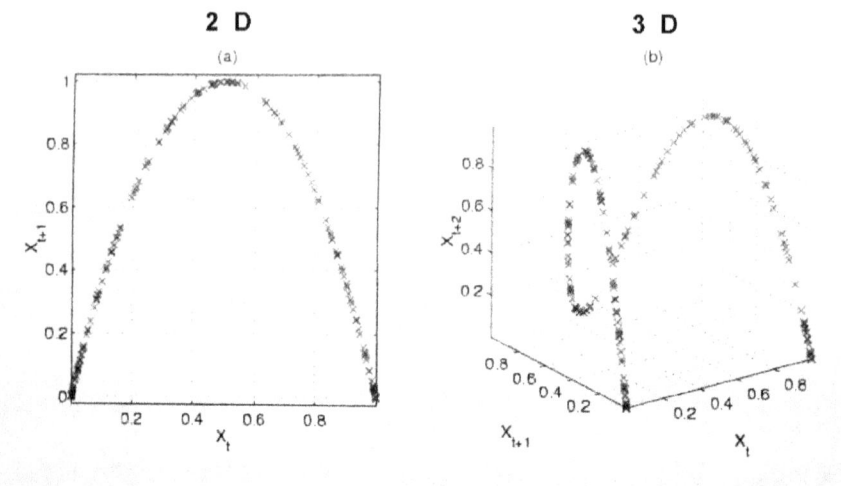

F 12 Logistic Map Later Bifurcations

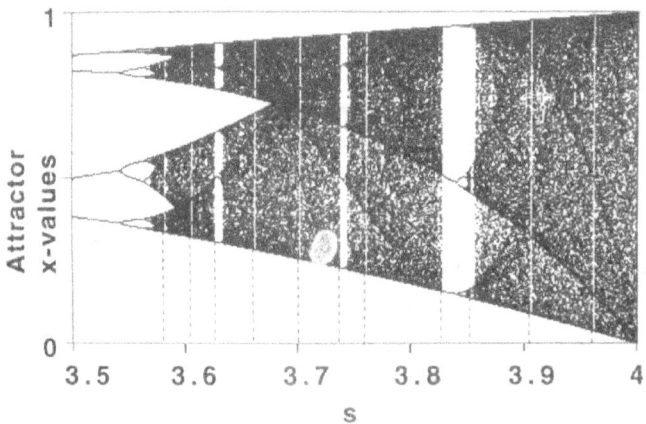

Burrowing down into narrower and narrower ranges of S as it increases we see the same form of bifurcation pattern repeated again and again. This is a pattern common in much more complex non-linear systems. The logistic map is a very much simplified model of population dynamics in biology but it serves to warn us of what might be lurking in real systems and the need for humility and caution when we interfere in ecosystems.

In section 5.11 we look at recent research on probable bifurcation structures in the climate system in response to the Croll-Milankovitch orbital / axis forcing signals which seem to control the current series of ice ages …and for more than a million years.

It may be instructive to look briefly at one more classical non-linear system which can teach us important lessons. Edward Lorenz was one of the pioneer explorers of real world non-linear systems behaviour. His particular field of study is of some relevance to this book. Lorenz was a student of weather systems. He did a wonderful job of reducing the partial differential equations (of Navier-Stokes) for fluid flow and heat transfer, into a small set of non-linear ordinary differential equations. If this is done sensibly much of the original *qualitative* behaviour of the more complex system is preserved and can be explored. He ended up with three equations representing the movement of gas in a finite 'box' heated from below:

$$dx/dt = P.(y - x)$$

$$dy/dt = R.x - y - x.z$$

$$dz/dt = x.y - B.y$$

P is the Prandtl number which measures the ratio of fluid viscosity to thermal conductivity. The lower the viscosity and the higher the thermal conductivity the greater the convection in the box. R is the difference in temperature between the top and base of the system. The higher R is, the greater is the driving force in the system. B is just the aspect ratio of the box or width / height. Lorenz used the values $P = 10$, $R = 28$, $B = 8/3$. The nonlinearities here seem modest: the two simple product terms x.z and x.y but this is enough to make the system conform to a strange attractor. Here it is showing the path through time of the three system variables x, y, z. We see that the attractor has a definite form with a complex surface to which the path of the system conforms but that path may flow anywhere on that surface making the future path generally unpredictable beyond a limited number of steps ahead. But in certain places on the attractor, if we know its form in detail, we can guarantee that longer term forecasts will be possible…for a while.

Remember this is a simplified model of a small piece of a weather system. We can see now why weather forecasting is difficult. Current short term forecasts are based on very detailed (but incomplete) atmospheric models driven by masses of real time weather data from satellites and so on. The author doubts if the forecasters have a formal 'strange attractor' model to consult but they do the next best thing.

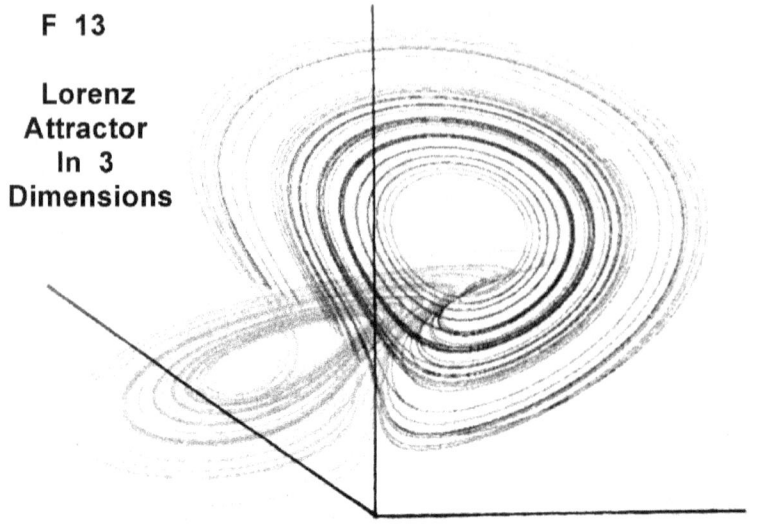

F 13

Lorenz Attractor In 3 Dimensions

They have a library of past weather situations through historic seasons which they match to the current position and then ask: what happened last time? If you like: where are we on the weather 'attractor'? They are well aware that some situations are intrinsically uncertain: are we on the wrong part of the 'attractor'? Similar models are used to make long term climate predictions. Is the weather system less chaotic, or less complex, on these longer time scales? I suspect not and many factors are not taken into account in the models. Just for fun consider figure F 14 which shows a 2 D projection of the Lorenz strange attractor onto variables x and y plane. Many have likened this projection to a butterfly and used it as an icon of chaos and the limits of prediction.
Famously it has been said that the landing of a butterfly on a leaf in the Brazilian rainforest could ultimately deflect the path of a storm over the Caribbean Sea. It is true that the paths beginning from two very, very close points on such an attractor will eventually diverge exponentially.

We should not forget the 'butterfly effect' when considering how apparently small changes caused by so-called 'negligible' external forces may change the output of a star or drive a planet into a new ice age or out of one (see section 5). If all this seems a little speculative we will end by looking at the Sun again but in the light of what we have learned about non-linear dynamic systems. Reference (3) looks at the classic sunspot number and sunspot area time series using smoothed monthly data from 1874 to 2012.

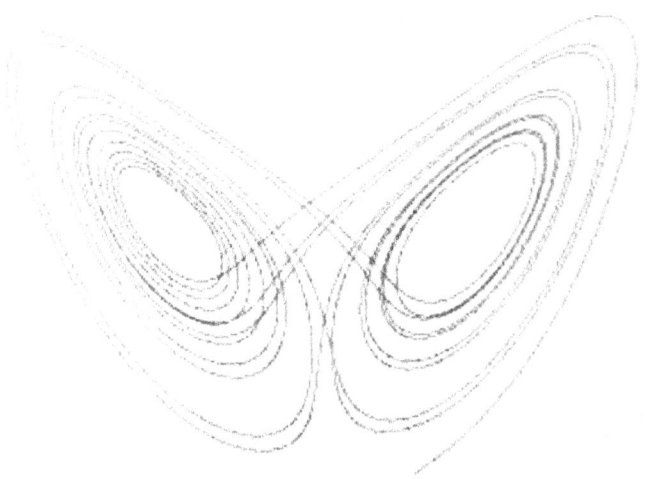

F 14   Lorenz Attractor 2 D Projection

Sunspot activity follows an ~11 year cycle but in addition the amplitude of the cycle changes over longer periods of time and sometimes the cycle appears to turn off completely for periods of decades. It is agreed that the spinning Sun contains a dynamo which generates a time varying magnetic field which, through convection and differential rotation, feeds knots of disturbed field to the surface which become our visible sunspots. The dynamo appears to be intrinsically oscillatory with limit cycle like behaviour. The question is: does the variation in the cycle reflect some random, stochastic process or is the Sun actually a chaotic oscillator? Several attempts have been made to test this by trying to reconstruct a possible underlying strange attractor like that of the Lorenz system we discussed above.

These early attempts failed, probably because of high frequency noise in the time series. The present authors (3) chose a long data window in which the Sun's behaviour was thought to be homogenous and the data of good quality. They found that smoothing the noise was also important. By forming sets of time delayed values of sunspot numbers and areas they discovered that three dimensional phase spaces created from $x(t)$, $x(t + T)$, $x(t + 2T)$ data sets showed well defined strange attractors with fractal dimensions of ~1.2. In 3D we can see solar activity winding around loops of different magnitude across a more or less flat attractor surface. In the book section on solar activity we will also show that the solar dynamo, despite meanderings in cycle amplitude, accurately preserves the phase of the

22 year Hale magnetic cycle over three centuries. This is also not reflective of stochastic behaviour.

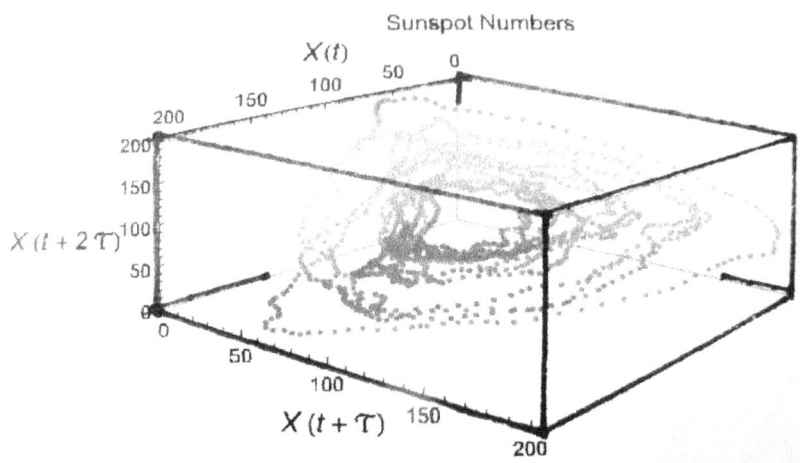

F 15   Sunspot Strange Attractor Reconstruction

This analysis seems to settle the matter of solar activity variation as presented through sunspot numbers and areas. These output variables of the Sun are chaotic. However we still do not know, given only this information, whether this behaviour is completely intrinsic to the solar dynamo and convection / differential rotation properties of the Sun or whether that dynamo is being forced by external influences i.e by planetary torque effects and tides. In the main book we will argue on the basis of spectral analysis that the movements of the planets and their gravitational 'influences' cause such forcing and discuss new and compelling evidence for that case.

In sections 2.5, 3.8 and 4.4 we looked at several variables of interest and showed that their various properties and relationships could be understood in terms of 'self organised critical' processes. These have the remarkable property that small inputs to them across a wide range of frequencies can produce large responses in say, energy release on longer time scales. This potentially explains how small inputs such as planetary tidal and torque effects on the Sun can have such large consequences. Similarly such small planetary forces are able to affect a system of critically tuned faults in the earth's crust to apparently 'cause' earthquakes to the extent we can pick up clear patterns of planetary dynamics in earthquake and volcanic energy spectra.

We have tried to simply introduce in this section the ideas behind non-linear systems of various kinds so it seems only fair to do the same with SOC systems. Self-organisation is often defined as the evolution of a dynamic system into an organised state in the absences of external forces explicitly guiding the local behaviour of sub-systems. 'Self' does not imply any conscious desire of components or the global system to organise itself. All we need to get 'order' is a set of components that interact in certain ways under certain constraints which turn out to be quite common. We noted in section 2 for example how the particles and gases in a rotating proto-planetary disk around a new star would form aggregations of all sizes in random orbits. The larger bodies would attract more matter clearing some paths in the disk. Initially many bodies would interact under gravity with nearby bodies and resonance effects between pairs or triplets or multiplets would alter distances and periods until the surviving large bodies reached some kind of 'stable' mutual arrangement. This is why we see certain common harmonic orbital period ratios in many exo-planetary systems. However we also noted in our solar system that the planetary system is only quasi-stable on long time scales. It is critically tuned and poised for change. Most changes will be small but eventually large changes will occur. Nobody designed the solar system: it evolved under natural forces.

The idea of self organisation began with the study of a simple system : the sand pile. Consider a sand pile sitting on a raised circular disk with a slow input of sand particles dropping on to the centre. Eventually a conical pile will build up with a nominal balance between particles falling from the edge of the disk and the new particles added at the centre. Depending on the size and roughness of the particles the pile will attain a certain natural (self organised) angle of repose.

From: 25 Years of Self-Organized Criticality: Solar and Astrophysics

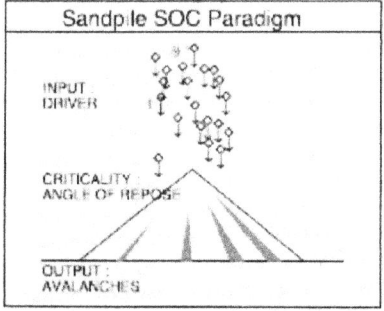
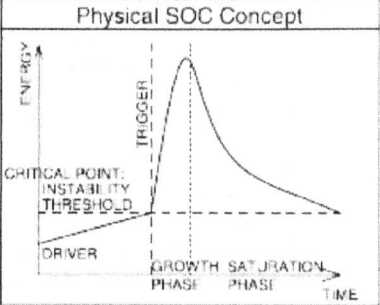

So the question is how do the sand particles leave the pile? We might expect the outflow to match the small steady inflow. But the pile is sitting at the edge of stability. It turns out that avalanches occur in practice on a wide range of size scales from a single particle to hundreds of particles. Small avalanches are the most frequent and the largest avalanches are rare.

It turns out in practice and in computer models that the avalanche size follows a power law in frequency of occurrence so that the log of frequency is a linear function of the log of avalanche size over orders of magnitude. This is a remarkable result with remarkable consequences. This physical property means that very small random, external additions of energy say, can lead to a large response of the system and the release of a large amount of energy at longer time scales. SOC systems act like non-linear amplifiers with a high gain at low frequencies. We suspected that this has happened in planetary system structuring, in the generation or modulation of solar activity as represented by flares, sunspots and the solar wind by planetary forcing; the occurrence patterns of earthquakes and volcanic activity again in response to planetary forcing; the response of the climate system to solar output forcing and ultimately to planetary dynamics on time scales from months to hundreds of millennia (by various physical mechanisms).

The evidence is very suggestive, at least to an open mind. Two last pieces of evidence are worth noting in passing. The figure below presents the power spectral density of Total Solar Irradiance energy as a function of frequency (1 / years) over several orders of magnitude in both axes. We essentially have a power law relationship from periods of days to centuries with some interesting deviations. The deviation around 27 days picks up excess energy associated with the rotation period of the Sun. The deviation at 11 years marks excess energy at our old friend the main 11.16 year solar activity cycle.

The lower graph records the spectral power for variations in global surface temperature as a function of frequency...another power law relationship. In this case the curve shows a deviation at one year representing the annual seasonal variation. We may also have a step up beyond the power law background around 11 years and for longer periods. The power laws are not identical but are similar in slope.

TSI reflects all frequencies of solar energy but it is known that the UV (ultraviolet) flux varies more than TSI overall and has an immediate heating effect on the ionosphere but less so at lower levels. UV flux shows a 27 day (solar rotation) modulation in our upper atmosphere.

The longer periods we have seen in climate variables appear to have an origin in the longer solar activity variations 'indicated' by the sunspot cycle, such as the strength of the solar wind, but the probable presence of SOC processes reminds us that energy inputs over a wide frequency range can result in strong responses at low frequencies. We have identified the elements of a solution to the climate change mystery but not a finished story.

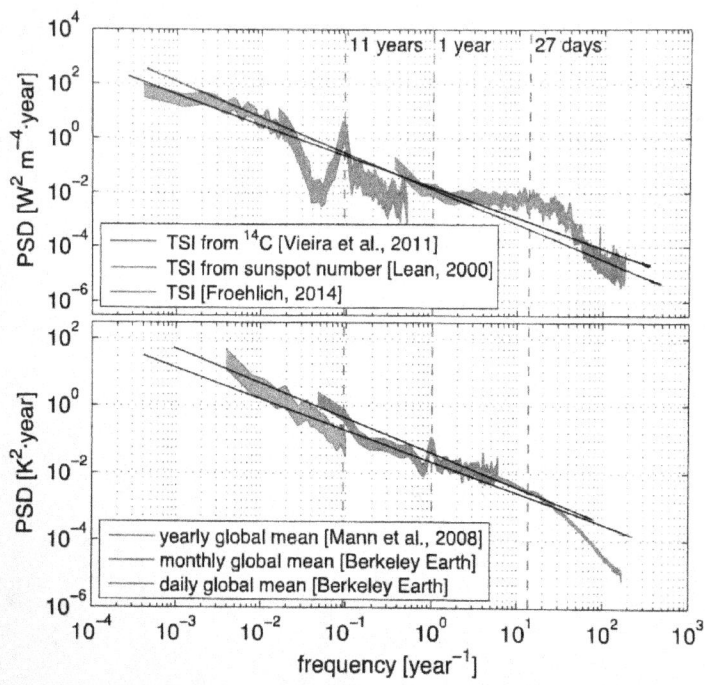

References Appendix 1

1. Billings S; 'Nonlinear system identification: NARMAX methods in the time, frequency and spatio-temporal domains', Wiley, September 2013.
2. Rial J A; 'Earth's orbital eccentricity and the rhythm of the Pleistocene ice ages: the concealed pacemaker', Global & Planetary Change 41 (2008) pp81-93.
3. Zhou S et al; 'Low-dimensional chaos and fractal properties of long term sunspot activity', Research in Astronomy & Astrophysics, 2014 Vol. 14, No. 1, pp 104-112.

Appendix 2
Solar-Terr. Predictions Workshop
Paris Obs. 18-22 June 1984

LONG TERM SOLAR CYCLE MODULATION

D. P. GREGG
Unilever Research Laboratory, Port Sunlight, Merseyside, UK

ABSTRACT

A long term model of solar cycle amplitude modulation which closely matches behaviour from ~ 1750 to the present has been developed. Despite this good performance certain features of the cycle suggest that "stable" quasi-periodic behaviour may break down from time to time. The Sun may have a much wider repertoire of dynamic behaviour up to and including chaos.

Long term forecasting of peak sunspot activity requires either stable regularities in the historical time series of this phenomenon which can be extrapolated or a stable leading indicator relationship between it and some other observable physical phenomenon. In the latter case of course the physical time lag between the phenomena must be of the same order as the desired forecast horizon or the leading indicator itself reliably predictable over appropriate periods.

ANALYSIS

A non-linear model which exploits an interactive, lagged relationship between sunspot activity and the occurrence of geomagnetic abnormal quiet days. AQD. (Brown 1974) has been developed (Gregg 1984). If this relationship remains stable it has potential for ~ 4 to 5 years ahead forecasting of activity. This basic model is complemented by a quasi-linear "statistical" model of the long term sunspot peak amplitude variation. This model accounts for 96% of peak height variance over the period 1859-1980. The combined model predicts a long decay for cycle 21 with a minimum in ~ 1989. Cycle 22 mean annual peak will be low at 36 (± 15) and may not occur until 1994-95. Cycle 23 should be more normal with a peak around 115.

Recent examination of sunspot data for cycle 21 (1975-1984) shows a very close similarity in shape to cycle 4 which would be expected from the model. If the decline of 21 follows the cycle 4 pattern from now on then relative skewness would be β = 0.63. However 21 was higher than 4 (peak 154 cf 130). Allowing for the positive relationship between peak height and skewness this suggests cycle 21 could be more skewed with β = 0.73. Using the Ramaswamy (1977) correlation between skewness and ratio of successive peak amplitudes indicates a cycle 22 peak height of 28 to 52, close to the non-linear model forecast range based on data up to 1980.

In statistical terms the model's historical performance is impressive. However the model is still structurally crude relative to the behavioural complexity apparent in longer term (i.e. 200 years +) sunspot and "related" proxy data. The model does not account for example for phenomena like the Maunder and Spörer minima. This longer term behaviour gives every appearance of being generated by a physical mechanism capable of a range of interesting dynamic phenomena including chaos and possibly havoc (exogenously modulated chaos). The implications of this possibility for long term forecasting are important. The peak amplitude variation over the period 1700 - 1980 appears to be dominated by quasi-cycles of ~ 45, 90 and 180 years (and some power at ~ 60 years). These periods are even sub-harmonics of the basic Hale, 22.35 year, magnetic cycle in the ratios 1:2:4:8. Examination of proxy C₁₄ data over the period 1000 to 1600 AD (Stuiver 1980) shows the presence of quasi-cycles of ~ 67, 135, 270 years, odd sub-harmonics of the Hale cycle in the ratios 1:3:6:12. Chinese large sunspot group records (back to 43 BC) also confirm the dominant presence of these sub-harmonics over long periods (ASRRG 1977).

Data series dominated by groups of even or odd sub-harmonics in these ratio patterns are characteristic of chaotic oscillatory systems whose repertoire of behaviour includes quasi-periodic episodes, large and sudden amplitude jumps and truly chaotic (aperiodic) behaviour. Small changes in the parameters of such mechanism can cause sudden flips between quite different dynamical regimes. If the sun hosts such an oscillatory mechanism there may be long periods of stable quasi- cyclical behaviour dominated by particular harmonic patterns as suggested by the author's amplitude model. However that stability may break down at any time precipitating a completely different pattern of

7

311

oscillation (including Maunder minimum like phenomena).

Such complex dynamic behaviour would limit our ability to make longer term forecasts with confidence whether based on statistical or physical models i.e. even fully deterministic physical models in this class can give rise to essentially random and therefore unpredictable behaviour. In the latter case however reliable prediction might still be possible if key parameters were modulated by predictable effects external to the sun. Many workers have suggested such external forcing related to either movements of the sun about the solar system centre of mass or direct planetary tidal effects.

Despite the problems, both physical and psychological, with this idea a remarkable similarity does exist between the spectrum of the sunspot time series and the spectrum of the sun - solar system centre of mass distance filtered by a simple non-linear device:

Table 1. Sunspot and Centre of Mass Distance Spectra

Sunspot Spectrum Key Periods Yrs	F(w1,w2)* Spectrum Key Periods Yrs	Non-linear Combination Frequencies
22.33	22.35	w1 - w2
17.8	17.7	w1 + w2
25.5	26	w1 - 2w2
16.2	16.3	w1 + 2w2
9.92	10.4	2w1 - w2
	9.92	2w1
	9.5	2w1 + w2
19.9	19.86	w1
7.4	7.44	3 (w1 - w2)

*The centre of mass distance series is dominated by the Jupiter-Saturn conjunction period of 19.86 yrs (1/w1) and a complex modulation pattern of 178.8 yrs (1/w2).

This degree of correspondence is surely too remarkable to be a coincidence. Given the critical problems of long term forecasting associated with the complex oscillatory mechanism apparently generating the solar cycle this possible relationship is worth serious examination given the long term predictability of planetary motions.

REFERENCES

Brown, G. M., A New Solar-Terrestrial Relationship, Nature, 251, 592, 1974.

Gregg, D. P., A Nonlinear Solar Cycle Model With Potential for Forecasting on a Decadal Time Scale, Solar Physics, 90, 185-194, 1984.

Ramaswamy, G., Sunspot Cycles and Solar Activity Forecasting, Nature, 265, 713, 1977.

Stuiver, M. and Quay, P. D., Changes in Atmospheric Carbon-14 Attributed to a Variable Sun, Science, 207, 11, 1980.

Ancient Sunspot Records Research Group, A Re-Compilation of Our Country's Records of Sunspots Through the Ages and an Inquiry into Possible Periodicities in their Activity, Chinese Astronomy, 1, 347, 1977.

Appendix 3 : Consensus Bullying & 'Cargo Cult Science'

In the book we touched briefly on the key facts of the current climate chaos. That is, the divergence between the upward global temperature forecasts from the IPCC approved climate models and real, observed temperatures for the last decade or decade and a half which show little or no warming, depending on the temperature series chosen. The model forecasts are way above the observations: full stop.

The UAH and RSS satellite based series are clear that there has been no warming from ~1998 to 21015. These temperature series give us direct, integrated measures of global surface temperature based on 14 on board, microwave instruments calibrated against precise laboratory standards before launch. There is no room for 'after the event' fiddling but sadly, recently, attempts have been made (1). The other 'standard' temperature series are summations of hundreds of local ground station records. This summation process presents many problems (and opportunities) including the relatively few stations in polar regions where 'simple' physics tells us any CO_2 generated warming would be concentrated. Other vast regions are also bereft of stations. So we have to use these few readings to 'infill' across vast areas and apply various weightings …so called 'homogenisation'. In looking back over many decades the environments of long standing stations have often changed e.g. because of urbanisation around them. Cities are well known heat islands. So again there is scope for legitimate adjustments in historical temperature records. However such adjustments become suspect unless carefully recorded and justified. There have been troublingly many claims that historical adjustments in some regions turn long term cooling trends into warming trends (2). We noted in the book similar 'adjustments' attempting to reduce the amplitude of the late 20th century solar activity 'grand maximum' by falsely claiming 'bad data' over some periods. Yet we saw that several independent proxy series for solar activity showed clearly that the late 20th century peak was unique in several thousand years of solar history.

Similarly attempts have been made to 'talk away' the 'Maunder Minimum' in solar activity and its coincidence with the Little Ice Age. We looked at the activity of other sun like stars and showed that they too exhibit grand minima with about the same frequency as the Sun. All this mischief is aimed at countering claims that the Sun affects climate significantly. 'Everybody', allegedly, knows that only CO_2 can claim that crown!

Yet our look at the spectral fingerprints of the Sun and the fingerprints of many regional climate variables showed strong and clear links...provided we allowed for possible non-linear relationships. We found cyclical connections on all time scales from months to thousands of years.

Considering all the above defines a situation where a scientific issue of great practical importance has taken on the mantle of a quasi-religious cult: there is no god but global warming and the IPCC is his prophet. The IPCC / western governments axis has grown so powerful that a consensus on man made global warming can be claimed which few have the ability or independence to challenge. This is still so despite the failure of the climate models to predict the long halt in warming post 1998... although CO_2 output levels continue to increase. In some cases instead of addressing the stark problem head on, the response has been to find reasons to 'adjust' global temperature series to 'find' the missing heat. Recent temperatures have been adjusted upwards; historical temperatures have been adjusted downwards ...suddenly we have a bigger trend or a trend where none existed before.

We saw that the other 'warmist' excuse was that some of the missing heat had gone into the deep oceans, but alas, recent measurements and analyses have refuted that. We also noted that 'natural cycles' such as El Nino or the AMO or the NAO were now claimed to be responsible for the post 2000 AD halt in warming : cycles not included in any of the expensive, supposedly definitive and sophisticated, climate models. Nor did the 'warmists' point to the inescapable corollary: these 'natural cycles' **must** then have been responsible for some (or all?) of the upward global temperature trend in the late 20th century.

With a few honourable exceptions, few scientific authorities have spoken out against all this nonsense. Why? Because anything the scientific establishment says against the global warming story may damage the wider reputation of science itself. The few who have spoken out always stress there is nothing wrong with the science 'underpinning' the theory of global warming. This is true: CO_2 is a strong green house gas. In isolation increasing CO_2 should cause warming – that is what the climate models, model. But we saw that the Earth has oceans and a wet atmosphere and clouds and a complex bio-sphere and negative feedbacks, not positive, appear to be in operation! The current models barely touch this complexity. Nor do they consider external forcing variables such as solar activity and cosmic ray variation. These stripped down models make various assumptions about CO_2 forcing and much else which is why the range of forecasts is so embarrassingly wide. The range is so wide that it is ludicrous to talk of a consensus on warming.

The forecasting range is so wide that taking an average does not constitute a meaningful result. The debate should be about why the forecast **range** is so wide: about the key areas of ignorance reflected in the model results.

Eventually, if the warming halt continues, some authority must step forward and address the elephant in the room. The longer this is delayed, the more the global reputation of science will be damaged. The consequences could spread far beyond the failure of 'climate science'.

This 'climate chaos' situation is not the first time crude 'consensus bullying' has damaged science. We explored in the book the Continental Drift controversy begun by Alfred Wegener, whose accurate and observation rich evidence was ridiculed or ignored because geological theory provided *no explanation* for those observations. *The correct procedure would have been to review current theoretical assumptions about the physics of the Earth.* Certainly by the 1950s it was obvious that radioactive decay in the mantle could supply the required energy. Instead it was only in the late 1960s, in the face of overwhelming evidence, that the paradigm, reluctantly, shifted. Continental Drift was accepted under the less embarrassing name of Plate Tectonics. Congratulations all round for how wonderful science is!

In climate change debate we are told that any planetary tidal or torque forces acting on the Sun are so small they cannot have an effect on solar activity. Solar variation itself is also small. There is no plausible mechanism they say. Yet the small Croll-Milankovitch forcings of the Earth's orbit control the sequence of ice ages of the last million years. The mechanism is still not fully understood but the climate data is unequivocal. We have shown the strong planetary signals in solar activity, earthquake energy release and many climate time series. The data is equally unequivocal. We do not have a complete 'theory' for the sequence of mechanisms involved **but observation trumps theory or lack of theory**. We have pointed to the properties of highly non-linear dynamic systems where small inputs can be magnified into strong output responses over some frequency ranges. Self Organised Criticality will probably form the basis of the much needed real climate change theory.

A similar story can be told about evolutionary theory. The Punctuated Equilibrium hypothesis proposed that the fossil records of given species often showed long periods of stability in size, or other defining characteristics, and occasional 'sudden' changes, or punctuations, for no apparent reason. This was contrary to the orthodox view of the slow accumulation of small changes under natural selection.

The heretics were still talking of changes occurring over tens of thousands of years or more, not overnight, but were accused by the establishment of giving support to the US Creationist movement...a capital offence! The proper course was the open, public, scientific debate of the issues and evidence...however messy and inconvenient...however 'stupid' the laymen ...and women, and not the (ultimately unsuccessful) suppression of new ideas. Attempted suppression of discourse as in the climate change case, is what damages the reputation of science...not the heretics.

There are endless examples which provide object lessons which are repeatedly ignored. Let's consider the distorting power of authorities. The leading authority on human intelligence, through much of the 20th century, was Sir Cyril Burt. Burt was so certain (from his socially prejudiced observations of the UK underclass, shared by many in the educated elite) that IQ was totally genetically determined, he notoriously not only fiddled and invented many of his field observations over decades, but also invented two non-existent field workers to collect that data. After 40 years Burt was exposed in a Sunday Times article and later in a detailed 1979 biography (4). As in the case of climate change, Burt's work had a profound influence on public policy making. If IQ is purely determined by genetics why waste money on trying to educate the underclass? Should we not do all we can to prevent the low IQ underclass from breeding? We still recall with horror where crude eugenics arguments took us in the 20th century.

I say 'we' because many US states had a legal sterilisation policy for undesirable, low IQ misfits into the 1970s. In the UK, under New Labour, a youngster on the autistic spectrum was 16.6 x more likely than a 'normal' youngster to become a young offender. Under Blair 60% of those in court under the ASBO process had *diagnosed* learning disability and mental health problems. Such people are 9x more likely to die in police custody than 'normal' prisoners (5). But, hey, it's just 'bad seed' so good riddance. If the reader thinks I am exaggerating consider what Tony Blair said, as recently as September 2006, based on bad social science data, misinterpreted to suit his prejudices...and to win votes.

'There is no point pussyfooting...If we are not prepared to predict and intervene more early...pre-birth even...these [underclass] kids, a few years down the line, are going to be a menace to society...'

He is talking about his ASBO kids, from poor lone mother families, with significant mental disorders. What is all this to do with you and your family? Well Blair, having already precipitated the Iraq disaster, to justify a mass of repressive law, also said

> 'Civil liberties arguments are not so much wrong,
> just made for a different time.'

Bad science has dangerous political and social consequences. Nor has physics been immune. Richard Feynman, one of the giants of 20th century science, condemned what he called 'cargo cult science' (6). At Caltech his predecessor, Professor Robert Millikan, a famous physicist, precipitated a scientific scandal, inadvertently, because of his prestige.

'It's this kind of care, not to fool yourself, that is missing to a large extent in much of the research in cargo cult science. One example: Robert Millikan measured the charge on the electron by an experiment with falling oil drops and got an answer which we now know was not right. It's a little bit off because he had the incorrect value for the viscosity of air. It's interesting to look at the history of measurements of the charge on the electron after Millikan. If you plot them as a function of time you find that one is a little bit bigger than Millikan's [value], and the next one's a little bit bigger than that, and the next one's a little bit bigger than that, until finally they settle down to a number which is higher [and correct].

Why didn't they discover that the real number was higher right away? It's a thing that scientists are ashamed of – this history – because it's apparent that people did things like this: when they got a number that was too high above Millikan's they thought something must be wrong – and they would look for and find a reason why something might be wrong. When they got a number closer to Millikan's value they didn't look so hard. And so they eliminated the numbers that were too far off, and did other things like that. We've learned those tricks nowadays, and now we don't have that kind of a disease.'

Sadly, looking at the 'climate chaos' we are still infected with, at best, self-delusion and wishful thinking. Millikan's pioneering but incorrect result, because of his eminence, exerted a psychological pull on the other experimenters. The measurements were delicate and various adjustments were needed to eliminate possible biasing effects. Experimenters would stop making adjustments when the result looked 'right'. I see the same problem with, for example, the standard global temperature time series. There *must* be CO_2 generated warming : the basic physics says so. Maybe historic temperature records were too high; maybe recent temperatures were too low: let's look at all the raw data, weightings and infillings again. The standard temperature histories now exist in a bewildering array of 'versions'.

Perhaps some adjustments were valid, perhaps not. Now it seems that nobody is happy to be an outlier in terms of upward temperature trend, except the UAH team. Recently even the RSS satellite series has been changed to increase its trend, for disputed reasons. It seems the RSS team got fed up with 'deniers' using their series to 'prove' the decade and a half halt in warming (1). That could arguably affect their commercial forecasting business. As we saw with the climate models, the remarkably wide initial range of forecasts, from ~1 to 6 degrees C of temperature increase over this century, has been whittled down to a 'consensus' 3 to 4 degrees C range (by the IPCC). But taking an average of a vast range is not the same as having a consensus. And even a genuine consensus may be badly wrong...as in the case of the rejection of Continental Drift.

Given the consequences of global warming (or not), and the current, panicky, knee-jerk, political policy responses, we now have a serious problem of our own making. The reader may nevertheless feel that the author is overly concerned about the suppression of serious debate on climate change. Not so: there is more to fear.

In 2015 a conference of judges and lawyers took place in London, entitled 'Climate Change & the Law'. *It took place in the UK Supreme Court building (7).* It was funded by the UK government and the United Nations Environmental Programme, UNEP. N.b. UNEP is strong supporter of the man made global warming hypothesis. The aim of the conference was two fold. Firstly, to discuss how to enforce in international law a commitment to the infamous, +2 degree C, global warming limit. Secondly, to do something about 'scientifically qualified, knowledgeable and influential individuals' who still deny global warming. The International Court of Justice in fact, should act to 'finally scotch' denialist claims. That is, it should be made **illegal** for a 'government, corporation or individual' to publically question the consensus 'science' on climate change. No doubt this book would be burned and the author arrested. These proposals seem unlikely to succeed if only because I believe the world leaders of science (outside the climate change bubble) would strongly condemn them. However, the author is chilled to the marrow that such proposals are being seriously debated by some leading, legal professionals. It is time to ventilate all this dangerous nonsense in public and fumigate climate 'science'.

When the crap hits the fan, as it will, and the truth emerges, I fear scientists will join politicians as people not to be trusted. On a practical level what will the UK public do when they discover that all the additional energy costs they have endured were unnecessary...with much more to come?

But a failure of trust in 'science' opens the door to frightening social consequences including a mass return to irrational models of belief and behaviour. We have already seen the consequences of the discrediting of politicians and elites in this country in the recent referendum on Europe. Let us hope that somebody with genuine authority faces up to the 'climate chaos bomb' and defuses it, by open and honest debate, before it is too late.

Appendix 3   References

1. Watts A; 'The 'Karlisation' of global temperatures – this time RSS makes a massive upwards adjustment' ; 2nd March 2016.
   www. wattsupwiththat.com

2. Booker C; 'Climategate, the sequel: how we are STILL being tricked with flawed data on global warming'
   www. telegraph.co.uk/comment/11367272

3. Clette F, et al ; 'Revisiting Sunspot Numbers'.
   A 400 year Perspective on the Solar Cycle.
   Arxiv.org/ftp/arxiv/papers/1407/1407.3231.pdf

4. Eysenk H & Kamin L; 'Intelligence: The Battle for the Mind', Pan Psychology series; Pan Books, 1981.

5. Gregg D P; 'Troubled Families: state lies, demonization & voodoo social engineering', Green Man Books, 2015.

6. Feynman R; 'Surely You're Joking Mr. Feynman', Vintage Books; 1981.

7. Booker C; 'Judges plan to outlaw climate change denial', The Sunday Telegraph; 11.10.2015.

www.ingramcontent.com/pod-product-compliance
Lightning Source LLC
Chambersburg PA
CBHW071349210526
45465CB00001B/31